STUDIES IN GEOPHYSICS

Effects of Past Global Change on Life

Board on Earth Sciences and Resources
Commission on Geosciences, Environment, and Resources
National Research Council

NATIONAL ACADEMY PRESS
Washington, D.C. 1995

NATIONAL ACADEMY PRESS • 2101 Constitution Avenue, N.W. • Washington, DC 20418

NOTICE: The project that is the subject of this report was approved by the Governing Board of the National Research Council, whose members are drawn from the councils of the National Academy of Sciences, the National Academy of Engineering, and the Institute of Medicine. The members of the committee responsible for the report were chosen for their special competencies and with regard for appropriate balance.

This report has been reviewed by a group other than the authors according to procedures approved by a Report Review Committee consisting of members of the National Academy of Sciences, the National Academy of Engineering, and the Institute of Medicine.

The National Academy of Sciences is a private, nonprofit, self-perpetuating society of distinguished scholars engaged in scientific and engineering research, dedicated to the furtherance of science and technology and to their use for the general welfare. Upon the authority of the charter granted to it by the Congress in 1863, the Academy has a mandate that requires it to advise the federal government on scientific and technical matters. Dr. Bruce Alberts is president of the National Academy of Sciences.

The National Academy of Engineering was established in 1964, under the charter of the National Academy of Sciences, as a parallel organization of outstanding engineers. It is autonomous in its administration and in the selection of its members, sharing with the National Academy of Sciences the responsibility for advising the federal government. The National Academy of Engineering also sponsors engineering programs aimed at meeting national needs, encourages education and research, and recognizes the superior achievements of engineers. Dr. Robert M. White is president of the National Academy of Engineering.

The Institute of Medicine was established in 1970 by the National Academy of Sciences to secure the services of eminent members of appropriate professions in the examination of policy matters pertaining to the health of the public. The Institute acts under the responsibility given to the National Academy of Sciences by its congressional charter to be an adviser to the federal government and, upon its own initiative, to identify issues of medical care, research, and education. Dr. Kenneth I. Shine is president of the Institute of Medicine.

The National Research Council was organized by the National Academy of Sciences in 1916 to associate the broad community of science and technology with the Academy's purposes of furthering knowledge and of advising the federal government. Functioning in accordance with general policies determined by the Academy, the Council has become the principal operating agency of both the National Academy of Sciences and the National Academy of Engineering in providing services to the government, the public, and the scientific and engineering communities. The Council is administered jointly by both Academies and the Institute of Medicine. Dr. Bruce Alberts and Dr. Robert M. White are chairman and vice chairman, respectively, of the National Research Council.

Support for this activity was provided by the Department of Energy, the National Science Foundation, the U.S. Geological Survey, and the National Oceanic and Atmospheric Administration.

Library of Congress Cataloging-in-Publication Data
Effects of past global change on life / Board on Earth Sciences and
   Resources, Commission on Geosciences, Environment, and Resources,
   National Research Council.
      p.   cm.
   Includes bibliographical references and index.
   ISBN 0-309-05127-4
   1. Paleoecology. 2. Paleoclimatology. I. National Research
Council (U.S.). Board on Earth Sciences and Resources.
QE720.E32   1994
560'.45—dc20   94-38695

Copyright 1995 by the National Academy of Sciences. All rights reserved.

Printed in the United States of America

Panel on Effects of Past Global Change on Life

STEVEN M. STANLEY, The Johns Hopkins University, *Chairman*
JAMES P. KENNETT, University of California, Santa Barbara, *Co-Vice Chairman*
ANDREW H. KNOLL, Harvard University, *Co-Vice Chairman*
ROSEMARY A. ASKIN, University of California, Riverside
ERIC J. BARRON, Pennsylvania State University
WILLIAM B. BERRY, University of California, Berkeley
DAVID C. CHRISTOPHEL, University of Adelaide, Australia
WILLIAM A. DIMICHELE, National Museum of Natural History, Smithsonian Institution
BENJAMIN P. FLOWER, University of California, Santa Barbara
ERLE G. KAUFFMAN, University of Colorado
GERTA KELLER, Princeton University
WILLIAM F. RUDDIMAN, Lamont-Doherty Earth Observatory
ROBERT A. SPICER, Oxford University, U.K.
S. DAVID WEBB, University of Florida
THOMPSON WEBB III, Brown University

Staff

THOMAS M. USSELMAN, Associate Director
JUDITH ESTEP, Administrative Assistant

Board on Earth Sciences and Resources

FREEMAN GILBERT, Scripps Institution of Oceanography, La Jolla, Calif., *Chair*
GAIL M. ASHLEY, Rutgers University, Piscataway, N.J.
THURE CERLING, University of Utah, Salt Lake City
MARK P. CLOOS, University of Texas at Austin
NEVILLE G.W. COOK, University of California, Berkeley
JOEL DARMSTADTER, Resources for the Future, Washington, D.C.
DONALD J. DePAOLO, University of California, Berkeley
MARCO EINAUDI, Stanford University, Stanford, Calif.
NORMAN H. FOSTER, Independent Petroleum Geologist, Denver, Colo.
CHARLES G. GROAT, Louisiana State University, Baton Rouge
DONALD C. HANEY, Kentucky Geological Survey, Lexington
ANDREW H. KNOLL, Harvard University, Cambridge, Mass.
PHILIP E. LaMOREAUX, P.E. LaMoreaux and Associates, Inc., Tuscaloosa, Ala.
SUSAN LANDON, Thomasson Partner Associates, Denver, Colo.
MARCIA K. McNUTT, Massachusetts Institute of Technology, Cambridge
J. BERNARD MINSTER, University of California, San Diego
JILL D. PASTERIS, Washington University, St. Louis, Mo.
EDWARD C. ROY, JR., Trinity University, San Antonio, Tex.

Staff

JONATHAN G. PRICE, Staff Director
THOMAS M. USSELMAN, Associate Staff Director
WILLIAM E. BENSON, Senior Program Officer
KEVIN CROWLEY, Senior Program Officer
ANNE LINN, Program Officer

Staff (continued)

LALLY A. ANDERSON, Staff Assistant
JENNIFER T. ESTEP, Administrative Assistant
JUDITH L. ESTEP, Administrative Assistant
SHELLEY MYERS, Project Assistant

Commission on Geosciences, Environment, and Resources

M. GORDON WOLMAN, The Johns Hopkins University, Baltimore, Md., *Chairman*
PATRICK R. ATKINS, Aluminum Company of America, Pittsburgh, Pa.
JAMES P. BRUCE, Canadian Climate Program Board, Ottawa, Canada
WILLIAM L. FISHER, University of Texas at Austin
EDWARD A. FRIEMAN, Scripps Institution of Oceanography, LaJolla, Calif.
GEORGE M. HORNBERGER, University of Virginia, Charlottesville
W. BARCLAY KAMB, California Institute of Technology, Pasadena
PERRY L. McCARTY, Stanford University, Stanford, Calif.
RAYMOND A. PRICE, Queen's University at Kingston, Ontario, Canada
THOMAS C. SCHELLING, University of Maryland, College Park
ELLEN SILBERGELD, Environmental Defense Fund, Washington, D.C.
STEVEN M. STANLEY, The Johns Hopkins University, Baltimore, Md.
VICTORIA J. TSCHINKEL, Landers and Parson, Tallahassee, Fla.
EDITH BROWN WEISS, Georgetown University Law Center, Washington, D.C.

Staff

STEPHEN RATTIEN, Executive Director
STEPHEN D. PARKER, Associate Executive Director
MORGAN GOPNIK, Assistant Executive Director
JEANETTE SPOON, Administrative Officer
SANDI FITZPATRICK, Administrative Associate

Studies in Geophysics*

ENERGY AND CLIMATE
 Roger R. Revelle, *panel chairman*, 1977, 158 pp.
ESTUARIES, GEOPHYSICS, AND THE ENVIRONMENT
 Charles B. Officer, *panel chairman*, 1977, 127 pp.
CLIMATE, CLIMATIC CHANGE, AND WATER SUPPLY
 James R. Wallis, *panel chairman*, 1977, 132 pp.
THE UPPER ATMOSPHERE AND MAGNETOSPHERE
 Francis S. Johnson, *panel chairman*, 1977, 168 pp.
GEOPHYSICAL PREDICTIONS
 Helmut E. Landsberg, *panel chairman*, 1978, 215 pp.
IMPACT OF TECHNOLOGY ON GEOPHYSICS
 Homer E. Newell, *panel chairman*, 1979, 136 pp.
CONTINENTAL TECTONICS
 B. Clark Burchfiel, Jack E. Oliver, and Leon T. Silver, *panel co-chairmen*, 1980, 197 pp.
MINERAL RESOURCES: GENETIC UNDERSTANDING FOR PRACTICAL APPLICATIONS
 Paul B. Barton, Jr., *panel chairman*, 1981, 119 pp.
SCIENTIFIC BASIS OF WATER-RESOURCE MANAGEMENT
 Myron B. Fiering, *panel chairman*, 1982, 127 pp.
SOLAR VARIABILITY, WEATHER, AND CLIMATE
 John A. Eddy, *panel chairman*, 1982, 104 pp.
CLIMATE IN EARTH HISTORY
 Wolfgang H. Berger and John C. Crowell, *panel co-chairmen*, 1982, 198 pp.

*Published to date.

FUNDAMENTAL RESEARCH ON ESTUARIES: THE IMPORTANCE OF AN INTERDISCIPLINARY APPROACH
L. Eugene Cronin and Charles B. Officer, *panel co-chairmen*, 1983, 79 pp.

EXPLOSIVE VOLCANISM: INCEPTION, EVOLUTION, AND HAZARDS
Francis R. Boyd, *panel chairman*, 1984, 176 pp.

GROUNDWATER CONTAMINATION
John D. Bredehoeft, *panel chairman*, 1984, 179 pp.

ACTIVE TECTONICS
Robert E. Wallace, *panel chairman*, 1986, 266 pp.

THE EARTH'S ELECTRICAL ENVIRONMENT
E. Philip Krider and Raymond G. Roble, *panel co-chairmen*, 1986, 263 pp.

SEA-LEVEL CHANGES
Roger Revelle, *panel chairman*, 1990, 217 pp.

THE ROLE OF FLUIDS IN CRUSTAL PROCESSES
John D. Bredehoeft and Denis L. Norton, *panel co-chairmen*, 1990, 170 pp.

MATERIAL FLUXES ON THE SURFACE OF THE EARTH
William W. Hay, *panel chairman*, 1994, 170 pp.

EFFECTS OF PAST GLOBAL CHANGE ON LIFE
Steven M. Stanley, *panel chairman*, Andrew H. Knoll and James P. Kennett, *panel co-vice-chairmen*, 1995, 250 pp.

Preface

This report is part of a series, *Studies in Geophysics*, that has been carried out over the past 15 years to provide (1) a source of information from the scientific community to aid policymakers on decisions on societal problems that involve geophysics and (2) assessments of emerging research topics within the broad scope of geophysics. An important part of such reports is an evaluation of the adequacy of current geophysical knowledge and the appropriateness of current research programs in addressing needed information.

The study "Effects of Past Global Change on Life" is designed to help provide a scientific framework to assist the evaluation of the possible impacts of present and future global changes on the biosphere. Such a framework is based on the geologic record, which provides a unique, long-term history of changes in the global environment and of the impact of these changes on life. Because organisms are intimately related to their environment, we can infer that environmental changes of the past will have molded the history of life, and the geologic record confirms this inference for a wide range of temporal and spatial scales. The geologic record also reveals how particular kinds of environmental change have caused species to migrate, become extinct, or give rise to new species. More generally, it shows that many kinds of species and ecosystems are naturally fragile, and therefore transient, whereas other kinds are inherently more stable.

The topic was initiated by the Geophysics Study Committee in consultation with the liaison representatives of the federal agencies that support the committee, relevant boards and committees within the National Research Council, and members of the scientific community. While this report was being completed, the Geophysics Study Committee ceased operations and its parent Board on Earth Sciences and Resources assumed the responsibility for the completion of this report.

The preliminary scientific findings of the authored background chapters were presented at a symposium during October 1989 meeting of the Geological Society of America. In completing their chapters, the authors had the benefit of discussions at this symposium as well as the comments of several scientific referees. Ultimate responsibility for the individual chapters, however, rests with the authors.

The Overview of the study draws from the scientific materials presented in the authored chapters and from other materials available in the traditional scientific literature to summarize the subject. The Overview also formulates conclusions and recommendations. In preparing the Overview, the panel chairmen had the benefit of meetings that took place at the symposium, comments of the panel, and the comments of scientists, who reviewed the report according to procedures established by the National Research Council's Report Review Committee. Responsibility of the Overview rests with the Board on Earth Sciences and Resources and the chairman and two co-vice chairmen of the panel.

Contents

OVERVIEW AND RECOMMENDATIONS 1

BACKGROUND

1. Oxygen and Proterozoic Evolution: An Update 21
   Andrew H. Knoll and *Heinrich D. Holland*

2. Impact of Late Ordovician Glaciation-Deglaciation on
   Marine Life... 34
   W. B. N. Berry, M. S. Quinby-Hunt, and P. Wilde

3. Global Change Leading to Biodiversity Crisis in a Greenhouse
   World: The Cenomanian-Turonian (Cretaceous) Mass Extinction 47
   Earle G. Kauffman

4. Cretaceous-Tertiary (K/T) Mass Extinction: Effect of
   Global Change on Calcareous Microplankton...................... 72
   Gerta Keller and Katharina v. Salis Perch-Nielsen

5. Terminal Paleocene Mass Extinction in the Deep Sea:
   Association with Global Warming 94
   James P. Kennett and Lowell D. Stott

6. Tropical Climate Stability and Implications for the
   Distribution of Life 108
   Eric J. Barron

7. Neogene Ice Age in the North Atlantic Region: Climatic Changes, Biotic Effects, and Forcing Factors 118
   Steven M. Stanley and William F. Ruddiman

8. The Response of Hierarchically Structured Ecosystems to Long-Term Climatic Change: A Case Study Using Tropical Peat Swamps of Pennsylvanian Age 134
   William A. DiMichele and Tom L. Phillips

9. The Late Cretaceous and Cenozoic History of Vegetation and Climate at Northern and Southern High Latitudes: A Comparison 156
   Rosemary A. Askin and Robert A. Spicer

10. The Impact of Climatic Changes on the Development of the Australian Flora 174
    David C. Christophel

11. Global Climatic Influence on Cenozoic Land Mammal Faunas......... 184
    S. David Webb and Neil D. Opdyke

12. Biotic Responses to Temperature and Salinity Changes During Last Deglaciation, Gulf of Mexico 209
    Benjamin P. Flower and James P. Kennett

13. Pollen Records of Late Quaternary Vegetation Change: Plant Community Rearrangements and Evolutionary Implications 221
    Thompson Webb III

14. Climatic Forcing and the Origin of the Human Genus 233
    Steven M. Stanley

INDEX 245

EFFECTS OF PAST GLOBAL CHANGE ON LIFE

Overview and Recommendations

OVERVIEW

The geologic record provides a unique, long-term history of changes in the global environment and of the impact of these changes on life. From the fact that organisms are intimately related to their environment, we can infer that environmental changes of the past have molded the history of life. The geologic record contains the paleontological evidence that confirms this inference for a wide range of temporal and spatial scales. Study of this record is providing a framework for evaluating the impact of present and future global change on the biosphere—a framework that is urgently needed for the formulation of public policy.

What can be expected to happen to biotic communities when climatic zones shift or habitats shrink? As trends of global change progress, what thresholds may trigger sudden shifts between environmental states or cause catastrophic destruction of life? Lessons of the past will serve us well as we confront the future. The geologic record reveals how particular kinds of environmental change have caused species to migrate, become extinct, or give rise to new species. More generally, it shows that many kinds of species and ecosystems are naturally fragile, and therefore transient, whereas other kinds are inherently more stable.

Many advances in the understanding of ancient ecosystems are interdisciplinary in nature. Accurate plate tectonic reconstructions are essential for the evaluation of circulation patterns for ancient atmospheres and oceans. Geochemical data help us to understand ancient atmospheric and oceanic compositions, as well as climates. Functional morphology and studies of fossil preservation reveal modes of life of extinct species, and knowledge of the ecological requirements of fossilized organisms complements sedimentological analyses in the reconstruction of ancient environments. Fossil plants are among the most important indicators of ancient terrestrial climates, and studies of microfossil assemblages and stable isotopes are critical for reconstructing the three-dimensional structure of ancient oceans. Information about biotic productivity and other aspects of ancient ecosystems contributes to the understanding of secular changes in geochemical cycles. In all large-scale studies of ancient ecosystems, high-resolution stratigraphy is essential for establishing time scales. (Table 1 offers a simplified geologic time scale, which is designed to assist readers who are nongeologists.)

INTRODUCTION

The past few years have seen the emergence of a new interdisciplinary field of earth science that addresses the impact of large-scale environmental changes on ancient life. Exemplifying this development has been the maturation of the overlapping disciplines of

TABLE 1 Simplified Geologic Time Scale

Era	Period	Epoch	Time (m.y. ago)[a]
Cenozoic	Neogene	Holocene	Past 10,000 years
		Pleistocene	1.6-0.01
		Pliocene	5.3-1.6
		Miocene	23.7-5.3
	Paleogene	Oligocene	34-23.7
		Eocene	55-34
		Paleocene	65-57.8
Mesozoic	Cretaceous		144-65
	Jurassic		208-144
	Triassic		245-208
Paleozoic	Permian		286-245
	Carboniferous	Pennsylvanian	320-286
		Mississippian	360-320
	Devonian		408-360
	Silurian		438-408
	Ordovician		505-438
	Cambrian		544-505
Precambrian	Proterozoic		2,500-544
	Archean		Prior to 2,500

NOTE: The time scale was initially devised based on paleontologic evidence, with each period and epoch representing a significant paleontologic change. Each of the epochs can be further subdivided (e.g., the Cenomanian age that is in the Cretaceous Period, with ages ranging from about 97.5 to 91 million years (m.y.) ago).

[a]The relative numerical ages, based largely on radiometric determinations, are mostly from the Decade of North American Geology (1983) time scale issued by the Geological Society of America, with more recent modifications for the Cenozoic part of the record and for the Cambrian-Precambrian boundary reconstruction. Diverse new techniques have also fostered progress—improved methods for dating strata, for example, and new techniques for studying rates of evolution and extinction, as well as innovative ways of using isotopes to evaluate changes in environments, biological activity, and biogeochemical cycles.

paleogeography, paleoceanography, and paleoclimatology. The recent surge of interest in mass extinctions has helped to promote these developments, but their roots go much deeper. The success of the Deep Sea Drilling Program and its successor, the Ocean Drilling Program, has opened new opportunities for research, as has recent progress in plate tectonic reconstruction. Diverse new techniques have also fostered progress—improved methods for dating strata, for example, and new techniques for studying rates of evolution and extinction, as well as innovative ways of using isotopes to evaluate changes in environments, biological activity, and biogeochemical cycles.

Progress in all these areas has created a new framework for paleobiology, which entered a renaissance in the 1960s and is now well positioned to study the history of life in the context of a dynamic global environment. Patterns of evolution and extinction derived from fossil data are taking on new meaning in this context and have major implications for evolutionary biology and for studies of human-induced biotic change. The geologic record shows not only how the modern biosphere emerged in association with past global change, but also which kinds of species and biotic communities are most vulnerable to environmental change and which are most resilient.

In this Overview, we offer examples of research that is emerging in the study of past changes in the global ecosystem and recommend fruitful areas for research. Detailed discussions of recent advances in understanding ancient environments, the life those environments supported, and reasons for the changes in both biotas and environments appear in the authored chapters that follow this Overview. The wide variety of methods employed to study the dynamics of ancient ecosystems illustrates the interdisciplinary nature of the subject.

METHODS

Functional morphology provides key biological information. For example, dental morphology reflects a mammal's diet, and the morphology of fossil leaves is an excellent indicator of ancient climatic conditions. Because these features reflect basic laws of physics, their testimony is as powerful for the past as for the future. Terrestrial pollen spectra and marine plankton assemblages offer pictures of climatic conditions that are especially detailed for the past several million years. Preservation of key materials is also of special value, as in the use of coal balls to study the fabric and composition of Pennsylvanian peat, or the use of deep-sea deposits to obtain nearly continuous records of oceanic life and environments.

Stable isotopes and other geochemical signatures have been used in a variety of ways to investigate the dynamics of the oceans and climates. Oxygen isotopic composition of marine microfossils is the best indicator available for estimating ocean temperatures for the past 150 m.y. The oxygen isotope ratio ($^{18}O/^{16}O$) increases in the secreted skeletons with decreasing water temperature. Estimated ocean temperatures need to take into account that ^{16}O evaporates from the sea surface more readily than ^{18}O and accumulates preferentially in glacial ice. This information further allows the estimation of the volume of Cenozoic ice sheets. Oxygen isotopic ratios differ between summer-dwelling species and those representing other seasons and between surface-water dwellers and forms that occupy deep, cool waters. Carbon isotopic ratios in deep-sea sediments shed light on productivity and rates of carbon burial. Concentrations of iron and manganese in deep water shales reflect degree of oxidation and, hence, ventilation of the deep sea. Carbon isotopic ratios in paleosols appear to provide a proxy for past CO_2 levels, as does the isotopic composition of specific biomarker organic molecules preserved in marine sediments.

General circulation models have opened new possibilities for studying past changes in atmospheres and oceans. Models that couple oceans and atmospheres are especially valuable. Even imperfect global models can assist in simulating consequences of regional perturbations, such as the tectonic elevation of mountain systems and ocean barriers.

Often such consequences can be tested against key features of the geologic record. Thus, certain empirical data may constrain models, whereas others test model results. At present, the effective utilization of models is often limited by a paucity of pertinent geologic data. In addition, some models fail, in detail, to square with geologic data.

Any study of past changes in ecosystems demands a certain level of stratigraphic and chronological information. Such information is needed, for example, to determine whether similar events in widespread areas were contemporaneous. It is also essential for documenting global trends—for effectively collapsing data from many regions onto a single time line. Even at a single locality, one must know the approximate length of time separating two different conditions in order to calculate the rate of change that produced the second condition from the first. In general, chronological accuracy increases with decreasing geologic age. Special advantages are gained, for example, by working within the ranges of ^{14}C dating, well-preserved glacial varves, and extant species. Farther back in the record, errors in correlation are frequently smaller than errors in actual dates. High-resolution stratigraphy based on widespread events of brief duration can yield correlations one or two orders of magnitude more accurate than conventional biostratigraphy. For example, chemical marker beds and changing isotopic ratios of carbon and oxygen, which reflect events that spanned less that 10^5 years, have contributed to a detailed global chronology for rapid environmental change and mass extinction at the Cenomanian-Turonian boundary, about 91 m.y. ago (see Kauffman, Chapter 3). Some events, such as volcanic eruptions and accumulations of chemical fallout from extraterrestrial impacts, have deposited widespread stratigraphic markers within less than a year (Toon *et al.*, 1982). Quantitative statistical methods based on first and last stratigraphic occurrences of species are also yielding improved correlations.

Calibration of sedimentation rates allows for estimation not only of rates of extinction but also of rates of biotic recovery. High rates of deposition yield an expanded stratigraphic record and therefore often improve the quality of both the record and its temporal resolution. Thus, for the terminal Cretaceous event at about 65 m.y. ago, the shallow (middle neritic) deposits exposed at El Kef, Tunisia, seem to offer a more accurate picture of the sequence of events than do deep-sea cores (see Keller and Perch-Nielsen, Chapter 4). For the terminal Ordovician crises about 440 m.y ago, intervals of biotic change have been estimated by using the numbers (including fractions) of graptolite zones that they span (see Berry *et al.*, Chapter 2). The average duration of a zone (on the order of a million years) is estimated from radiometric ages for the boundaries of longer stratigraphic intervals.

SHIFTS BETWEEN ENVIRONMENTAL STATES

Many of the changes that have altered the global ecosystem in the course of Earth history can be viewed as shifts between environmental states. The most important shifts to affect the course of biotic evolution and the nature of the biosphere have been ones that are unique and unidirectional. Others have been components of episodic or periodic cycles, some of which have been superimposed on long-term trends.

Periodic Cycles

The controversial issue as to whether mass extinctions have occurred at equally spaced intervals has stimulated interest in the periodicity of geologic events. The most striking examples of periodic oscillations between environmental states in the recent past are those between glacial maxima and glacial minima during the past 2.5 m.y. These transitions, which have affected sea level, climates, and biotas, have been linked to periodic changes in the Earth's axial and orbital rotations—the so-called Milankovich cycles. These cycles are best documented by foraminiferal fossils from deep-sea deposits, which exhibit relative enrichment in ^{18}O when polar glacial expansion preferentially sequesters ^{16}O from the hydrosphere. For reasons not yet known, periodicities of ~41,000 years, reflecting the tilt cycle of the Earth's axis, dominated until about 0.8 m.y. ago, when periodicities of ~100,000 years, reflecting the shape of the Earth's orbit, began to prevail.

Cycles in some pre-Neogene marine successions appear to reflect minor changes in sea level or biotic productivity that were forced by Milankovich controls mediated by factors that remain poorly understood but may not always have entailed changes in ice volume. Certain Mesozoic lake deposits also contain evidence for Milankovitch-driven cyclicity, perhaps related to shifting monsoons or other patterns of rainfall and evaporation.

Nonperiodic Cycles

The most profound nonperiodic cycles of global change have been long-terms oscillations between what have been termed the "hothouse" and the "icehouse" states for oceans and atmospheres. The term hothouse is preferred to "greenhouse" because the conditions described may not always result from greenhouse warming; the hothouse states are, however, characterized by warm polar regions and warm deep oceans. In contrast, the icehouse state entails cold (usually glacial) polar conditions and a frigid deep sea that results from the descent of cold polar waters.

The geologic record spanning the Eocene-Oligocene boundary documents the transition between a hothouse state and the icehouse state that has persisted to the present (Kennett *et al.*, 1972). Much farther back in the geologic record, the interval spanning the Ordovician-Silurian boundary documents a similar transition, as well as the subsequent melting and retreat of glaciers and return of warmer conditions across broad regions (see Berry *et al.*, Chapter 2).

The Eocene-Oligocene Transition

The recent ice age in the Northern Hemisphere constitutes only an intensification of the icehouse state that our planet entered about 34 m.y. ago, at the end of the Eocene Epoch. Fossil floras and vertebrate faunas reveal that early in Eocene time, subtropical conditions extended north of the Arctic Circle and that southeastern England and the Paris Basin (45 to 50°N) supported tropical rain forests. Fossil floras are, in fact, the most valuable indicators of terrestrial climates for the past 100 m.y. Not only does the taxonomic composition of fossil floras reflect climatic conditions, but so does leaf morphology, especially the percentage of species with smooth, as opposed to jagged or lobed, leaf margins; this percentage varies linearly in the modern world with mean annual temperature (Figure 1). Leaf morphology and cuticular structure also provide a guide to precipitation conditions. Fossil floras show that the Eocene-Oligocene climatic shift was profound at middle and high latitudes in both hemispheres. As warm-adapted floral elements disappeared from these regions, other types of vegetation, adapted to colder and drier conditions, expanded (see Christophel, Chapter 10).

Climates actually did not undergo a simple shift between Early Eocene and Early Oligocene time. The tropical flora of England began to disappear at the end of Early Eocene time, as global temperatures began to cool, especially at high latitudes. By Late Eocene time, woodland savanna had already become the dominant vegetation of mid-continental North America (see Webb and Opdyke, Chapter 11). Whether the particular temporal pattern observed for North America characterized other continents remains uncertain, in part because of uncertain dating and in part because in some regions, such as Australia, a floral record is missing for much of the Eocene.

It is now widely agreed that the plate tectonic separation of Australia from Antarctica was a primary trigger of climatic changes near the end of the Eocene (and continuing separation caused further climatic changes after Eocene time). This event created the

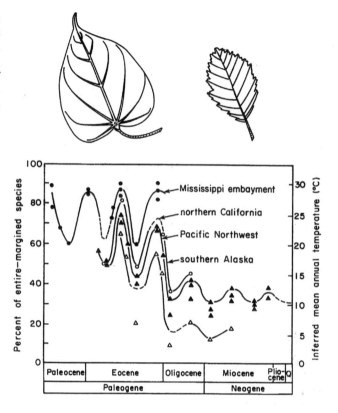

FIGURE 1 Estimated changes in temperature in four areas over the course of Cenozoic time, based on percentages of species or fossil terrestrial plants having smooth leaf margins. Especially evident is dramatic cooling near the end of the Eocene Epoch (after Wolfe, 1978).

incipient circum-Antarctic current, which began to isolate the Antarctic continent from warm waters flowing from the north. The resulting cooling of surface waters led to the formation of cool deep waters. Enrichment of ^{18}O in both planktonic and deep-sea benthic foraminifera and an influx of ice-rafted sediments indicate a significant, although temporary, expansion of the East Antarctic ice sheet at this time (see Kennett and Stott, Chapter 5). North Atlantic deep water (NADW), which is less dense than South Atlantic deep water (SADW), began to form slightly later, when rifting separated Greenland from Europe, permitting Arctic waters to descend into the North Atlantic (Schnitker, 1980). There is no question that climates cooled at middle and high latitudes, but four major questions remain:

1. Fossil floras indicate that temperate conditions extended to high latitudes in both hemispheres during Early Eocene time. How was so much heat transported from the equator toward the poles?
2. During this very warm interval, were the tropics warmer than, cooler than, or comparable to the tropics today?
3. How did the oceanographic changes during the Eocene-Oligocene transition shut down the heat transport system that had existed previously?
4. How did biotas throughout the world respond to the environmental changes?

Strong wind-driven currents cannot account for most of the meridional transport during the Eocene because, being dependent on steep thermal gradients, such currents are self-limiting. Most likely, a primary transport mechanism was the poleward flow of warm, saline subsurface water masses that formed at low latitudes. Whether the flux was sufficient to depress tropical temperatures below their modern levels remains a major question. Another is how the new system of thermohaline circulation thwarted this heat flux. Also at issue is the role of the greenhouse effect in producing the widespread warmth of the Early Eocene. If mean global temperature was well above that of the modern world, greenhouse warming is perhaps the most likely cause. If warmer high latitudes were accompanied by cooler equatorial conditions, then heat transport may have been the dominant control on the latitudinal temperature gradient.

More generally, one can ask how the secular changes in the greenhouse capacity of the atmosphere have interacted with increasing solar luminosity, continental geographies, and orogenic uplift to produce the significant climatic oscillations recorded throughout geologic time.

The Younger Dryas Cooling

Events in the North Atlantic during the earliest stages of deglaciation between 14,000 and 11,400 years ago represent another example of change in thermohaline circulation. The North Atlantic is part of a circuit that extends to the Pacific—the so-called great conveyor belt (Broecker and Denton, 1989).

North Atlantic deep water flows southward and is entrained in the Antarctic Current. It then passes into and through the Indian Ocean and the South Pacific to the North Pacific, where it upwells and returns by surface currents to the North Atlantic. There it loses heat that warms the climate of northern Europe and sinks again. During the emergence from the most recent glacial maximum, there was a brief expansion of glaciers between 11,400 and 10,200 years ago known as the Younger Dryas event. The cause of this reversal was in part a change in the flow of meltwater from North America to the ocean: Surface salinities declined in the North Atlantic because meltwater was diverted from the Mississippi drainage eastward to the St. Lawrence. Mixture with these buoyant waters reduced the density of the waters flowing northward into the North Atlantic and interrupted the formation at the surface of relatively dense NADW until the meltwater pulse had largely ended. Shifts in the isotopic composition of fossil planktonic foraminifera document this sequence of events, especially in the Gulf of Mexico region (see Flower and Kennett, Chapter 12). Plankton assemblages responded to changes in the salinity of surface waters. The Younger Dryas event illustrates how even a relatively small-scale perturbation of thermohaline circulation can have global oceanographic and climatic consequences.

The Terminal Ordovician Transition

Studies of events at the close of the Ordovician Period, about 440 m.y. ago, illustrate how it is also possible to interpret general causes and consequences of major climatic fluctuations for pre-Cenozoic intervals (see Berry *et al.*, Chapter 2). These events are associated with one of the largest mass extinctions of all time. The stratigraphic record offers evidence that continental glaciation began at this time in the supercontinent of Gondwanaland and that the buildup of glaciers lowered sea level by at least 50 m. Paleomagnetic data and environmental reconstructions reveal that Gondwanaland was moving over the South Pole. Marine fossils indicate a transition to a hothouse state: Hirnantian fauna, which had previously been restricted to cool water masses of the deep sea and to high latitudes, expanded over broad regions of the ocean, replacing warm-adapted taxa that became extinct.

Shifts to Hothouse Intervals A shift in the opposite direction, from the icehouse to the hothouse state, is associated with the development of warmer polar regions and deep ocean waters that are warm, sluggish, and dysaerobic to anoxic. Such a shift will automatically eliminate much life in the deep sea. During the mid-Cretaceous highstand of sea level, anoxic conditions extended upward into the deep portions of epicontinental seas, and it was during this interval that environmental perturbations produced pulses of biotic destruction that constitute the Cenomanian-Turonian mass extinction (see Kauffman,

Chapter 3). Near the end of the Ordovician Period, upward advection of toxic anoxic waters associated with expansion of the oxygen minimum zone may have caused major extinctions of midwater planktonic graptolites, while cooling also eliminated marine taxa (see Berry *et al.*, Chapter 2).

Value of High-Latitude Biotas Stratigraphic records at high latitudes are often critical to understanding patterns and causes of global climatic change. They also document the radiation of cold-adapted biotas during icehouse intervals and the extinction of these biotas during transitions to the hothouse state. It is enlightening to compare the middle to late Cenozoic transition from warm-adapted to cold-adapted terrestrial biotas at northern and southern high latitudes (see Askin and Spicer, Chapter 9). Apparently because higher taxa in the north spread from seasonally arid middle-latitude regions, angiosperms that occupied the new cold climates of the Northern Hemisphere were all deciduous or capable of dormancy. Antarctica, in contrast, became a center of evolutionary innovation as it grew increasingly isolated with the rifting apart of Australia and South America. Here evergreen rain forests prevailed in relatively warm coastal areas. Both northern and southern polar regions suffered drastic declines in floral diversity during the climatic cooling trend that began in the Middle Eocene. Today, only mosses and lichens grow along ice-free margins of Antarctica. In cold temperate and boreal zones of the Northern Hemisphere, the mixed coniferous forest that is widespread today became well established early in the Miocene at a time of widespread nonglacial climates. Grasses assumed a prominent role in floras of northern high latitudes near the Miocene-Pliocene transition (about 5.3 m.y. ago), when the taiga and tundra expanded dramatically.

Unidirectional Shifts

Whereas most large-scale environmental transitions in Earth history have been reversed after an interval of time, others have represented unreversed net secular trends. The composition of ancient soils, for example, points to a buildup of atmospheric oxygen from about 1% of the present atmospheric level (PAL) at about 2200 m.y. ago to about 15% PAL at about 1900 m.y. ago (see Knoll and Holland, Chapter 1). This shift, which may have been affected by a complex feedback system involving the marine geochemistry of iron and phosphorus, must have dramatically increased the production of nitrates and thus permanently altered patterns of productivity in the oceans.

Carbon-13 enrichment of carbonates and buried organic matter during the interval between 850 and 580 m.y. ago probably reflects accelerated burial of organic carbon, especially near the end of this interval. Thus, it may also reflect an increase in the partial pressure of atmospheric oxygen, as may a contemporaneous shift in the isotopic composition of marine sulfates. New data continue to support the hypothesis that atmospheric oxygen levels increased both at the beginning and at the end of the long Proterozoic Eon and had important consequences for biological evolution.

RATES OF TRANSITION

During the past few years, new evidence from high-resolution stratigraphy has revealed that many important environmental changes and biotic responses were more sudden than previously believed. The most notable example is the group of events that ended the Cretaceous Period. Most of these appear to have occurred in a crisis, perhaps measured in months rather than years, that many experts believe resulted from the impact of a comet or a meteorite—or from two or more related crises of this type. Evidence of this event, and of other sudden (though less dramatic) changes in the global ecosystem, has led to a resurgence of catastrophism as a paradigm to explain some fraction of the change in the earth system through time. How large a shift toward catastrophism is justified is a matter of current debate. On a smaller scale, ice core data are now revealing that during the past 40,000 years the climate of Greenland has changed dramatically for intervals of just 10 to 20 years (Taylor *et al.*, 1993).

Sudden Shifts and Gradual Trends

Two issues of timing are especially difficult to resolve. One is whether important events were protracted. The other is whether protracted events were pulsatile. An incomplete geologic record can give the false appearance of suddenness for an event that was actually protracted. Similarly, an imperfect record can give the false appearance of simultaneity for physical events, such as onsets or terminations of glaciation in two or more regions. A worldwide chronology for the multiple glaciations near the end of Proterozoic time has yet to be established, for example. Events of severe extinction that appear to have been protracted or to have occurred in multiple steps warrant statistical scrutiny. A key issue is the completeness of the records of taxa before their final disappearance. Imperfect records can produce an illusion of gradual or multistep extinction for a group of taxa that actually died out simultaneously: the so-called Signor-Lipps effect.

For some major events, however, the geologic record is of sufficiently high quality to document a stepwise or pulsatile pattern. Eight steps of extinction, for example, have been identified for the Cenomanian-Turonian crisis (about 91 m.y. ago). The Ordovician crisis (about 440 m.y. ago) had two principal phases, each possibly lasting hundreds of thousands of years (see Berry *et al.*, Chapter 2). The first pulse, at or near the Rawtheyan-Hirnantian stage boundary, coincided with glacial expansion. The second occurred within the Hirnantian (latest Ordovician) age, during the glacial maximum.

The geologic record documents numerous changes in the global ecosystem that spanned many millions of years. For these trends, the record, though imperfect, is too extensive to be masking a single dramatic event. We may nonetheless have difficulty in distinguishing between gradual and stepwise patterns for such trends. The classic example of this kind of trend is the climatic transition toward cooler and drier conditions on many continents between Eocene and Pleistocene times. During this time, prevailing biomes over broad regions shifted from tropical forest through savanna to grassland and steppe (see Christophel, Chapter 10; Webb and Opdyke, Chapter 11). While the terminal Eocene transition described earlier was a major early step in this trend (and was itself a complex event), the degree to which later changes were stepwise is not well established. It is, however, clear that net rates of change varied from place to place. One of the difficulties in resolving the details is in distinguishing between global changes and regional changes that resulted from such events as tectonic uplift in the American West or the Himalayan region.

The most recent geologic record offers special opportunities to establish temporal resolution for events on very short time scales. Studies of glacial varves suggest that the Younger Dryas cooling episode that interrupted deglaciation in the Northern Hemisphere between about 11,400 and 10,200 years ago developed during an interval of less than 300 years and may have ended during an interval of less than 20 years (Dansgaard *et al.*, 1989).

The Nature of Thresholds

Even a gradual environmental change can result in a sudden change of state when a threshold is crossed. The growth and contraction of glaciers are inherently unstable processes because glaciers have a higher albedo than land. One result is that the birth of a relatively small glacier can plunge a high-latitude region into an interval of widespread glaciation. The development of the Antarctic cryosphere during the Eocene-Oligocene transition is such an episode. Cooling during the Late Eocene eventually proceeded to a point where the cryosphere expanded (and it has never returned to its previous state). Another example is the abrupt shift of meltwater drainage and the resulting change in thermohaline circulation patterns that may have triggered and terminated the Younger Dryas.

Ecological limitations of organisms yield thresholds of biotic response to environmental change. For example, global circulation models suggest that during the Cretaceous Period, surface water temperatures and salinities across much of the Tethyan Ocean may have come to exceed the tolerance of most modern reef-building corals (30°C and 3.7% salinity). A shift past critical limits may explain the corals' relatively sudden loss of dominance to rudist bivalves in the central Tethyan reefs during Albian time, about 113 to 98 m.y. ago (see Barron, Chapter 6).

At the end of the Westphalian age of the Pennsylvanian Period, drier climates swept across North America and Europe, with profound consequences for vegetation (see DiMichele and Phillips, Chapter 8). As loss of swamp habitat reached a critical threshold, the arborescent lycopods that had been a conspicuous feature of landscapes for tens of millions of years disappeared rapidly. When climates amenable to swamp formation returned, ferns and tree ferns rose to dominance in these environments. The paleobotanical record shows repeated evidence for the adaptation of plants to particular environments, followed by extinction when habitats were disrupted.

Thresholds appear to have been crossed for antelopes, micromammals, and members of the human family close to 2.5 m.y. ago in Africa, with the rapid shrinkage of forests during the onset of the modern ice age. Forest-dependent species within all three groups disappeared over a broad area during an interval on the order of 10,000 years, and many species adapted to open, grassy habitats made their first appearances (Vrba, 1985). Within the human family, it appears that gracile australopithecines died out because they had depended on forests for food and refuge (see Stanley, Chapter 14). In the manner of modern chimpanzees, these animals presumably slept in trees and fled into them when threatened by predators. The characterization of paleofloras at fossil hominid sites using carbon isotopic ratios of paleosols reveals an acceleration at about 2.5 m.y. ago in the long-term Neogene shift from closed forests toward grassy habitats (Cerling, 1992). Apparently no fossil site supported a closed canopy forest after this time.

Although the preceding examples highlight the role of environmental thresholds in promoting extinction, at times environmental change has opened up new evolutionary possibilities. For example, increases in the partial pressure of oxygen must have crossed crucial thresholds for the evolution of life during Archean and Proterozoic time (see Knoll and Holland, Chapter 1). Increased partial pressure of oxygen to 1, 10, and essentially 100% of present-day levels would have cleared the environmental path for the evolution of aerobic bacteria and mitochondria-bearing eukaryotes, obligately photosynthetic eukaryotes, and large animals, respectively. In addition, an increase to levels much closer to those of today may have permitted the evolution of large animals. (As animals evolved complex circulatory and respiratory systems, they would have been able to expand into less oxygen-rich regions of the ocean and to enclose their tissues within thick mineralized shells.)

PATTERNS OF BIOTIC RESPONSE

An adverse change in the environment can cause species to migrate, or if migration to a suitable habitat is impossible, it can lead to their extinction. Patterns of migration and extinction for ancient biotas are of particular interest because they yield predictions as to how modern communities may respond to future global change. Environmental changes of the past have also had positive effects on certain surviving forms—especially ecological opportunists—or have triggered adaptive radiations within an impoverished ecosystem or a newly expanding habitat. Interactions between species intensify biotic responses to environmental change by producing chain reactions of extinction. On the other hand, interactions promote the diversification of certain taxa when others on which they depend diversify.

Migration

The intercontinental migration of Cenozoic land mammals has produced numerous natural experiments of faunal mixing. In interpreting the fossil record of these events it is often difficult to trace the causes of dispersal in detail. Correlation of the deep-sea oxygen isotopic record with pulses of mammalian migration between Eurasia and North America implicates land bridges produced by glacially controlled eustatic lowering of sea level (see Webb and Opdyke, Chapter 11). The subsequent spread of taxa throughout new regions may have been influenced by climatic or other environmental changes. In addition, it is not always clear whether the excessive rates of extinction that have typified regions being invaded by new species have resulted from habitat change or adverse species interactions. What is clear at present is that during the Cenozoic there had been a strong correlation between mammalian turnover and changes in sea level and climate.

The behavior of plant associations during floral migration has recently attracted much attention, partly because of its implications for floral changes during future global warming and partly because new evidence has contradicted the traditional view that modern biomes are ancient, coadapted associations. It appears that during the glacially induced climatic and eustatic fluctuations of the Pennsylvanian Period, coal swamp floras retained their ecological structure through many cycles of expansion and contraction (see DiMichele and Phillips, Chapter 8). Perhaps this can be attributed in part to the discrete character of the moist coal swamp environment, which did not easily exchange species with neighboring habitats. On the other hand, the pollen record of the past 20,000 years reveals that modern forest biomes of the temperate zone are transitory associations, not long-standing ones (see Webb, Chapter 13). Today in eastern North America, for example, *Pinus* (pines) and *Quercus* (oaks) have largely complementary geographic distributions outside the coastal plain, but this pattern has developed since the rapid contraction of glaciers about 10,000 years ago. During the most recent glacial maximum, pines and oaks were both largely restricted to a small region of the southeastern United States. Independent migration of plant species during future climatic changes could have important consequences for negative interactions between species of both plants and animals. Additional evidence of biotic mixing comes from small areas of Australia, where the present blending of floral provinces seems to have resulted from mid-Miocene warming.

Exactly what happened to tropical rain forests during Pleistocene glacial maxima remains unclear. Limited palynological and paleogeomorphological data suggest that the Amazon rain forest was considerably reduced in area during the last glacial maximum. Also, present geographic occurrences of certain taxa of plants and animals have been interpreted as relict distributions produced by fragmentation of rain forests during glacial maxima. This possibility needs further study, as does the more general question of coherence of rain forest communities during the past 2.5 m.y.

In the modern marine realm, many species that lived together in shallow-water Pliocene environments of eastern North America are now confined to separate depth zones. Increased seasonality (especially colder winter temperatures) since the onset of the recent ice age has driven thermally intolerant forms into deeper waters (see Stanley and Ruddiman, Chapter 7).

Extinction

In recent years, numerous biological patterns—biases against certain kinds of taxa—have been detected for particular episodes of extinction. Commonalities among victims often point to causes. In global mass extinctions such as the terminal Ordovician, Cenomanian-Turonian, and terminal Cretaceous events, tropical taxa, including reef communities, suffered preferentially. This is consistent with the idea that climatic cooling played a major role in extinction, but it may also reflect the typically narrow niche breadth of tropical taxa and the high degree of interdependence among species.

High-resolution stratigraphic and paleoenvironmental studies are crucial for understanding mass extinctions. Such studies reveal that a major extinction in deep-sea benthic assemblages at the end of the Paleocene, the most profound of the last 90 m.y. for this habitat, resulted from rapid warming of the deep oceans in conjunction with global warming (see Kennett and Stott, Chapter 5). Extinctions that removed between 35 and 50% of deep-sea taxa occurred in less than 2000 years, equal to the time required for the deep water to circulate through its basins. For a few thousand years, ocean circulation underwent fundamental changes that affected the deep-sea biota. High-resolution study of this event has illustrated, first, how events that are geologically brief but not instantaneous can strongly alter the ecosystem and, second, how such changes can be largely decoupled from events in other segments of the biosphere.

Patterns of extinction can point to particular causes of mass extinction. For example, the severe extinction of western Atlantic bivalve mollusks during the onset of the modern ice age seems to have eliminated all strictly tropical species of southern Florida; all survivors have broad thermal tolerances, ranging well beyond the tropics today. Here, a thermal filter seems clearly to have operated (see Stanley and Ruddiman, Chapter 7). Similarly, that climatic change was the ultimate cause of the previously discussed severe extinction of African mammals at the start of the modern ice age (~2.5 m.y. ago) is supported not only by the evidence that forest habitats shrank at this time but also by the fact that forest-adapted species were the primary victims (Vrba, 1985).

Some patterns of extinction have characterized higher taxa in more than one mass extinction. A striking aspect of the terminal Cretaceous extinction of planktonic foraminifera was the disappearance of species with large, complex, highly ornamented skeletons (see Keller and Perch-Nielsen, Chapter 4). Survivors were inherently small species or species that became dwarfed during the crisis. Deep-water planktonic species also died out first and in the largest numbers. These patterns must be taken into account in any analysis of the proximate causes of extinction. Most species of planktonic foraminifera that became extinct during the Late Eocene to Early Oligocene extinction were also complex, highly ornamented species. These were also largely warm-adapted taxa, which is compatible with evidence that climatic changes were the primary cause of this global crisis.

If there is one general pattern for extinctions, it is the *rate* of environmental change and not necessarily its *magnitude* that places most populations in jeopardy. This consideration is highly relevant to global changes predicted for the next century. If current models are correct, the magnitude of change will not be unusual on a geological time scale, but the rate of change may be.

Evolutionary Turnover

Evidence of causation also comes from the nature of species that immigrate into a region or originate within it during or soon after a pulse of extinction. In other words, the disappearance of some species and the appearance of others during a brief episode of evolutionary turnover should offer compatible testimony about environmental change. Thus, not only did forest-adapted species of antelopes preferentially die out in Africa about 2.5 m.y. ago, but newly appearing species were virtually all adapted to grassy habitats. Simultaneously, the apparently semiarboreal gracile australopithecines gave way to early *Homo*, which had helpless infants and could not have climbed trees habitually for refuge (see Stanley, Chapter 14).

Not only during the Pliocene but throughout much of the Cenozoic Era, mammals experienced pulses of evolutionary turnover that produced stepwise net increases in the relative number of species adapted to savannas, grasslands, or steppes. The first pulse of evolutionary turnover came at the end of the Eocene, when the extinction event that also affected the marine realm removed numerous browsers, including the huge, rhino-like titanotheres. New mammalian taxa included numerous taxa adapted to eating coarse fodder (see Webb and Opdyke, Chapter 11).

By mid-Miocene time, the diversification of taxa—adapted to grassy habitats—had produced the greatest North American land mammal diversity of all time, in savannas that were the biotic equivalents of those in Africa today. Continuation of the trend produced drier grasslands and steppes with lower mammalian diversities later in the Neogene.

Although post-Eocene pulses of turnover for Cenozoic mammals have not yet been shown to correlate well with particular floral shifts, they have been correlated with isotopic evidence of glacial expansion. Efforts to associate turnover with global climatic changes are complicated by regional trends produced by major tectonic events, such as the uplift of the Sierra Nevada, the Colorado Plateau, and the Himalayan Plateau. Similarly, although the spectacular diversification of grasses and other plants adapted to dry, seasonal habitats clearly resulted from the general post-Eocene climatic trends, intervals of diversification have not as yet been associated with pulses of extinction of moist-adapted forms.

Delayed Recovery

Severe extinctions that are not largely offset by simultaneous immigration or speciation result in impoverished ecosystems that sometimes persist for millions of years. Several factors can contribute to delayed recovery. Sometimes a delay results from a dearth of taxa capable of responding to the opportunity created by severe extinction. A striking example is the absence throughout Mississippian and Pennsylvanian times of a framework-building reef community to replace the tabulate-stromatoporoid community that had been devastated in the Late Devonian mass extinction (James, 1984). Contrasting with this situation was the rapid diversification of sclerophyllous terrestrial plants (forms with reduced leaves and thickened cuticles) in Australia after the Eocene (see Christophel, Chapter 10). These taxa seem to have originated in nutrient-poor soils at the margins of Paleogene rain forests and were, in effect, poised for rapid evolutionary response when, because of aridification, soils deteriorated over a broad area of the continent. Similarly, the mammals' evolutionary recovery from severe Late Eocene extinction in the Northern Hemisphere was accelerated by the fact that a variety of mammalian taxa with high-crowned teeth adapted for grazing on coarse vegetation had already evolved during the Eocene, prior to the severe climatic change (see Webb and Opdyke, Chapter 11).

For reasons that remain to be explained, small brachiopods that occupied the chalky seafloor of western Europe attained their former diversity within about 1 m.y. after the terminal Cretaceous extinction (see Figure 2). In general, delayed recovery from severe extinction has typified the marine realm. After the severe extinction of Late Eocene and Early Oligocene times, for example, marine faunas remained relatively impoverished throughout the Oligocene. Delayed marine recovery appears to have two primary causes. One is the inherently slow rate of adaptive radiation that characterizes many taxa of marine animals. The other is the typical failure of postcrisis conditions in the marine realm to stimulate the adaptive radiation of new kinds of taxa adapted to these conditions—to provide a new resource base comparable to productive savannas on the land (see Stanley and Ruddiman, Chapter 7). Even as overall mammalian diversity declined in North America after mid-Miocene time, certain mammalian and other taxa favored directly or indirectly by aridification underwent spectacular adaptive radiations: songbirds and Old World rats and mice—two groups that included many species that fed on the seeds of the

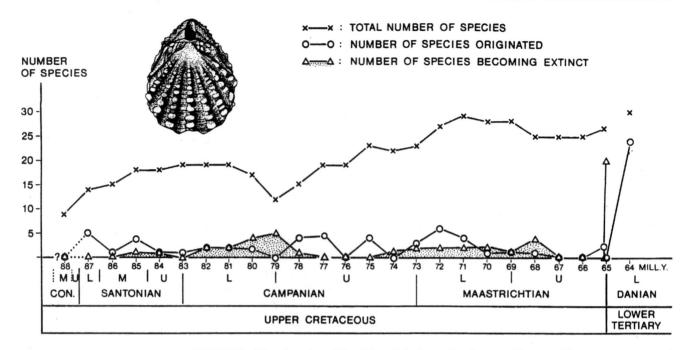

FIGURE 2 Diversity of small brachiopods in the chalk of western Europe. About three-quarters of the species died out suddenly in the terminal Cretaceous extinction, but a larger number of new species then originated very rapidly, during the next million years (from Johansen, 1988).

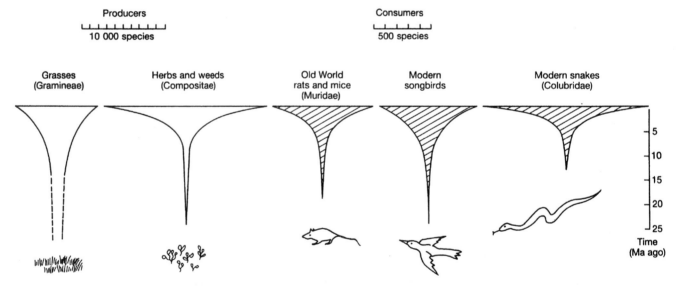

FIGURE 3 Proliferation of adaptive radiation upward from the base of the food web in terrestrial ecosystems that were favored by the global trend toward aridification during the past 25 m.y. (from Stanley, 1990). Many songbirds and Old World rats and mice feed on the seeds of taxa of grasses, herbs, and weeds that diversified dramatically during the spread of grasslands. The large majority of modern snake species belong to the family Colubridae, and many of these feed on rats and mice or on songbird eggs and chicks.

rampantly radiating herbs and grasses—as well as colubrid snakes, which prey on rats and mice and songbird chicks and eggs (see Figure 3).

Pre-Cenozoic marine faunas offer many additional examples of delayed recovery. The few small planktonic foraminifera that survived the terminal Cretaceous extinction died out sequentially during the first 200,000 years or so of Paleocene time, while new species evolved (see Keller and Perch-Nielsen, Chapter 4). The new forms were initially small, simple, and unornamented. Planktonic foraminifera did not recover their pre-extinction level of diversity until late in the Early Paleocene. Very low $\delta^{13}C$ values and low vertical $\delta^{13}C$ gradients indicate that the crisis produced low productivity in surface waters of the ocean. Drastic reduction in the abundance of calcareous nannofossils offers similar testimony, although a small number of opportunistic species experienced regional blooms. The planktonic realm began to recover only after 250,000-300,000 years. Exactly when shallow water benthic invertebrates began to rebound from the terminal Cretaceous crisis is unknown, but their recovery occupied most of Paleocene time (Hansen, 1984).

Recovery of marine life from the Cenomanian-Turonian crisis, earlier in the Cretaceous, was also slow, in part because of the loss of basic elements of the ecosystem, such as numerous taxa of reef-building rudists (see Kauffman, Chapter 3). During the terminal Ordovician mass extinction, the cool-adapted Hirnantian fauna expanded geographically but did not diversify appreciably (see Berry *et al.*, Chapter 2). Reradiation of graptolites and decimated benthic faunas was slow during the early phases of deglaciation. Few graptolite species survived to initiate radiations after the crisis, and brachiopods began to diversify only after sea level began to rise and climates became warmer. Trilobite faunas remained impoverished until late in the Early Silurian, with most species tolerating a wide range of environmental conditions.

RECOMMENDATIONS

The issues and examples cited in previous pages demonstrate the phenomenological richness of past environmental changes and biological responses to it. Earth scientists are now presented with opportunities and needs to reconstruct and interpret these interactions in unprecedented temporal and spatial detail. Some of their findings will shed light on future global change. Special attention should be given to the most recent segment of the geologic record, because it can be studied in great detail and reveals how present conditions have developed; however, older intervals that document key events also warrant study. The results will benefit evolutionary biology by bringing to light fundamental aspects of evolution and extinction, and will provide a perspective for anticipating the environmental and biotic consequences of future global change scenarios.

We present the following specific recommendations.

1. Expand interdisciplinary research that elucidates the geologic history of the biosphere in the context of earth system science—research that reveals how environments have changed on a global scale and how life has responded.

Key intervals that warrant attention are the following:

• intervals marked by major transitions between environmental states, some of which have dramatically transformed the biosphere;
• intervals marked by very rapid environmental change;
• intervals characterized by warmer conditions than those of the present—conditions that may resemble those produced by future global warming; and
• events that have produced the modern world since the latest glacial maximum in the Northern Hemisphere, about 20,000 years ago.

Features of special importance include:

- distribution of landmasses, shallow seas, deep oceans, and biogeographic provinces;
- for the terrestrial realm: the location of climatic zones and mountain belts; and
- for the oceans: three-dimensional structure, including major currents, thermohaline circulation, patterns of upwelling, and the global influence of polar regions.

2. Identify secular changes in biogeochemical cycles, including reservoir sizes and fluxes, and evaluate the consequences of these changes that are of particular importance to the documentation of past environmental change.

Topics deserving high priority include the following:

- the history of photosynthetic productivity in both terrestrial and marine environments;
- the history of atmospheric CO_2 and other greenhouse gases; and
- the Precambrian history of the atmosphere.

3. Identify and interpret patterns of extinction, migration, and evolution of life during intervals of environmental change.

Taxonomic patterns are critically important, but so are patterns based on functional and ecological groupings of organisms.

4. Construct conceptual and numerical models that portray the earth system as it existed during key geologic intervals.

Emphasis should be given to the following:

- causal explanations for changes between environmental states (crossing of environmental thresholds) that affected the biosphere;
- environmental consequences of changes in terrestrial topography and in land-sea configurations;
- modeling that couples the ocean and the atmosphere;
- synergistic interactions between building of models and gathering of the data required to constrain and test these models; and
- factors that amplify the influence of Milankovich cycles.

5. Improve existing, and develop new, techniques for characterizing ancient environments and for determining the ecological roles of species in these environments.

Approaches of special importance include the following:

- improved methodologies for characterizing the environmental tolerances of fossilized taxa;
- synthetic studies that focus on both plants and animals, for example, or both macrobacteria and microbacteria; and
- innovative isotopic, elemental, and organic geochemical techniques for environmental reconstruction.

6. Apply high-resolution stratigraphy and develop new techniques for dating and correlation in order to improve the chronological framework for studying ancient ecosystems.

Those areas deserving increased emphasis include the following:

- new or improved isotopic approaches to dating and correlation;
- dating of widespread events that were sudden, cyclical, or of great biotic consequence;
- quantitative correlation;
- refined biostratigraphic techniques; and
- studies that integrate physical and biological approaches.

REFERENCES

Broecker, W. S., and G. H. Denton (1989). The role of ocean-atmosphere reorganizations in glacial cycles, *Geochimica et Cosmochimica Acta 53*, 2465-2501.

Cerling, T. E. (1992). Development of grasslands and savannahs in East Africa during the Neogene, *Palaeogeography, Palaeoclimatology, Palaeoecology 97*, 241-247.

Dansgaard, W., W. C. White, and S. J. Johnson (1989). The abrupt termination of the Younger Dryas climatic event, *Nature 339*, 532-534.

Hansen, T. A. (1984). Early Tertiary radiation of marine molluscs and the long-term effects of the Cretaceous-Tertiary Boundary, *Paleobiology 14*, 37-51.

James, N. (1984). Reefs, in *Facies Models*, R. G. Walker, ed., Geoscience Canada Reprint Series 1, 2nd edition, pp. 229-244.

Johansen, M. B. (1988). Brachiopod extinctions in the Upper Cretaceous to lowermost Tertiary chalk of northwest Europe, Revista Espanola de Paleontologia, n° Extraordinario, 41-56.

Kennett, J. P., R. E. Burns, J. E. Andrews, M. Churkin, T. A. Davies, P. Dumitricia, A. R. Edwards, J. S. Galehouse, G. H. Packham, and G. J. Van der Lingen (1972). Australian-Antarctic continental drift, paleo-circulation changes, and Oligocene deep-sea erosion, *Nature 239*, 51-55.

Schnitker, D. (1980). West Atlantic circulation during the past 120,000 years, *Annual Reviews of Earth and Planetary Sciences 8*, 343-370.

Stanley, S. M. (1990). Adaptive radiation and macroevolution, *Systematics Association Special Volume 42*, 1-16.

Taylor, K. C., *et al.* (1993). The "flickering switch" of late Pleistocene climatic change, *Nature 361*, 432-436.

Toon, O. B., J. B. Pollack, T. P. Ackerman, R. P. Turco, C. P. McKay, and M. S. Liu (1982). Evolution of an impact-generated dust cloud and its effects on the atmosphere, *Geological Society of America Special Paper 190*, 187-200.

Vrba, E. S. (1985). African Bovidae: Evolutionary events since the Miocene, *South African Journal of Science 81*, 263-266.

Wolfe, J. A. (1978). A paleobotanical interpretation of Tertiary climates in the Northern Hemisphere, *American Scientist 66*, 694-703.

BACKGROUND

Oxygen and Proterozoic Evolution: An Update

ANDREW H. KNOLL and HEINRICH D. HOLLAND
Harvard University

Complex events can rarely be reconstructed from single lines of evidence, even where the record is well preserved.

Preston Cloud (1983)

ABSTRACT

Many authors, most notably Preston Cloud, have argued that major events in early evolution were coupled to changes in the oxygen content of the Precambrian atmosphere. Interest has focused on events close to 2000 million years ago (Ma) and 600 Ma, when increases in PO_2 are thought to have stimulated the radiations of aerobically respiring eubacteria and (via endosymbiosis) protists, and macroscopic metazoans, respectively. Acceptance of these hypotheses requires (1) geochemical evidence of environmental change; (2) paleontological evidence of coeval evolutionary innovation; and (3) physiological, ecological, and phylogenetic reasons for linking the two records. New data from Paleoproterozoic weathering profiles are providing increasingly quantitative constraints on the timing and magnitude of an early Proterozoic PO_2 increase. The emerging environmental history correlates well with the known phylogeny of protists and is consistent with the fossil record. Quantitative data on possible Neoproterozoic PO_2 changes remain elusive, but new geochemical data strongly indicate that the interval just prior to the Ediacaran radiation was a time of marked environmental change. An increase in PO_2 may well have accompanied this event, providing oxygen levels sufficient to support the metabolism of large heterotrophs. Geochemical, paleontological, and biological data support the hypothesis that atmospheric composition significantly constrained and at times provided important opportunities for early biological evolution.

INTRODUCTION

The coevolution of life and its environment has long been a principal theme in interpretations of Earth's early history. In particular, molecular oxygen has frequently been singled out as a major factor in Precambrian evolution, a view argued eloquently by Preston Cloud (1968a,b, 1972, 1983).

Why this coevolutionary view should be so popular is clear enough. Although the present day atmosphere contains 21% O_2, there is widespread agreement that prior to the emergence of oxygenic cyanobacteria, PO_2 must have been extremely low. Recent models by Canuto et al. (1983) and Kasting (1987) suggest that the prebiotic atmosphere contained no more than about 10^{-10} bar of molecular oxygen—enough to make hematite stable, but far too little to provide an effective ozone screen or to support aerobic metabolism. Indeed, a standard tenet of chemical evolution is that prebiotic chemistry could not have proceeded in environments containing significant amounts of O_2. Increasingly well resolved phylogenetic trees (e.g., Woese, 1987) complement this perspective. These trees indicate that anaerobic organisms diverged earlier than aerobes, and that aerobes requiring high PO_2 (i.e., large animals) appeared later than aerobes able to function in less oxic environments. It is, therefore, attractive to link biological to environmental history; however, the entire pattern of biological evolution can potentially be explained quite differently. The facts noted in the previous paragraph really only require a specified initial condition: that is, that the Earth's prebiotic atmosphere was essentially anoxic and that the first organisms were, therefore, anaerobic. There is no a priori reason why an early radiation of cyanobacteria could not have engendered an early and rapid increase in PO_2 approximating or even exceeding today's levels. Very different controls would then have to be sought for the observed evolutionary patterns.

Acceptance of what might fairly be called the Cloud model requires that three criteria be satisfied:

1. geochemical documentation of environmental change;
2. independent paleontological evidence for coeval evolutionary innovation; and
3. physiological, phylogenetic, and ecological reasons for linking criteria 1 and 2.

In the following pages, we evaluate the rapidly accumulating data on oxygen and biological evolution during two intervals often inferred to have been critical junctures in the history of life: (1) early in the Proterozoic Eon (ca. 2000 Ma), when increases in PO_2 above the Pasteur point are thought to have made possible the evolution of aerobic prokaryotes and mitochondria-bearing protists; and (2) the latest Proterozoic (ca. 600 Ma), when another substantial increase in PO_2 may have made possible the initial evolution of macroscopic animals.

THE EARLY PROTEROZOIC EON

Geochemical Evidence for Atmospheric Change

The Paleoproterozoic Era (2500 to 1600 Ma) was a time of profound environmental change (Cloud, 1968a, 1972; Holland, 1984). Two independent sedimentological observations have long been cited in support of the hypothesis that the atmosphere first accumulated significant amounts of oxygen during this interval. Banded iron formations (BIF), quintessentially Precambrian sediments composed of iron-bearing minerals and silica, are abundant in successions older than ca. 1900 Ma, but are rare in younger sequences (Figure 1.1). Continental red beds display an inverse distribution. The origin of marine iron

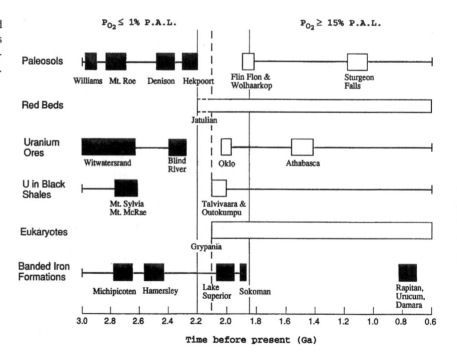

FIGURE 1.1 A summary of geochemical and paleobiological data relevant to considerations of Paleoproterozoic evolution and environmental change (PAL = present atmospheric level).

formations probably requires anoxic mid- and deep oceans for the storage and transportation of ferrous iron, while it is likely that red beds can form only when terrestrial or nearshore marine sediments come in contact with atmospheric oxygen. Thus, it has been reasoned that the BIF-red bed transition marks the rise of atmospheric oxygen. Complementary information comes from detrital uraninite in Archean and earliest Proterozoic alluvial rocks. Because this uranium mineral can survive prolonged transport only in media containing little or no oxygen, the lack of detrital uraninite deposits younger than ca. 2300 Ma also points toward a significant environmental transition (Figure 1.1; Roscoe, 1969; Grandstaff, 1980; Holland, 1984).

Not all scientists have accepted the validity of these observations or of their interpretation (see, for example, Dimroth and Kimberley, 1976; Clemmey and Badham, 1982; Windley *et al.*, 1984). It has been argued repeatedly that at least some red beds antedate the end of BIF deposition, that Archean granites have paleoweathering profiles indicative of oxic environments, and that oxidized sulfur minerals (sulfates) occur in some of the oldest known sedimentary successions. All of these observations are correct, and we must ask whether they preclude the interpretation of Archean and earliest Proterozoic environments as oxygen poor. The answer appears to be no. The formation of red beds and oxidized weathering profiles on granitic substrates requires oxygen, but only in minute quantities (see, for example, Holland, 1984; Pinto and Holland, 1988)—considerably less than is needed for aerobic metabolism. Marine sulfate does not require free oxygen at all—H_2S can be photooxidized anaerobically to SO_4^{2-} by photosynthetic bacteria, while the photochemical oxidation of volcanogenic S and SO_2 to sulfate was probably a steady source of oxidized sulfur in the Archean oceans (Walker, 1983).

Towe (1990) has specifically argued for the development of aerobic respiration early in the Archean and, therefore, for the presence of 1 to 2% PAL (present atmospheric level) O_2 in the atmosphere since that time. The possibility that oxygen levels reached this physiologically important threshold so early is not contradicted by the sparse geochemical data available for early Archean rocks (see below); however, Towe's model suffers from the absence of Archean O_2 sinks other than Fe^{2+}. We believe that the neglect of volcanic gases in his model casts significant doubts on the validity of his analysis.

Other arguments against the Cloud model posit that the geochemical indicators of low PO_2 during the Archean and earliest Proterozoic could be the result of burial digenesis, which generally acts to reduce minerals. Equally, it has been argued that oxide facies iron formations are themselves diagenetic replacements of carbonates. Neither of these views can sustain critical scrutiny. Although both oxidation and reduction can occur during diagenesis, there is ample evidence that at least some detrital uraninite and most iron formations have a primary sedimentary origin.

New data from paleosols add quantitative rigor to arguments for Paleoproterozoic environmental change (Holland and Zbinden, 1988; Holland *et al.*, 1989; Holland and Beukes, 1990). All paleosols younger than 1900 Ma that have been studied to date are highly oxidized. Fe^{2+} in the parent rocks of these paleosols was oxidized quantitatively, or nearly so, to Fe^{3+}, and was retained in the paleosols as a constituent of Fe^{3+} oxides or hydroxides. This is demonstrated by the near constancy of the ratio of total Fe to Al_2O_3 and of total Fe to TiO_2 within these paleosols and their parent rocks. Paleosols older than 1900 Ma that were developed on basaltic rocks have lost nearly all iron from their upper sections. Some of this lost iron was reprecipitated in the lower sections of the paleosols. There is some evidence that iron loss from pre-1900 Ma paleosols developed on granitic rocks was much less pronounced. These observations suggest that the O_2 content of the atmosphere prior to 1900 Ma was insufficient to oxidize more than a small fraction of the iron developed in soils on basaltic rocks, but was sufficient to oxidize a good deal of the much smaller amount of iron in soils developed on granitic rocks (Figure 1.2). More detailed studies of paleosols are needed to confirm the generality of these observations. If confirmed, they can be used to assign a rough value of ca. 1% PAL to the O_2 content of the atmosphere between about 2700 and 2200 Ma (Pinto and Holland, 1988). Pre-

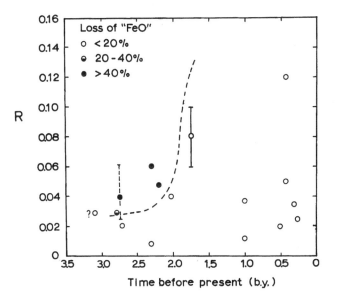

FIGURE 1.2 Iron retention in Precambrian paleosols plotted in terms of R—the ratio of the oxygen to CO_2 demand in the weathering of parent rocks—and geological age. Parent rocks with $R < 0.025$ are granitic; those with higher R are basaltic.

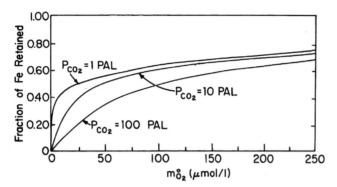

FIGURE 1.3 The fraction of iron retained during the weathering of siderite of composition $(Fe_{0.78}Mg_{0.22})CO_3$ as a function of the initial concentration of O_2 in groundwater and the CO_2 pressure in the atmosphere with which the groundwater equilibrated before reacting with siderite.

liminary data for the distribution of the rare earth elements (REEs) in paleosols suggest that Eu in paleosols has been present in the +3 valence state since at least 2750 Ma, and that the valence state of Ce in paleosols changed from +3 to +4 between 2750 and 1800 Ma. These results are consistent with those for Fe and indicate that the REEs may join iron as useful indicators of oxygen evolution in the Precambrian atmosphere.

The data for paleosols developed on igneous rocks have been supplemented recently by information on a paleoweathering profile developed on carbonate facies Kuruman Iron Formation in Griqualand West, South Africa (Holland and Beukes, 1990). The profile was probably developed ca. 1900 Ma. It is highly oxidized, and the high degree of iron retention during weathering can be used to show that PO_2 was probably in excess of 15% PAL (Figure 1.3). This is a higher minimum for PO_2 than that set by the behavior of iron in paleosols developed in igneous rocks, and indicates that the O_2 content of the atmosphere rose from about 1 to >15% PAL between 2200 and 1900 Ma (Figure 1.1).

The transition inferred from paleosol data is consistent with that inferred from the time distribution of iron formations and postdates the last known occurrence of detrital uraninite ores by several hundred million years (Knoll, 1979; Walker et al., 1983). Isotopically very light organic matter in late Archean and earliest Proterozoic sedimentary rocks has also been interpreted in terms of an early appearance of environments capable of sustaining aerobic metabolism, at least locally (Hayes, 1983). These data suggest that PO_2 may have increased in at least two steps: an initial rise from extremely low oxygen tensions to levels about 1 to 2% PAL, and a later increase to levels >15% PAL approximately 2100 Ma (see also Walker et al., 1983).

Why oxygen levels should have increased in this manner is not clear. The origin of oxygenic cyanobacteria is poorly constrained in time, but it certainly occurred before 2100 Ma. Fossils morphologically diagnostic for the group are known only from about 2000 Ma (Golubic and Hofmann, 1976), but plausibly cyanobacterial remains have been found in early Archean cherts (e.g., Schopf and Packer, 1987). Buick (1992) has argued on sedimentological and geochemical grounds that stromatolites in lacustrine carbonates of the 2800 Ma Fortescue Group, Australia, must have been built by oxygenic photoautotrophs. If Hayes' interpretation of the carbon isotope record is correct, cyanobacteria radiated 2800 Ma or earlier. The pre-2800 Ma sedimentary record has been sampled too poorly to establish whether anomalously light carbon was widespread in early Archean lacustrine environments.

Increases in atmospheric oxygen were probably occasioned by increases in primary productivity and/or decreased rates of oxygen consumption. The increase from very low O_2 levels to 1 to 2% PAL may have been related to productivity increases associated with rapid continental growth and stabilization during the late Archean/earliest Proterozoic (Knoll, 1979, 1984; Cameron, 1983). In contrast, the later increase to >15% PAL does not seem to be related to a major tectonic event. The high oxygen level in today's atmosphere must be related to the role of PO_2 in maintaining the redox balance of the atmosphere-biosphere-ocean-lithosphere system. However, the nature of the connection is still in dispute. Atmospheric PO_2 determines the concentration of O_2 in surface ocean water, but the influence of the O_2 concentration in seawater on the burial efficiency of organic matter within marine sediments seems to be slight (see, for instance, Betts and Holland, 1991). Nutrients are a more likely link between PO_2 and the burial rate of organic matter, and hence between PO_2 and rates of long-term O_2 generation. A plausible argument can be made that links the marine geochemistry of PO_4^{3-} to that of iron and hence to the O_2 content of the atmosphere today. If this argument turns out to be valid, then the history of atmospheric O_2 may have been controlled by a complicated feedback system involving the marine geochemistry of iron and phosphorus. The rapid increase in PO_2 ca. 2100 Ma may have marked the passage of the system across a threshold from one steady state to another.

Paleontological Evidence for Evolutionary Innovation

At first glance, the fossil record appears to provide strong support for the linkage of environmental and biological evolution. The oldest known fossils of probable eukaryotic origin are spirally coiled, megascopic remains in 2100 Ma shales from Michigan (Han and Runnegar, 1992). Microfossils of probable eukaryotic origin first become widespread in rocks 1800 to 1600 Ma (Figure 1.4A-D), and molecular biomarkers for eukaryotes are similarly known from rocks ≤1760 Ma (Summons and Walter, 1990). Unfortunately, paleobiological documentation of Paleoproterozoic evolutionary change is hampered by a serious problem. At about 1800 Ma the fossil record improves markedly (e.g., Schopf, 1983), so it is not clear that paleontological first appearances necessarily record evolutionary innovations.

Paleontological evidence is certainly *consistent* with a model of linked early Proterozoic environmental and biological evolution, but at the present time, fossils do not provide strong, independent confirmation of such a linkage. Documentation of microfossil assemblages from a number of pre-2100 Ma localities representing diverse paleoenvironments is needed to strengthen or reject the conclusion that fossilizable protists radiated about the time the deposition of iron formations ceased.

Biological Reasons for Linkage

Bearing in mind the unsatisfactory state of paleontological evidence, let us ask *why* evolutionary change might have attended the atmospheric transitions of the Paleoproterozoic Era. At oxygen levels less than about 1% PAL (a relatively poorly defined number; see Schopf, 1983),

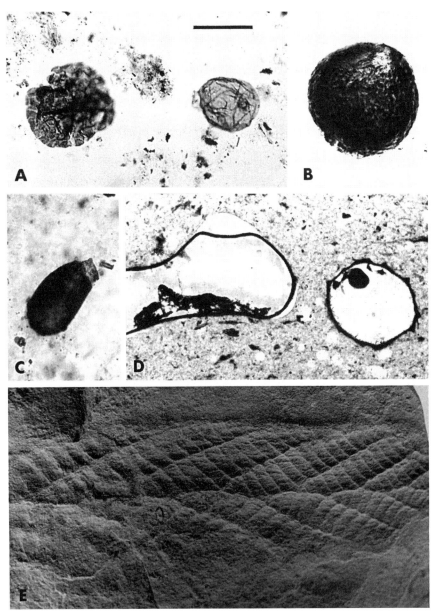

FIGURE 1.4 A: Probable eukaryotic microfossils from the Mesoproterozoic Roper Group, northern Australia (bar = 50 μm); B: a weakly ornamented protistan cyst from the Neoproterozoic Visingsö Beds, Sweden (bar = 25 μm); C: a vase-shaped protistan microfossil from the Neoproterozoic Elbobreen Formation, Spitsbergen (bar = 50 μm); D: large process-bearing protistan microfossils from the Neoproterozoic Draken Conglomerate Formation, Spitsbergen (bar = 200 μm); E: Ediacaran metazoan from the White Sea, USSR (bar = 1.5 mm).

aerobic metabolism is impossible, and there is limited atmospheric protection against ultraviolet (UV) radiation that destroys DNA (Kasting, 1987). The rate of nitrate production in the atmosphere by lightning may well have been an order of magnitude lower than today (Yung and McElroy, 1979; Levine et al., 1982); however, the difference between the present-day rate of nitrate production (Borucki and Chameides, 1984) and the rate 2500 to 2000 Ma depends on the CO_2 pressure in the Paleoproterozoic atmosphere. Kasting (1990) proposed that at a PCO_2 of 0.2 atmosphere (atm), the NO production rate in the absence of O_2 is only about a factor of two lower than at present. At a CO_2 pressure of 0.02 atm (i.e., 60 PAL), the NO fixation rate would probably be only one-tenth of the present rate. Denitrification was probably more widespread and intense in a low-O_2 ocean than in the present ocean. Therefore, NO_3^- was almost certainly in very short supply prior to 2100 Ma, and biological N_2 fixation must have been the principal source of usable nitrogen for primary producers. H_2O_2 may have been an important oxidant on the early anoxic Earth (Kasting et al., 1987); it is possible that biochemical defenses against molecules generally regarded as reactive intermediates in oxygen biochemistry evolved before O_2 itself became a significant constituent of the atmosphere (McKay and Hartman, 1991).

The conditions described in the previous paragraph certainly apply to the biota that existed before the evolution of cyanobacterial photosynthesis. How long such conditions persisted after the advent of oxygenic photosynthesis is unclear. As noted above, the antiquity of oxygenic cyanobacteria is poorly constrained, although it could easily be as great as 3500 Ma, the age of the oldest negligibly metamorphosed sedimentary rocks (Knoll, 1979; Schopf and Packer, 1987). As noted previously, a PO_2 of approximately 1 to 2% PAL appears likely for an extended period prior to 2100 Ma (see also Towe, 1990). This is an oxygen level of both biological and environmental significance. At about 1% PAL, aerobic metabolism by single-celled organisms becomes possible, while an effective ozone screen expands the ecological possibilities of life. When PO_2 rose to 1 to 2% PAL, aerobic metabolism probably followed quickly in organisms already protected against oxygen toxicity. In particular, bacteria capable of aerobic respiration, with its tremendous energetic advantage over fermentation, probably radiated rapidly (and polyphyletically) from photosynthetic ancestors. Flavin-based oxygen-utilizing pathways evolved in archaebacteria and in amitochondrial eukaryotes. At this O_2 level, nitrate production levels in the atmosphere may well have been significantly lower than today's (see above). Nitrogen fixers, therefore, could have retained a considerable advantage in primary production.

Ancestral eukaryotes formed endosymbiotic associations with purple bacterial aerobes, gaining the benefits of aerobic respiration. Symbioses with photosynthetic prokaryotes may not have formed concurrently, however. Nitrogen fixation is unknown in plastids and appears to be prohibited (Postgate and Eady, 1988); the reasons for this are not clear, but may involve oxygen toxicity. Although some early algae might have obtained nitrogen heterotrophically, obligately photosynthetic eukaryotes (including all extant megascopic algae) are unlikely to have occurred in the absence of significant quantities of nitrate in the environment.

These considerations suggest a different biological focus for the 2100 Ma oxygen event. It is not that fundamentally new metabolisms were made possible, but rather that as oxygen increased to levels above 10% PAL, nitrate availability may well have increased dramatically (see above). Obligately photosynthetic eukaryotes would then have become feasible. With their ability to avoid formation of nutrient-depleted boundary layers adjacent to cells, eukaryotic primary producers would soon have become ecologically important as primary producers. Thus, it is not surprising that 2100 Ma shales contain megascopic algae or that slightly younger rocks contain abundant acritarchs whose morphology and distribution are similar to those of younger eukaryotic phytoplankton.

How does this environmental scenario compare with the known phylogeny of eukaryotes? Figure 1.5 shows evolutionary relationships among living eukaryotes as determined by Sogin et al. (1989). At the base of the tree is *Giardia*, a common pathogen in the digestive system of vertebrates. Biochemically, *Giardia* shares more features with prokaryotes than any other known eukaryote. Ultrastructurally, however, it is clearly a true eukaryote; it contains a membrane-bounded nucleus, undulipodia (9+2 flagella), and a cytoskeleton (albeit a biochemically very simple one). On the other hand, *Giardia* has no mitochondria and no well-developed ER or Golgi apparatus. These organisms are heterotrophic, engulfing particulate food (phagocytosis) and absorbing dissolved organic molecules. Food is metabolized by the classic Embden-Meyerhoff pathway of glycolysis. *Giardia* cells are not capable of classical aerobiosis, but can use oxygen as a terminal acceptor of reducing equivalents. This system uses flavins and iron-sulfur proteins, and does not include cytochromes; it appears that the cells derive little energetic benefit from this reaction (Müller, 1988).

The next branches in Figure 1.5 are occupied by the microsporidia, trichomonads, and related protists. Both groups have clearly become specialized as obligate parasites (microsporidians are apparently dependent on an external source of ATP), but they retain features that complement the picture of early eukaryotes developed from *Giardia*.

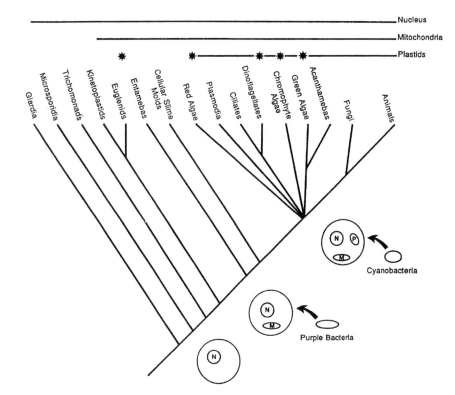

FIGURE 1.5 Summary of eukaryotic phylogeny as determined by comparisons of small subunit ribosomal RNA sequences (redrawn as a Hennigian comb from Sogin et al., 1989).

Microsporidians have prokaryote-like features of ribosomal organization and lack mitochondria; unlike *Giardia* they have a well-developed endomembrane system. Trichomonads also lack mitochondria, and free-living species live as aerotolerant anaerobic heterotrophs (Margulis et al., 1989). Some contain small organelles called hydrogenosomes, which are thought to be anaerobic equivalents of mitochondria. Others harbor intracellular bacterial symbionts that confer specific metabolic capabilities such as cellulose catabolism (Müller, 1988).

To date, RNA sequence data are available for relatively few protists, and it is entirely possible that organisms will be recognized that branch even earlier than *Giardia*. Nonetheless, the significant features shared by *Giardia*, microsporidians, and trichomonads suggest that these organisms can provide important clues to the nature of the earliest eukaryotes. They appear to have been anaerobic but aerotolerant heterotrophs, motile (using typically eukaryotic undulipodia), and endowed with a cytoskeleton and membrane system capable of endocytosis. Whether or not the first eukaryotes could have lived in the essentially O_2-free environments in which life began is uncertain. *Giardia* is unable to synthesize most of its lipids and so must incorporate lipids from its environment (Jarroll et al., 1989). The growth environment of modern *Giardia* is the small intestine of vertebrate hosts, and the lipids taken up from this environment include sterols. This raises an important issue, because sterol synthesis requires molecular oxygen at concentrations of ca. ≥0.2% PAL (Chapman and Schopf, 1983). If sterol synthesis is a primitive feature of eukaryotes subsequently lost by *Giardia*, the eukaryotic cell could not have arisen until at least low levels of oxygen had accumulated in the atmosphere. This requires that the origin of cyanobacteria predate the divergence of eukaryotes, a scenario of rapid early diversification consistent with recent phylogenies that root the universal tree between the eubacteria and an archaebacterial/eukaryote clade (Iwabe et al., 1989; Woese et al., 1990), but not with those in which all three kingdoms are viewed as diverging from a simple common ancestor (Woese, 1987). Alternatively, sterol synthesis could be a later innovation of eukaryotes, with sterol *incorporation* by *Giardia* being a relatively recent phenomenon, perhaps related to its specialized habitat. The important point is that regardless of the phylogeny preferred, organisms such as *Giardia* and trichomonads could have existed in Archean environments containing significantly less than 1 to 2% PAL PO_2. As PO_2 rose to the 1 to 2% PAL level, their aerotolerance and ability to phagocytize and maintain intracellular symbionts would have positioned them well for continued evolution.

All branches above the level of trichomonads are occupied principally by mitochondria-containing organisms. The kinetoplastids include both free-living bacteriophagous forms and obligate parasites (the notorious trypanosomes).

Often coprozoic, they are in general most common in organic-rich environments of low oxygen content (Margulis et al., 1989). Kinetoplastids are closely related to a much better known group, the euglenids. Euglenids are commonly characterized by the green photosynthetic protist *Euglena*; however, most organisms in this group are heterotrophic, and there is reason to believe that the euglenid acquisition of plastids was a relatively recent event involving the incorporation of a green algal symbiont or chloroplast (Gibbs, 1981; Whatley, 1981). Therefore, for the purposes of this argument, undue weight should not be accorded to the euglenid plastid. Kinetoplastids and euglenids both provide a perspective on early mitochondria-bearing eukaryotes as aerobic, bacteriophagous heterotrophs capable of engulfing potential symbionts and able to thrive at relatively low PO_2. Such organisms are plausible candidates for the types of eukaryotes that might have radiated during the period when PO_2 stood at 1 to 2% PAL. Eukaryotes may have gained ecological prominence as micropredators and scavengers well before they became important as photoautotrophs.

Other sequenced organisms that branch earlier than the main algal limbs include amebomastigotes (common soil and water organisms that live as flagellates under low-nutrient conditions but as amebas in nutrient-rich environments); entamebas (often parasitic, some without mitochondria, others with bacterial symbionts); and cellular slime molds.

The "crown" of the eukaryote tree is studded with photosynthetic members (Sogin et al., 1989). As argued by Cavalier-Smith (1987), there is no ultrastructural reason why organisms capable of engulfing mitochondrial precursors could not also have incorporated cyanobacteria. The diversity of photosynthetic eukaryotes certainly indicates that once plastid acquisition became feasible, a number of protistan lineages acquired them. The barrier to protoplastid acquisition could have been environmental, and the relatively late rise of PO_2 from 1 to 2% PAL to >15% PAL provides a plausible explanation for the observed phylogeny. Until this increase in PO_2, the availability of NO_3^- may have been severely limited, giving ecological advantage to free-living, nitrogen-fixing cyanobacteria. Once oxygen levels increased, however, nitrogenase activity was inhibited (Towe, 1985) and odd-nitrogen availability increased by more than an order of magnitude. Eukaryotic phytoplankton and benthos could radiate to become important parts of most surficial ecosystems, except for stressed environments such as the upper intertidal zone of restricted seaways—environments well represented in the Proterozoic fossil records. Surprisingly, increasing paleontological data suggest that the "big bang" of higher eukaryotic evolution did not occur until 1200 to 1000 Ma (Knoll, 1992b). Earlier algae, including those that formed the 2100 Ma fossils, apparently belonged to extinct lineages.

Summary of the Paleoproterozoic Earth

Given the requirement that three independent criteria must be satisfied, we cannot unequivocally accept the Cloud hypothesis as it relates to early Proterozoic evolution. The geochemical and paleontological record is improving rapidly. It now appears to fit well with the molecular phylogenetic record, and it is at least consistent with paleontological observation. A modified Cloud model provides the best framework for the available biological and geochemical data.

THE END OF THE PROTEROZOIC EON

In the Cloud model, the other principal period of linked environmental and biological evolution is the end of the Proterozoic Eon, when further increases in PO_2 are thought to have allowed the evolution of macroscopic animals. Here the relative strengths of the three lines of evidence are reversed. The paleontological data are quite extensive; it is the geochemical evidence that is consistent and suggestive rather than compelling. Details of latest Proterozoic Earth history are presented in Knoll (1992a) and Derry et al. (1992); therefore, only a brief synopsis is provided here.

Paleontological Data

The radiation of macroscopic animals was the cardinal evolutionary event of the Neoproterozoic Era. In 1968, Cloud argued that no unequivocal animal remains are present in rocks older than the great Varangian (ca. 610 to 590 Ma) ice age, and in the ensuing 20 years, a great deal of detailed stratigraphic research has strengthened this conclusion. Hofmann et al. (1990) have reported small, simple disks of probable metazoan origin in immediately sub-Varanger strata from northwestern Canada, but macroscopic animal remains and traces are otherwise conspicuously absent from pre-Varanger successions. On six continents, large and diverse, but structurally simple, animals—the so-called Ediacaran fauna (Figure 1.4C)—first appear in strata that lie above Varangian glaciogenic rocks (Runnegar, 1982a; Glaessner, 1984). Metazoan trace fossils have a parallel history of appearance and diversification (Crimes, 1987).

It is important to note the terms "macroscopic" and "large." Ancestral *microscopic* metazoans may well have evolved significantly earlier, but left no fossil record. Indeed, it is biologically appealing to posit some sort of metazoan prehistory. The significant point, however, is that the fossil record of *macroscopic* animals begins about 600 Ma.

Biological Reasons for Linkage to Environmental Change

Early hypotheses linking metazoan evolution to increases in atmospheric oxygen stressed the ozone screen and its consequences for UV absorption (e.g., Nursall, 1959: Berkner and Marshall, 1965; Cloud, 1968b). Such scenarios now seem unlikely in that an essentially complete ozone screen was probably in place by the time PO_2 reached 1% PAL (Kasting, 1987), a threshold attained long before 600 Ma (see above). Nonetheless, oxygen is important to metazoan evolution for physiological reasons (Raff and Raff, 1970; Towe, 1970; Runnegar, 1982b). Although tiny animals, perhaps ecologically similar to extant millimeter-scale nematodes and other meiofauna, can live at a PO_2 of a few percent PAL, macroscopic animals require higher oxygen concentrations to ensure the oxygenation of multiple cell layers. Large animals also require a higher PO_2 to support collagen manufacture, exercise metabolism, and organismic function within skeletons. Values of 6 to 10% PAL appear to be minimum values for the support of large, unskeletonized animals that have a circulatory system or, in the case of coelenterates, a system of finely divided mesentery folds (essentially thin, flat animals folded into a three-dimensional architecture; Raff and Raff, 1970; Runnegar, 1982b). Much higher oxygen concentrations, approaching present-day levels, are necessary to support macroscopic animals without circulatory systems (Runnegar, 1982b).

Thus, there are two potential ways in which increasing PO_2 might correlate with metazoan evolution. If tiny ur-metazoans developed circulatory systems, then a PO_2 increase to more than 6 to 10% PAL would remove this environmental barrier against the evolution of macroscopic animals. On the other hand, if macroscopic size preceded the efficient internal circulation of fluids, PO_2 increases to nearly modern levels would be necessary for large animal metabolism. Most discussions of oxygen and early animal evolution have tacitly or explicitly assumed the former case (e.g., Cloud, 1968b, 1976; Towe, 1970; Runnegar, 1982a); however, this is by no means demonstrated. It is not at all clear that Ediacaran animals had well-developed circulatory systems or finely divided mesenteries. Although many present-day animals live in relatively oxygen-poor waters (e.g., Thompson *et al.*, 1986), this may tell us little about the ancestral habitat of macroscopic animals. Once circulatory and respiratory systems were invented, large animals would have been able to inhabit oxygen-poor environments previously closed to them.

Geochemical Data

In the preceding section it is argued that oxygen levels greater than 6 to 10% PAL may have been necessary for the evolution of large animals, but we argue earlier that PO_2 probably equaled or exceeded 15% PAL as early as 2100 Ma. We, therefore, have three choices. We can reject the Cloud hypothesis; we can suggest that oxygen levels actually decreased during the Meso- or early Neoproterozoic; or we can suggest that the first macroscopic animals did not possess well developed circulatory systems and, therefore, required oxygen levels substantially greater than 15% PAL. It is certainly premature to choose the first option, and we know little about the history of atmospheric oxygen between 1900 and 600 Ma or the internal architecture of Ediacaran animals. There is no a priori reason to believe that the secular curve for atmospheric oxygen has been monotonic, although it has often been drawn that way.

In the absence of convincing data, we can ask whether or not the geological record contains any evidence suggestive of immediately pre-Ediacaran environmental change. The answer is clearly yes. At least four independent lines of evidence indicate that the period immediately preceding the Ediacaran radiation was a highly distinctive epoch in Earth history, certainly involving marked environmental change and plausibly including biologically significant variations in PO_2 (Figure 1.6; Derry *et al.*, 1992; Knoll, 1992a).

The distinctive nature of ca. 850-600 Ma sedimentary successions is clearly seen in two lithologies. After a hiatus of more than 1000 million years (m.y.), iron formations—some of them thick and laterally extensive—reappear on five continents (Young, 1976). It is difficult to envision their formation in the absence of extensive deep ocean anoxia. In general, Neoproterozoic iron formations are associated with the other distinctive lithologies of this period: tillites and related glaciogenic rocks. At least four ice ages punctuated late Proterozoic history; the Sturtian and Varanger glaciations were arguably the most severe in our planet's history (Hambrey and Harland, 1985; Kirschvink, 1992).

Independent evidence of environmental change comes from the isotopic composition of Neoproterozoic sedimentary rocks. Over the past few years, stratigraphic variations in the marine Neoproterozoic carbon, sulfur, and strontium isotopic records have been detailed. Carbon in ca. 850 to 600 Ma carbonates and organic matter is isotopically unusual in two respects—these materials are often anomalously enriched in ^{13}C ($\delta^{13}C >+5\permil$ PDB), and within this interval there are several negative $\delta^{13}C$ excursions of 6 to 8‰, at least in part associated stratigraphically with glaciogenic rocks (Knoll *et al.*, 1986; Kaufman *et al.*,

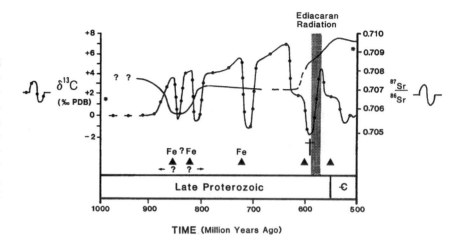

FIGURE 1.6 Summary of geochemical and paleobiological data relevant to considerations of Neoproterozoic evolution and environmental change. Filled triangles indicate ice ages; Fe indicates iron formation deposition; the cross marks a major extinction of large, morphologically complex protistan fossils; asterisks indicate present-day values for the carbon isotopic composition of diagenetically stabilized carbonates (left margin) and the strontium isotopic composition of seawater (right margin).

1990). During this same interval, the $^{87}Sr/^{86}Sr$ ratio of carbonates is anomalously low ($\Delta^{87}Sr$ as low as –500; Veizer et al., 1983; Derry et al., 1989; Asmeron et al., 1991).

The positive carbon isotope anomalies indicate a substantial increase in the proportional rate of organic carbon burial in the Neoproterozoic oceans. Insofar as oxygen generation is dependent on the burial of photosynthetically produced organic matter, this may signal a significant Neoproterozoic oxygen increase. This is not necessarily the case, however, because increased introduction of H_2, CO, and/or other reduced materials at midocean ridges and/or from terrestrial volcanoes could have balanced the oxygen generated by the burial of excess organic carbon. Interpretation of the Sr isotopic composition of seawater remains problematic; however, the extremely low $\Delta^{87}Sr$ values of carbonates that also exhibit anomalous ^{13}C enrichment may well indicate a high rate of hydrothermal input into seawater during this interval. Recent models of Neoproterozoic environmental change by Knoll and Walker (1990) and Derry et al. (1992), although differing in assumptions and procedures, both suggest that during the ca. 850 to 600 Ma interval of unusual carbon and strontium isotopic signatures, PO_2 remained relatively low. Both models further suggest that during latest Proterozoic time, when Sr isotopic ratios in seawater increased from their lowest to nearly their highest values in the past 1000 m.y. PO_2 may have increased significantly. The relatively rapid change in the isotopic composition of Sr occurred just prior to the Ediacaran radiation.

The Knoll/Walker and Derry et al. models indicate that a latest Proterozoic PO_2 increase is plausible, but not that it is proven. There are as yet no direct quantitative data indicating a change in PO_2. One indirect line of evidence that supports the idea of latest Proterozoic oxygen increase in the sulfur isotopic record. The sulfur isotopic composition of 850 to 600 Ma sulfates does not move antithetically to the carbon curve, as during much of the Phanerozoic. Antithetic movement was established in latest Proterozoic times—during the brief but eventful interval when the isotopic composition of Sr in seawater shifted; Ediacaran metazoans radiated; and intriguingly, most of the large, morphologically complex protists that characterize the 850 to 600 Ma microplankton record disappeared (Figure 1.4D; Knoll and Butterfield, 1989; Zang and Walter, 1989). The substantial shift in the isotopic composition of marine sulfate recorded at this time indicates a marked shift of the sedimentary sulfur reservoir toward pyrite (Claypool et al., 1980; François and Gerard, 1986)—a shift that contributed further to the production rate of O_2.

Summary of the Latest Proterozoic Record

The Cloud model links the diversification of macroscopic animals to new evolutionary opportunities attendant on increasing PO_2. Certainly, there can no longer be any doubt that the period immediately prior to the Ediacaran radiation was a time of marked environmental fluctuation. This may have included a PO_2 increase, although quantification of Neoproterozoic oxygen levels and even the O_2 levels at which Ediacaran-grade animals were able to function remains uncertain. As in the case of Paleoproterozoic atmospheric change and evolution, we can claim to have only two of the three required pieces of the puzzle in place; however, specific attention can now be focused on paleosols and other features of the Neoproterozoic rock record in a concentrated effort to understand the pattern of atmospheric change at or just before the first appearance of macroscopic metazoans.

CONCLUSIONS

In the Phanerozoic geological record, environmental change is often associated with extinction. Its link to evolutionary innovation is usually indirect and depends on the removal of pre-existing ecological dominants to provide evolutionary opportunity. The thesis evaluated in this chapter is that the Archean and the Proterozoic Earth were different—that major environmental changes early in Earth history directly facilitated evolutionary innovation.

The ecological specificity of many eubacteria and archaebacteria indicates that at some level this must surely be true for prokaryotic organisms (e.g., Knoll and Bauld, 1989). The Cloud model suggests that environmental-biological coevolution also applies to fundamental aspects of eukaryotic evolution, specifically to the profoundly important radiations of aerobic protists and animals. Proof of the relationship remains elusive, but accumulating evidence lends new support to the model's basic tenets. Continued research is needed to strengthen the paleontological and geochemical bases on which the empirical evidence for Proterozoic evolution and environmental change rests. Specifically, paleontological and organic geochemical research on Paleoproterozoic shales and other subtidal facies is needed to document in a more satisfactory fashion the early fossil record of eukaryotic photoautotrophs. Geochemical research on Neoproterozoic paleosols and additional indicators of PO_2 are required to document the possible role of changing oxygen concentrations in latest Proterozoic environmental and evolutionary events. These outstanding questions provide an agenda by means of which we may finally be able to document what many scientists have long believed—that two of the most significant radiations in the history of life are linked closely to secular variations in atmospheric oxygen.

ACKNOWLEDGMENTS

We acknowledge our deep debt to Preston Cloud for his articulation of fundamental problems in biological and environmental history. We thank Joseph Montoya, Mitchell Sogin, and John Postgate for useful discussions and advice. James Kasting and Kenneth Towe provided helpful reviews of the manuscript. Our research on problems of Precambrian paleontology and geochemistry is supported by NSF Grant BSR 88-17662 and NASA Grants NAGW-893 (A.H.K.) and NAGW-599 (H.D.H.).

REFERENCES

Asmeron, Y., S. Jacobesen, and A. H. Knoll (1991). Sr isotope variations in Late Proterozoic sea water: Implications for crustal evolution, *Geochimica et Cosmochimica Acta 55*, 2883-2894.

Berkner, L. V., and L. C. Marshall (1965). On the origin and rise of oxygen concentration in the Earth's atmosphere, *Journal of Atmospheric Science 22*, 225-261.

Betts, J. N., and H. D. Holland (1991). The oxygen content of ocean bottom waters, the burial efficiency of organic carbon, and the regulation of atmospheric oxygen, *Palaeogeography, Palaeoclimatology, Palaeoecology 97*, 5-18.

Borucki, W. J., and W. L. Chameides (1984). Lightning: Estimates of the rates of energy dissipation and nitrogen fixation, *Reviews of Geophysics and Space Physics 22*, 363-372.

Buick, R. (1992). The antiquity of oxygenic photosynthesis: Evidence from stromatolites in sulphate-deficient Archaean lakes, *Science 255*, 74-77.

Cameron, E. M. (1983). Sulphate and sulphate reduction in early Precambrian oceans, *Nature 296*, 145-148.

Canuto, V. M., J. S. Levine, T. R. Augustsson, C. L. Imhoff, and M. S. Giampapa (1983). UV radiation from the young Sun and oxygen and ozone levels in the prebiological atmosphere, *Nature 296*, 816-820.

Cavalier-Smith, T. (1987). The simultaneous origin of mitochondria, chloroplasts, and microtubules, *Annals of the New York Academy of Sciences 503*, 55-71.

Chapman, D. J., and J. W. Schopf (1983). Biological and biochemical effects of the development of an aerobic environment, in *Earth's Earliest Biosphere: Its Origin and Evolution*, J. W. Schopf, ed., Princeton University Press, Princeton, N.J., pp. 302-320.

Claypool, G. E., W. T. Holser, I. R. Kaplan, H. Sakai, and I. Zak (1980). The age curves of sulfur and oxygen isotopes in marine sulfate and their mutual interpretation, *Chemical Geology 28*, 199-260.

Clemmey, H., and N. Badham (1982). Oxygen in the Precambrian atmosphere: An evaluation of geological evidence, *Geology 10*, 141-146.

Cloud, P. (1968a). Atmospheric and hydrospheric evolution on the primitive Earth, *Science 160*, 729-736.

Cloud, P. (1968b). Pre-metazoan evolution and the origins of the metazoa, in *Evolution and Environment*, T. Drake, ed., Yale University Press, New Haven, Conn., pp. 1-72.

Cloud, P. (1972). A working model of the primitive Earth, *American Journal of Science 272*, 537-548.

Cloud, P. (1976). The beginnings of biospheric evolution and their biogeochemical consequences, *Paleobiology 2*, 351-387.

Cloud, P. (1983). Banded iron-formation—A gradualist's dilemma, in *Iron-Formation: Facts and Problems*, A.F. Trendall and R. C. Morris, ed., Elsevier, Amsterdam, pp. 401-416.

Crimes, T. P. (1987). Trace fossils and correlation of late Precambrian and early Cambrian strata, *Geological Magazine 124*, 97-119.

Derry, L., L. S. Keto, S. Jacobsen, A. H. Knoll, and K. Swett (1989). Sr isotopic variations of Upper Proterozoic carbonates from East Greenland and Svalbard, *Geochimica Cosmochimica Acta 53*, 2331-2339.

Derry, L., A. J. Kaufman, and S. Jacobsen (1992). Sedimentary cycling and environmental change in the Late Proterozoic: Evidence from stable and radiogenic isotopes, *Geochimica et Cosmochimica Acta 56*, 1317-1329.

Dimroth, E., and M. M. Kimberley (1976). Precambrian atmospheric oxygen: Evidence in the sedimentary distributions of carbon, sulfur, uranium, and iron, *Canadian Journal of Earth Sciences 13*, 1161-1185.

François, L. M., and J.-C. Gerard (1986). A numerical model of the evolution of ocean sulfate and sedimentary sulfur during the last 800 million years, *Geochimica Cosmochimica Acta 50*, 2289-2302.

Gibbs, S. P. (1981). The chloroplasts of some algal groups may have evolved from endosymbiotic eukaryotic algae, *Annals of the New York Academy of Sciences 361*, 193-208.

Glaessner, M. (1984). *The Dawn of Animal Life*, Cambridge University Press, Cambridge, 244 pp.

Golubic, S., and H. J. Hofmann (1976). Comparison of modern and mid-Precambrian Entophysalidaceae (Cyanophyta) in stromatolitic algal mats: Cell division and degradation, *Journal of Paleontology 50*, 1074-1082.

Grandstaff, D. E. (1980). Origin of uraniferous conglomerates at Elliot Lake, Canada, and Witwatersrand, South Africa: Implications for oxygen in the Precambrian atmosphere, *Precambrian Research 13*, 1-26.

Hambrey, M. B., and W. B. Harland (1985). The Late Proterozoic glacial era, *Paleogeography, Palaeoclimatology, Palaeoecology 51*, 255-272.

Han, T.-M., and B. Runnegar (1992). Megascopic eukaryotic algae from the 2.1 billion-year-old Negounee iron formation, Michigan, *Science 257*, 232-235.

Hayes, J. M. (1983). Geochemical evidence bearing on the origin of aerobiosis: A speculative hypothesis, *Earth's Earliest Biosphere: Its Origin and Evolution*, J. W. Schopf, ed., Princeton University Press, Princeton, N.J., pp. 291-301.

Hofmann, H. J., G. M. Narbonne, and J. D. Aitken (1990). Ediacaran remains from intertillite beds in northwestern Canada, *Geology 18*, 1199-1202.

Holland, H. D. (1984). *The Chemical Evolution of the Atmosphere and Oceans*, Princeton University Press, Princeton, N.J., 582 pp.

Holland, H. D., and N. J. Beukes (1990). A paleoweathering profile from Griqualand West, South Africa: Evidence for a dramatic rise in atmospheric oxygen between 2.2 and 1.9 BYBP, *American Journal of Science 290A*, 1-34.

Holland, H. D., and E. A. Zbinden (1988). Paleosols and the evolution of the atmosphere, Part I, in *Physical and Chemical Weathering in Geochemical Cycles*, A. Lerman and M. Meybeck, eds., Kluwer Academic Publishers, Dordrecht, pp. 61-82.

Holland, H. D., C. R. Feakes, and E. A. Zbinden (1989). The Flin Flon paleosol and the composition of the atmosphere 1.8 BYBP, *American Journal of Sciences 289*, 362-389

Iwabe, N., K. Kuma, M. Hasegawa, S. Osawa, and T. Miyata (1989). Evolutionary relationship of archaebacteria, eubacteria, and eukaryotes inferred from phylogenetic trees of duplicated genes, *Proceedings of the National Academy of Sciences USA 86*, 9355-9359.

Jarroll, E. L., P. Manning, A. Berrada, D. Hare, and D. G. Lindmark (1989). Biochemistry and metabolism of *Giardia*, *Journal of Protozoology 26*, 190-197.

Kasting, J. F. (1987). Theoretical constraints on oxygen and carbon dioxide concentrations in the Precambrian atmosphere, *Precambrian Research 34*, 205-229.

Kasting, J. F. (1990). Bolide impacts and the oxidation state of carbon in the Earth's early atmosphere, *Origins of Life 20*, 199-231.

Kaufman, A. J., J. M Hayes, A. H. Knoll, and G. J. B. Germs (1990). Isotopic compositions of carbonates and organic carbon from Upper Proterozoic successions in Namibia: Stratigraphic variation and the effects of diagenesis and metamorphism, *Precambrian Research 49*, 301-327.

Kirschvink, J. L. (1992). Late Proterozoic low-latitude global glaciation: The snowball Earth, in *The Proterozoic Biosphere*, J. W. Schopf and C. Klein, eds., Cambridge University Press, Cambridge, pp. 51-57.

Knoll, A. H. (1979). Archean photoautotrophy: Some limits and alternatives, *Origins of Life 9*, 313-327

Knoll, A. H. (1984). The Archean/Proterozoic transition: A sedimentary and paleobiological perspective, in *Patterns of Change in Earth Evolution*, H. D. Holland and A. F. Trendall, eds., Springer-Verlag, Berlin, pp. 221-242.

Knoll, A. H. (1992a). Biological and biogeochemical preludes to the Ediacaran radiation, in *Origins and Early Evolutionary History of the Metazoa*, J. H. Lipps and P. W. Signor, eds., Plenum, New York, pp. 53-84.

Knoll, A. H. (1992b). The early evolution of eukaryotes: A global perspective, *Science 256*, 622-627.

Knoll, A. H., and J. Bauld (1989). The evolution of ecological tolerance in prokaryotes, *Transactions Royal Society of Edinburgh, Earth Science 80*, 209-223.

Knoll, A. H., and N. J. Butterfield (1989). New window on Proterozoic life, *Nature 337*, 602-603.

Knoll, A. H., and J. C. G. Walker (1990). The environmental context of early metazoan evolution, *Geological Society of America, Abstracts with Program 22(7)*, A128.

Knoll, A. H., J. M. Hayes, A. J. Kaufman, K. Swett, and I. M. Lambert (1986). Secular variation in carbon isotope ratios from Upper Proterozoic successions of Svalbard and East Greenland, *Nature 321*, 832-838.

Levine, J. S., G. L. Gregory, G. A. Harvey, W. E. Howell, W. J. Borucki, and R. E. Orville (1982). Production of nitric oxide by lightning on Venus, *Geophysical Research Letters 9*, 893-896.

Margulis, M., J. O. Corliss, M. Melkonian, and D. I. Chapman, eds. (1989). *Handbook of Protoctista*, Jones and Bartlett, Boston, 914 pp.

McKay, C. P., and H. Hartman (1991). Hydrogen peroxide and the evolution of oxygenic photosynthesis, *Origins of Life 21*, 157-164.

Müller, M. (1988). Energy metabolism of protozoa without mitochondria, *Annual Reviews of Microbiology 42*, 465-488.

Nursall, J. R. (1959). Oxygen as a prerequisite for the origin of the metazoa, *Nature 183*, 1170-1171.

Pinto, J. P., and H. D. Holland (1988). Paleosols and the evolution of the atmosphere, Part II, in *Paleosols and Weathering Through Geologic Time*, J. Reinhardt and W. R. Sigleo, eds., Geological Society America Special Paper 216, 21-34.

Postgate, J. R., and R. R. Eady (1988). The evolution of biological nitrogen fixation, in *Nitrogen Fixation: Hundred Years After*, H. Bothe, M. de Bruijn, and W. E. Newton, eds., Gustav Fischer, Stuttgart, pp. 31-39.

Raff, R. A., and E. C. Raff (1970). Respiratory mechanisms and the metazoan fossil record, *Nature 228*, 1003-1004.

Roscoe, S. M. (1969). Huronian rocks and uraniferous conglomerates in the Canadian Shield, *Geological Survey of Canada Paper 68-40*, 1-205.

Runnegar, B. (1982a). The Cambrian explosion: Animals or fossils? *Journal of Geological Society of Australia 29,* 395-411.

Runnegar, B. (1982b). Oxygen requirements, biology, and phylogenetic significance of the late Proterozoic worm *Dickinsonia,* and the evolution of the burrowing habit, *Alcheringa 6,* 223-239.

Schopf, J. W., ed. (1983). *Earth's Earliest Biosphere: Its Origin and Evolution,* Princeton University Press, Princeton, N.J., 543 pp.

Schopf, J. W., and B. Packer (1987). Early Archean (3.3-billion to 3.5-billion-year-old) microfossils from the Warrawoona Group, Australia, *Science 237,* 70-73.

Sogin, M. L., J. Gunderson, H. Elwood, R. Alonso, and D. Peattie (1989). Phylogenetic meaning of the kingdom concept: An unusual ribosomal RNA from *Giardia lamblia, Science 243,* 75-77.

Summons, R. E., and M. R.Walter (1990). Molecular fossils and microfossils of prokaryote and protists from Proterozoic sediments, *American Journal of Science 290-A,* 212-244.

Towe, K. M. (1970). Oxygen-collagen priority and the early metazoan fossil record, *Proceedings of the National Academy of Sciences USA 65,* 781-788.

Towe, K. M. (1985). Habitability of the earth Earth: Clues from the physiology of nitrigen fixation and photosynthesis, *Origins of Life 15,* 235-250.

Towe, K. M. (1990). Aerobic respiration in the Archaean? *Nature 348,* 54-56.

Veizer, J., W. Compston, N. Clauer, and M. Schidlowski (1983). $^{87}Sr/^{86}Sr$ in late Proterozoic carbonates: Evidence of a "mantle" event at 900 Ma ago, *Geochimica Cosmochimica Acta 47,* 295-302.

Walker, J. C. G. (1983). Possible limits on the composition of the Archean crust, *Nature 302,* 518-520.

Walker, J. C. G., C. Klein, J. W. Schopf, D. J. Stevenson, and M. R. Walter (1983). Environmental evolution of the Archean-Early Proterozoic Earth, in *Earth's Earliest Biosphere: Its Origin and Evolution,* J. W. Schopf, ed., Princeton University Press, Princeton, N.J., pp. 260-290.

Whatley, J. M. (1981). Chloroplast evolution: Ancient and modern, *Annals of the New York Academy of Science 351,* 154-165.

Windley, B. F., P. R. Simpson, and M. D. Muir (1984). The role of atmospheric evolution in Precambrian metallogenesis, *Fortschritte der Mineralogie 62,* 253-267.

Woese, C. R. (1987). Bacterial evolution, *Microbiology Review 51,* 221-271.

Woese, C. R., O. Kandler, and M. L. Wheelis (1990). Towards a natural system of organisms: Proposal for the domains Archaea, Bacteria, and Eucarya, *Proceedings of the National Academy of Sciences USA 87,* 4576-4579.

Young, G. M. (1976). Iron-formation and glaciogenic rocks of the Rapitan Group, Northwest Territories, *Precambrian Research 3,* 137-158.

Yung, Y., and M. B. McElroy (1979). Fixation of nitrogen in a prebiotic atmosphere, *Science 203,* 1002-1004.

Zang, W., and M. R. Walter (1989). Latest Proterozoic plankton from the Amadeus Basin in Central Australia, *Nature 337,* 642-645.

2
Impact of Late Ordovician Glaciation-Deglaciation on Marine Life

W. B. N. BERRY, M. S. QUINBY-HUNT, and P. WILDE
University of California, Berkeley

ABSTRACT

Sea level fell at least 50 m during Late Ordovician continental glaciation which centered on the South Pole. Oxygen isotope analyses indicate that ocean surface waters cooled during glaciation. As sea level fell and surface waters cooled, mass mortalities occurred among most marine benthic faunas, primarily brachiopods and trilobites. Carbon isotope analyses reveal a significant biomass loss at the time of the mass mortalities. The brachiopod-dominated *Hirnantia* fauna spread widely during glacial maximum. That fauna essentially became extinct during deglaciation. Cold, oxygen-rich deep ocean waters generated at the South Pole during glaciation drove a strong deep ocean circulation and ventilated the deep oceans. Potentially, waters bearing metal ions and other substances toxic to organisms were advected upward into ocean mixed layer during glacial maximum. Graptolite mass mortality apparently was a consequence. Mass mortalities took place among pre-*Hirnantia* brachiopod and trilobite faunas at the same time as the graptolite mass mortality. Reradiation among graptolites and benthic marine faunas followed after sea-level rise, and deep ocean circulation slowed as deglaciation proceeded. Initially, reradiation rates were slow as unstable environments persisted during the early phases of deglaciation. New colony organization developed among graptolites, but significant originations did not take place until habitats preferred by graptolites stabilized. Conodont mass mortality occurred at the onset of deglaciation. Originations of new taxa were slow initially, but the pace increased as shelf seas expanded and new environments became stable. Similarly, the pace of marine benthic faunal reradiation was slow at first but increased after shelf sea environments stabilized.

INTRODUCTION

Although many glaciations have occurred during the history of the Earth, the glaciation near the end of the Ordovician stands out because the environmental changes that took place then were accompanied by near extinction of a great number of organisms (Berry and Boucot, 1973; Sheehan, 1973). A massive continental ice cap centered near the South Pole covered a large part of a continent that may be called Gondwanaland for about 2 million years (m.y.) about 435 to 437 m.y. ago (Ma). During that time of maximum glaciation, sea level fell at least 50 m and perhaps as much as 100 m. The rock record suggests that sea-level fall may have been slower than its rise following glacial melting. These glacioeustatic sea-level changes created significant changes in marine environments. Extinctions among most marine organisms living at the time appear to have been related to the environmental changes.

The features typical of continental ice sheets, including glacial pavements, striated pebbles, esker-like ridges, and glacial ice-carried dropstones have been described by Beuf *et al.* (1971) and Rognon *et al.* (1972) in Saharan Africa and Saudi Arabia (see summaries in Brenchley, 1988; Vaslet, 1990). Evidence of glaciation appears to have extended as far north from the South Pole as about 40° of south latitude (Brenchley and Newall, 1984; Brenchley, 1988). Tilloids of glaciomarine origin have been recorded from many localities in Spain, Portugal, and France (Brenchley, 1988). The presence of large dropstones that deform finely spaced sediment laminae in deposits in Spain and France indicates that large icebergs floated many miles from the shores of Gondwanaland. The spread of continental ice and the icebergs that calved off from it are analogous to the development of ice and icebergs during Pleistocene glacial maximum (Brenchley and Newall, 1984). Brenchley *et al.* (1991) described Late Ordovician glaciomarine diamictites in stratal sequences in Portugal and in the Prague Basin, Czechoslovakia. They concluded that "the sequence in the Prague Basin suggests that cold climates with floating marine ice developed early in the Hirnantian, before the main glacioeustatic regression, whereas in Portugal, deposition from marine ice was somewhat later and postdated the regression."

Areas that had been sites of shallow marine deposition became lands into which rivers cut channels and across which sands spread in sites of nearby terrigenous materials or karst topography formed on lime rock deposits in tropical areas (Brenchley and Newall, 1984). The rock record suggests that the features formed during glacial maximum and lowstand of sea were covered rapidly as sea level rose during glacial melting. Postglacial environmental changes seemingly were rapid (Brenchley, 1988).

DATA SUMMARIES

Cocks and Rickards (1988) assembled an extensive body of data concerning environmental and faunal changes in the Ordovician-Silurian boundary interval. Those summaries, arranged by region and faunal group, provide precise information on environmental changes, near extinctions, and radiations of organisms in the postglacial interval. Brenchley and Newall (1984) summarized the evidence for sea-level changes and related marine environmental changes during and after glaciation. Detailed summaries of changes in specific groups of organisms during the interval of Late Ordovician faunal change may be found in Rickards *et al.* (1977), Briggs *et al.* (1988), Fortey (1989), and Sheehan and Coorough (1990). Summaries in Barnes and Williams (1991) enhance the faunal and stratigraphic data for the Late Ordovician glacial interval. These several summaries and the data on which they are based provide the information essential to this overview of faunal changes in the Late Ordovician glacial-postglacial sequence of environmental changes.

THE TIME FRAME

Six primary divisions, termed Series, have been recognized within the Ordovician Period. Each of the Series is typified by a unique fauna. The youngest Series, the Ashgill, has been divided into four stages based on associations of brachiopods and some trilobites. The youngest Ashgill division or stage, the Hirnantian, is succeeded by the earliest Silurian stage, the Rhuddanian (see Table 2.1). The stages are characterized by fossil faunas that were benthic in life and lived in shelf sea environments. Remains of planktic organisms are rare, with certain exceptions, among the shelly faunas. Graptolites—the remains of colonial, marine plankton—occur in abundance in sequences of thinly laminated black shales and mudstones in which shelly fossils do not occur, except in debris materials derived from environments in which shelly fossils live. Accordingly, graptolite zones have been recognized as divisions of the Ordovician in black shale sequences. Graptolite zone divisions of the Ashgill are shown in Table 2.1. Time synchronous correlation of the graptolite zones with the shelly fossil stages is imprecise because the occurrence of the two types of organisms in the same deposit is so rare. Two sets of graptolite zones are shown for the early part of the Ashgill because graptolites of that time interval were distributed in two faunal provinces.

The stages typified by brachiopod-trilobite associations and the graptolite zones provide a temporal context for analysis of patterns in mass mortality and reradiation. Glacial maximum occurred during the Hirnantian Stage, although glaciation probably commenced early in the Ashgill.

TABLE 2.1 Time Units for the Late Ordovician (Ashgill) and Earliest Silurian

Period	Age	Stage	Zone	
Silurian	Llandivery	Rhuddanian	*acuminatus*	
Ordovician	Ashgill	Hirnantian	*persculptus*	
			extraordinarius	
		Rawtheyan	*pacificus*	*bohemicus*
				uniformicus
			complexus	
				mirus
		Cautleyan		*typicus*
			complanatus	
		Pusgillian		

Deglaciation started in the graptolite *Glyptograptus persculptus* zone. At least part of that zone is coeval with the latter part of the Hirnantian. The Ordovician-Silurian boundary is at the base of the *Parakidograptus acuminatus* zone. The glaciation was thus entirely within the latter part of the Ashgill.

THE PALEOGEOGRAPHIC FRAMEWORK

The paleogeographic maps in Figures 2.1 and 2.2 were supplied by C. R. Scotese. The Ashgill and Early Silurian (Llandovery) paleogeography (i.e., the positions of the lands, shelf seas and open oceans of the time) has been developed from remnant magnetism data, glacially derived sedimentary materials and features, and the locales of carbonates and evaporites. The carbonates most likely formed in tropical shallow marine environments. Ocean water masses and ocean currents are proposed based on paleogeography and consideration of modern oceanic circulation patterns. Three primary features stand out (see Figure 2.1 and 2.2): (1) no record of land or shallow marine shelf environments in the Northern Hemisphere north of the tropics; (2) a significant number of plates bearing shallow marine environments aligned essentially east-west within the tropics; and (3) a large land mass, Gondwanaland, that covered the South Pole, with a portion of it extending northward into the equatorial region.

The positions of the plates bearing lands and shallow shelf seas, as well as the zoogeographic relationships of the faunas present in the rocks formed in these environments, allow the major features of ocean circulation to be proposed. Late Ordovician-Early Silurian ocean circulation (see Figures 2.1 and 2.2) is based on the assumption that similar insolation patterns existed in the Ashgill as at present. The presence of a large part of Gondwanaland in southern temperate and tropical latitudes suggests that ocean surface circulation near it was influenced by seasonal monsoons. The positions of pressure systems over land and nearby seas would have shifted seasonally, leading to seasonal reversals in surface ocean circulation, as is seen today in the Indian Ocean. Accordingly, surface ocean circulation patterns are proposed for winter and summer seasons (Figures 2.1 and 2.2).

Ocean surface circulation north of the Northern Hemisphere tropics would have been zonal. Surface circulation in the tropics would have been influenced greatly by the several plates bearing shallow marine environments. The relatively long north-south orientation of the Laurentian (North America), Baltoscania, and Gondwanaland plates would have resulted in major oceanic circulation that was essentially within a single ocean. The ocean bordered by Gondwanaland and Baltoscania (see Figures 2.1 and 2.2) would have had a unique surface circulation that included a relatively cold polar western boundary current that flowed northward on the west side of Gondwanaland. Upwelling would have been strong and continuous there as a result of Coriolis deflection. The modern analogue of that current is the Humboldt Current off Peru-Chile. A similar but weaker western boundary current probably flowed along the west side of Baltoscania.

Pre-Hirnantian Ashgill and Hirnantian brachiopod zoogeography provides clues to potential changes in surface circulation in the ocean between Baltoscania and Gondwanaland. Sheehan and Coorough (1990) recognized four unique zoogeographic faunas, prior to the Hirnantian, along the Baltoscanian-Gondwanaland coasts. A single fauna,

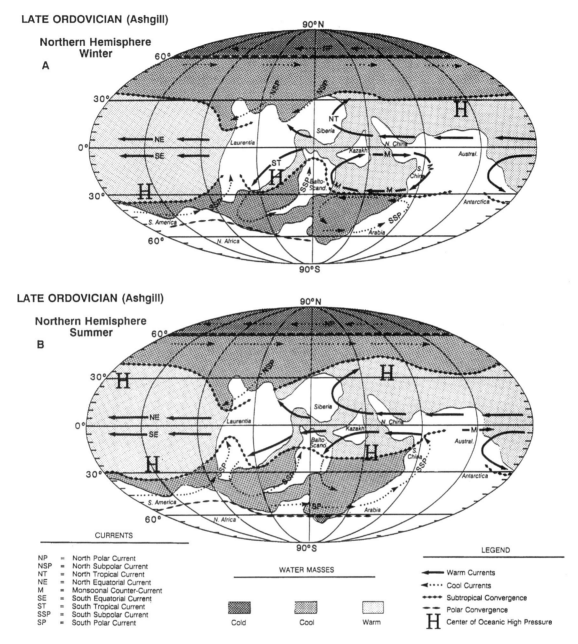

FIGURE 2.1 Paleogeographic maps for Ashgill time showing positions of lands, shelf seas, and open oceans. Ocean surface circulation is suggested. The size of Gondwanaland suggests that monsoonal conditions near it generated seasonal differences in ocean surface circulation. As sea level fell during the latter part of the Ashgill, large areas of the shelves were exposed and ocean surface currents may have been enhanced along shelf margins of the time. Base map provided by C. R. Scotese.

the *Hirnantian* fauna, replaced three of the four provincial faunas during the Hirnantian Stage. This zoogeographic change suggests that the current flowing from the tropics south along Baltoscania strengthened to such an extent that it had an influence on the Gondwanaland shelf. The Gulf Stream may be a somewhat analogous modern current. The southerly flowing current would have flowed from the tropics into high latitudes, carrying warm surface water into a cold air regime. When the Gulf Stream originated to carry warm water northward in the past 1.5 to 2 m.y., the infusion of warm water led to enhanced evaporation in a cold air regime and so was a factor in the development of glacial ice. Potentially, the ocean surface current circulation implied by the wide geographic spread

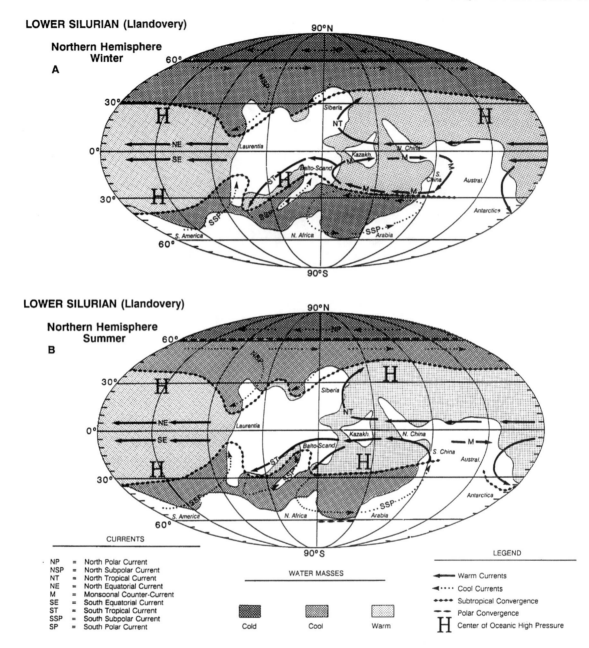

FIGURE 2.2 Paleogeographic maps for the Llandovery showing positions of lands, shallow shelf seas, and open oceans. Ocean surface water circulation currents are indicated. The size of Gondwanaland suggests that monsoonal conditions were generated near it. Seasonal monsoons would lead to seasonal differences in surface circulation. The shelf seas were sites of marine transgression that commenced in the latest Ordovician and continued into the early part of the Llandovery. Return of stable shelf sea environments during the Llandovery was followed by significant reradiation among marine communities, both benthic and planktic. Sheehan (1982) described development of Llandovery brachiopod communities in shelf sea environments that had become stable following deglaciation and transgression. Base map provided by C. R. Scotese.

of the *Hirnantia* fauna functioned as the Gulf Stream did during the onset of North Polar glaciation in the Pleistocene.

Glaciation during the Hirnantian Stage resulted in lower sea level. Surface water current impinged on only the outer margins of plates, and shelf seas were significantly narrower than prior to glaciation. Surface currents may have swept along shelf margins more strongly that when waters were widespread across the shelves. Currents may have slowed when sea level rose as a result of deglaciation.

More zoogeographic provinces among benthic faunas would have resulted as a consequence of slowed surface circulation. Sheehan and Coorough (1990) noted that the Early Silurian brachiopod faunas were characterized by more provinces than the Hirnantian Stage brachiopod faunas.

Deep ocean circulation probably increased significantly during glacial maximum. Deep ocean circulation would have been driven by both evaporation and cooling of large volumes of ocean water situated around the margins of a glaciated Gondwanaland. Cold, dense water formed in this manner sinks and flows along the ocean floor. Such water bears more oxygen than warmer surface water. Thus, during glacial maximum, the deep oceans would have received a significant and prolonged increase in oxygen. As deep water flowed north, the positions of the plates (aligned essentially east-west in the tropics and south of the tropics) would have created baffles against which deep ocean water would have advected upward. A consequence of that advection would have been a general toward-the-surface movement of the thermocline and of the zone of oxygen-depleted waters or oxygen minimum zone waters. Upward motion of oxygen-depleted water would have forced waters that contained metal ions and other substances toxic to many organisms into the mixed layer. Breaking of internal waves, as well as upwelling of waters bearing unconditioned nutrients (nutrients in proportions or in oxidation states such that organisms cannot take them in) or toxic trace metals, would have resulted in episodic incursion of water toxic to nearly all phytoplankton and zooplankton into those waters inhabited by most plankton and nekton, the mixed layer. If glacial development waxed and waned, as suggested by Vaslet (1990) and Brenchley et al. (1991), then these incursions may have had a greater or lesser effect on organisms, depending on the influence of bottom-water generation, which was related to glacial developments. Such episodic stronger and weaker incursions of toxic waters into plankton and nekton habitats could have had a greater impact on the long-term survival or extinction of many organisms than a single incursion. Wilde and Berry (1984) and Wilde et al. (1990) described how regional- to global-scale vertical advection of deep ocean waters into near-surface mixed layer water can create an environmental change crisis for many marine organisms. Such vertical advection into the mixed layer can result in the following (Wilde et al., 1990): (1) direct toxicity of mixed layer water; (2) modification or reduction of nutrients and food resources through inhibition of photosynthesis; (3) chronic debilitation through continued contacts with toxic waters; and (4) increased predation by more adapted organisms. Such environmental crises for most organisms could also result in new ecologic opportunities for organisms that had been ecologically suppressed under prior environmental conditions.

Whereas deep circulation was vigorous during glaciation, it slowed markedly with the onset of deglaciation. A characteristic of Pleistocene glacial to interglacial change is rapid development of deglaciation (W. Broecker, Lamont-Doherty Earth Observatory, oral communication, 1990). Relatively rapid deglaciation results in rapid change in deep ocean circulation. Marked vertical advection of the glacial maximum was followed by ocean conditions in which the zone of oxygen-depleted water expanded and descended somewhat in the ocean. Upward vertical advection from that zone diminished as a result. Mixed zone water expanded downward and rapidly became more hospitable to life. Sea level rose as a consequence of deglaciation. As sea level rose and oxygen-depleted waters expanded, these waters spread anoxia across the outer parts of shelves and platforms.

Vertical advection of toxic waters during glaciation would have created inhospitable environments not only for nektic and planktic organisms, but also for many benthic organisms. Reduction or absence of vertical advection of toxic waters during deglaciation would have reopened many environments in the mixed layers to resettlement by organisms.

GEOCHEMICAL EVIDENCE OF DEEP OCEAN VENTILATION

Although the ocean surface and deep circulation suggested for the Late Ordovician glacial and subsequent nonglacial interval is essentially speculative and derived from proposed paleogeographic reconstructions, some direct geochemical evidence has been developed to support the proposed model. Quinby-Hunt et al. (1989) summarized the results of approximately 300 neutron activation analyses of dark shales. Many of the samples came from the Late Ordovician-Early Silurian succession at Dob's Linn (Wilde et al., 1986; Quinby-Hunt et al., 1989). Dark shales in which the calcium concentration is less than 0.4% were selected for close scrutiny. The low calcium concentration in such samples allows the assumption that the Fe and Mn contained in them are in oxides and sulfides that reflect reducing conditions. The Fe and Mn in such low-calcium rocks are not bound in carbonates. Accordingly, Fe and Mn concentrations may be used as indicators of the intensity of reducing conditions.

Under oxic conditions (environments in which oxygen is relatively plentiful), Fe and Mn concentrations are relatively high. As oxygen availability diminishes to a condition in which it is no longer present, Mn is reduced before Fe and becomes more soluble. As a consequence, Mn concentration diminishes because it may form oxides and sulfides. Manganese diminishes in the early stages of onset of reducing conditions. As reducing conditions become somewhat more intense, Fe^{3+} is reduced to Fe^{2+}, possibly as a result of resolution of oxyhydroxides (Libes, 1992, p. 196). Much of the reduced Fe will form iron sulfide. These developments in Mn and Fe, which follow from a change from oxic to mildly reducing conditions in depositional environments, suggest the existence of three basic environmental situations that reflect the change from oxic to reducing conditions: (1) relatively high concentrations of Fe and Mn in oxic environments; (2) relatively low concentration of Mn and high concentration of Fe in mildly reducing conditions; and (3) relatively low concentrations of both Mn and Fe in slightly more highly reducing conditions.

Based on Fe and Mn concentrations reflective of the change from oxic to reducing environments, the low-calcium dark shale sample analyses were studied to ascertain if their Mn and Fe concentrations reflected depositional environment. The Mn and Fe concentrations in about 200 low-calcium dark shale samples revealed three clusters (see Quinby-Hunt et al., 1988, 1990). As indicated in Table 2.2, the three clusters appear to reflect the degree of oxidation in the above three environmental situations. Also indicated in Table 2.2, a fourth cluster is present in the samples analyzed. That cluster of samples is characterized by relatively low concentrations of manganese and iron but high concentrations of vanadium.

Samples from modern oxic and anoxic depositional environments were analyzed by using neutron activation to ascertain if Mn and Fe concentrations in them were closely similar to those characteristic of any of the clusters recognized among the ancient shales. Sediment samples from the Santa Barbara and Santa Monica Basins, housed at the University of Southern California, were analyzed. Oxygen content of waters approximately a meter above these sediment had been measured (D. Gorsline, oral communication, 1988.). The Mn and Fe concentrations of samples of basin sediment beneath waters in which the oxygen content varied from 0.1 ml/l to undetectable were comparable to cluster 2. Sediments beneath waters that had an oxygen content of 0.5 ml/l or greater had relatively high concentrations of Mn and Fe (similar to cluster 1).

Petroleum source-rock samples from the Miocene Monterey Formation recovered from a producing oil well in California also were examined with neutron activation. These source-rock samples had low Mn and Fe and high V concentrations, the unique geochemical signature of cluster 4. Kastner (1983) pointed out that the highly organic-rich, petroleum source-rock shales of the Monterey Formation formed under highly reducing, methanogenic conditions.

The analyses of modern sediments from the Santa Barbara and Santa Monica Basins and those of the organic-rich shales from the Monterey Formation suggest certain geochemical aspects of the depositional environments in which dark shales in clusters 1, 2, and 4 accumulated. Berner's (1981) discussion of a geochemical classification of sediments indicates possible geochemical conditions in the depositional environment of cluster 3. Berner (1981) described four primary geochemical categories of environmental conditions under which sediment accumulates: (1) oxic; (2) post-oxic, nonsulfidic; (3) sulfidic or sulfate reducing; and (4) methanogenic. Categories 2, 3, and 4 are indicative of an increasingly greater degree of reducing conditions in the depositional environment (Berner, 1981). The geochemical data from analyses of ancient dark shales and modern sediment are consistent with Berner's (1981) geochemical categories for clusters 1 and 2. If cluster 4 is indicative of Berner's methanogenic zone, then cluster 3 is likely, at least in part, to be a product of sediment accumulation in Berner's sulfidic category. In view of the lack of direct comparison with samples from modern environments, some cluster 3 samples could have been derived from sediment that accumulated in Berner's post-oxic, nonsulfidic interval. Bacterial sulfate reduction generates hydrogen sulfide that may react with iron to form pyrite (Berner, 1981). Pyrite occurs in many dark, graptolite-bearing shales that have Mn and Fe concentrations characteristic of cluster 3 (Quinby-Hunt et al., 1989).

Shale samples were taken from closely spaced stratigraphic intervals in the Late Ordovician-Early Silurian shales that comprise the Ordovician-Silurian boundary stratotype at Dob's Linn, southern Scotland. These samples were analyzed by neutron activation (Wilde et al., 1986). The Fe, Mn, and V concentrations (Wilde et al., 1986) in the Ordovician-Silurian boundary interval shales at Dob's Linn are consistent with the assignment of each sample to one of the four clusters reflective of oxic and the sequence of increasingly more reducing anoxic depositional environments (Figure 2.3). The Mn and Fe concentrations in gray shales that bear rare graptolites or are unfossiliferous are similar to the Mn and Fe concentrations in Santa Barbara and Santa Monica Basin sediments that accumulated

TABLE 2.2 Chemical Clustering of ~200 Low-Calcium Dark Shales

Cluster	Mn (ppm)	Fe (ppm)
1	1300	56,000
2	310	52,000
3	176	19,000
4	Similar to cluster 3 but V ranges from 350 to 1500 ppm	

SOURCE: Quinby-Hunt et al., 1988, 1990.

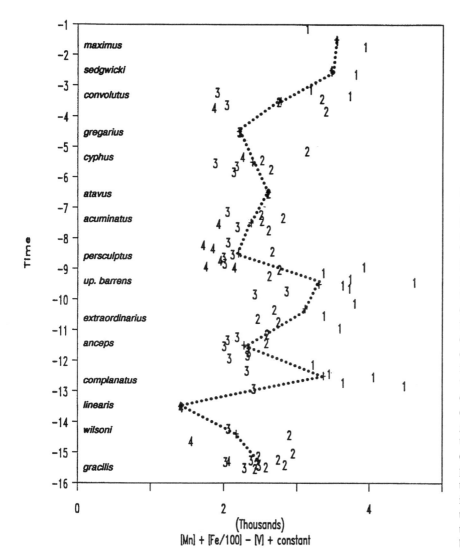

FIGURE 2.3 Diagram illustrating the results of the neutron activation analyses of shales in the Ordovician-Silurian boundary section at Dob's Linn, southern Scotland. The Mn and Fe concentrations of cluster 1 indicate relatively oxic depositional environments. Concentrations of Mn and Fe suggest that cluster 2 is mildly anoxic. Mn and Fe concentrations of clusters 3 and 4 are indicative of relatively highly anoxic depositional environments. Shales with Mn and Fe concentrations suggestive of oxic conditions formed during glaciation, an interval during which deep ocean circulation was strong and ocean water ventilation great. The data indicate that samples bearing extraordinarius zone graptolites accumulated under anoxic conditions, perhaps during an interglacial. Anoxic environments developed in the area during persculptus zone time, probably reflecting the onset of deglaciation and global warming. Oxic conditions appeared in the depositional environment during sedgwicki zone time, possibly reflecting shallowing of the basin and/or influx of currents with oxic water.

under waters with some oxygen. Dob's Linn section shales bearing numerous graptolites representative of numbers of different taxa commonly have Mn, Fe, and V concentrations characteristic of clusters 3 or 4 (Quinby-Hunt et al., 1989). Shales having Mn and Fe concentrations characteristic of cluster 2 contain fewer graptolites than those with Mn and Fe concentrations typical of clusters 3 and 4. Shales with Mn and Fe concentrations indicative of oxic depositional environments (cluster 1) contain only specimens of climacograptids of C. miserabilis and C. normalis groups (normalograptids) in the Dob's Linn Ordovician-Silurian boundary interval. These graptolites were survivors of the Late Ordovician mass mortality among graptolites (Berry et al., 1990). The shales with Mn and Fe concentrations indicative of oxic depositional environments, containing only normalograptids, were deposited during the Late Ordovician glaciation in the Southern Hemisphere. That is, they were deposited during the Hirnantian Stage of the post-D. anceps into the G. persculptus zone interval. Deep ocean water circulation would have been at its maximum during glacial maximum, and the oxygen content of that water would have been at its greatest.

Orth (see Wang et al., 1990) analyzed closely spaced black shale samples from the Ordovician-Silurian boundary interval at two localities in south China using neutron activation. The Mn, Fe, and V concentrations from the south China samples reflect a pattern of change from anoxic to oxic and a return to anoxic conditions in the depositional environment that is similar to the change seen in coeval rocks at Dob's Linn. The Mn and Fe concentrations in south China indicate that oxic waters were present in the area during the glacial maximum. The graptolite faunas obtained from several Ordovician-Silurian boundary sequences in south China suggest that environmental conditions favored by many graptolites persisted longer in south China than in Dob's Linn in the Late Ordovician (Mu, 1988; Berry et al., 1990). Furthermore, those waters preferred as habitats by many graptolites returned more rapidly to areas in south China than they did in southern Scotland (Berry et al., 1990).

The geochemical evidence from two sites that were distant from one another in the Late Ordovician appears to be consistent with significant deep ocean circulation and ventilation during glaciation. The geochemical data are also consistent with the hypothesis that only certain graptolites lived in near-surface and relatively oxic waters, and that many others lived at some depth close to oxygen-poor but nitrogen oxide rich waters (see Berry et al., 1987). The geochemical data indicate that graptolites disappeared from areas in which they had been plentiful during glacial maximum when the oxygen-poor, nitrogen oxide-rich waters were diminished and forced upward into locations nearer the surface than in nonglacial times. Lateral spread of oxygen-poor and possibly toxin-bearing waters across shelf-marginal sites during glacioeustatic sea-level fall may have been a factor in mass mortalities among certain benthic shell dwellers as well as graptolites living relatively low in the oceanic mixed layer.

PATTERNS IN EXTINCTION AND RERADIATION

Graptolites

The graptolite extinction and reradiation pattern is linked closely to the presence and absence of richly graptoliferous black shales. Such shales both at Dob's Linn and in south China bear Fe and Mn concentrations of cluster 3 or 4 (Quinby-Hunt et al., 1989). The most abundant and richly diverse graptolite faunas occur in sediments deposited under anoxic ocean water. Analogous conditions are found today in the eastern tropical Pacific (Berry et al., 1987). There, oxygen-poor, nitrogen oxide-rich waters are inhabited by zooplankton that may be modern ecologic analogues of many ancient graptolites (Berry et al., 1987, 1990). Sulfate-reducing bacteria have been recorded from sediment accumulating under the oxygen-poor, nitrogen oxide-rich waters (Gallardo, 1963). Stratigraphic study of Ordovician-Silurian boundary sections indicates that such conditions were diminished markedly during glacial maximum (see summaries in Cocks and Rickards, 1988). Koren (1991) pointed out that the Late Ordovician graptolite mass mortality occurred at the end of the P. pacificus zone, a time when black shales disappeared from nearly all Ordovician-Silurian boundary sequences. Commenting on the Late Ordovician graptolite mass mortality, Melchin and Mitchell (1988) pointed out that the "Late Ordovician graptolites experienced nearly total extinction," and that "post-extinction morphological radiation stemmed from only a few species, most of which had the same pattern in colony development."

Graptolite faunas during glacial maximum included mostly normalograptids (i.e., small climacograptids of the C. normalis and C. miserabilis types, glyptograptids of the G. persculptus group, and Climacograptus extraordinarius) as well as Diplograptus bohemicus and a few other rare diplograptid (biserial) species. All taxa except the climacograptids of the C. normalis and C. miserabilis type and glyptograptids of the G. persculptus group became extinct prior to the appearance of those glyptograptids characteristic of the G. persculptus zone.

Rickards et al. (1977) and Rickards (1988) drew attention to graptolite development at the onset of reradiation. Reradiation commences when black shales become more widespread during G. persculptus zone time. Species diversity is low, but a number of glyptograptids and climacograptids appear. The most significant innovation among graptolites in G. persculptus zone time is the appearance of the uniserial scadent or monograptid colony (Atavograptus ceryx and similar species). The base of the superjacent Parakidograptus acuminatus zone is characterized by the appearance of a number of biserial scandent graptolites (new genera) with a slender, elongate proximal region (initial part of the colony). The sicula in these taxa is a relatively long, slender cone, as are the first few thecae or zooidal cups. Commonly, the origin of the first zooidal cup is high enough on the conical sicula so that much of the sicula is exposed. Taxa with slender initial regions spread widely as sea level rose and platform areas flooded. Most genera and species are new. Their appearances are relatively slow, taking most of the time duration of three graptolite zones for a fauna that is relatively species rich to redevelop. The morphological innovation that is most striking among the reradiation faunas is that of the monograptid colony organization. At least three graptolite zones elapsed; however, that new style of colony development led to many new taxa.

Shelly Faunas

Brachiopods were numerous and widespread during the latter part of the Ordovician. Mass mortalities occurred among them in two phases in the Late Ordovician. The first phase took place at or near the Rawtheyan-Hirnantian Stage boundary. At that time, sea level was falling as glaciers expanded. Ocean surface water circulation and chemistry changed as well. Sheehan and Coorough (1990, p. 184) point out that "... of the 211 genera recorded in the early-middle Ashgill, 92 (44 percent) did not survive into the Hirnantian. A substantial portion of these genera probably died during the early-middle Ashgill rather than at the base of the Hirnantian." Of 124 brachiopod genera known from the Hirnantian Stage, only 21 originated during that stage (Sheehan and Coorough, 1990, p. 185). Approximately one-half the Hirnantian brachiopod genera become extinct by the end of the Stage; thus, the Hirnantian extinctions comprise the second phase of Late Ordovician brachiopod mass mortality. Taken together, approximately two-thirds of the early-middle Ashgill brachiopods became extinct by the end of the second phase of mass mortality.

Apparently, the shelf sea brachiopod fauna typified by members of the genus *Hirnantia* spread widely during glacial maximum conditions. Many members of the fauna may have been adapted to somewhat cooler ocean surface water temperatures than were present either before or after glacial maximum. Oxygen isotope determinations made from brachiopod shells in Sweden, which lay within the tropics in the Late Ordovician, indicate that surface water temperatures were cooler in the Hirnantian than earlier (Marshall and Middleton, 1990). Marshall and Middleton (1990) also note that the carbon isotope values from the same shells suggest "enhanced deposition of organic carbon, a process which would have decreased" the partial pressure of CO_2 in both the ocean and the atmosphere, contributing to rapid global cooling.

Sheehan and Coorough (1990) note that of 130 genera recorded from Early Silurian strata, about 40% are new and 60% range into the Silurian from the Ordovician. Cocks (1988) noted that the earliest Silurian brachiopod faunas have fewer species and fewer individuals in similar samples than Hirnantian Stage faunas, based on a precise review of Ordovician-Silurian boundary interval brachiopods. Sheehan (1982) pointed out that the Early Silurian brachiopod communities changed in species composition relatively rapidly. The time interval of the most rapid change in community organization was during sea-level rise and surface ocean water warming. After the transgression of seas across formerly emergent platforms had stabilized and surface waters warmed, brachiopod community organization stabilized as well.

Late Ordovician trilobite mass mortality took place during and at the Rawtheyan-Hirnantian Stage boundary (Briggs *et al.*, 1988). Hirnantian trilobites lived on the outer parts of marine shelves and were highly provincial (Lesperance, 1988). Shelf sea-dwelling trilobites generally had high rates of generic extinction as waters cooled with the onset of glaciation (Fortey, 1989). Survivors are those trilobites that either lived among reefs or "reef-like calcareous habitats, or were commoning inshore clastics around Ordovician Gondwana" (Fortey, 1989, p. 106). Lesperance (1988, p. 359) noted that Early Silurian trilobites appear to be holdovers from or survivors of the mass mortality at the end of the Rawtheyan with a few new taxa. Early Silurian originations were rare, with the result that Early Silurian trilobite faunas include primarily relatively long-ranging taxa with broad environmental tolerances. Trilobites adapted to particular habitats disappeared in the Late Ordovician mass mortality, and no taxa adapted to discrete environments appeared until the latter part of the Early Silurian.

Barnes and Bergstrom (1988) summarized conodont faunas in the Ordovician-Silurian boundary interval, noting striking differences between Ordovician and Silurian faunas. The position of most intense conodont faunal turnover seems to be within the *Glytograptus persculptus* zone, at a position about coeval with the beginnings of graptolite reradiation. The stratigraphic position coincides with the onset of rising sea level coincident with the commencement of deglaciation.

Late Ordovician conodont mass mortality "was not a sudden catastrophic event although only a few species survived into the Silurian; rather, during the Ashgill there was a gradual disappearance involving many characteristic and long-established stocks and the new taxa that appeared were considerable fewer than those that became extinct" (Barnes and Bergstrom, 1988, p. 334). Clearly, origination rates of conodont species were reduced markedly during the Late Ordovician and earliest Silurian. The extinction rate seemingly increased during the Hirnantian Stage and the early part of the *G. persculptus* zone (Barnes and Bergstrom, 1988, Figures 4 and 5). Demonstrably, conodonts survived the environmental changes coincident with glaciation that had so profound an impact on extinction and origination rates among other organisms. Interestingly, the environmental changes related to sea-level rises and global warming had the greatest influence on conodont mass mortality and the origination of new taxa.

Late Ordovician chitinozoans are primarily taxa with long stratigraphic ranges. Most chitinozoans in Ordovician strata disappear near the end of the Rawtheyan (Grahn, 1988). Chitinozoans are rare in the Hirnantian Stage-*G. persculptus* zone interval. New chitinozoan taxa appear in the Early Silurian, with reradiation commencing near the base of the Silurian (Grahn, 1988).

ENVIRONMENTAL-ORGANISMAL CHANGES: A SUMMARY

Late Ordovician glaciation involved several significant environmental changes, including the following: a lowering and then a rising sea level; marked deep and midocean ventilation; advection upward of waters potentially toxic to many organisms into the lower part of the ocean mixed layer; the cooling and then warming of ocean surface water temperatures; and changes in nutrient availability in many nearshore marine environments. The most prominent organismal (Table 2.3) mass mortality in the Late Ordovician took place at or close to the Rawtheyan-Hirnantian Stage boundary. Many graptolites, brachiopods, trilobites, corals, and chitinozoans became extinct or were markedly reduced in numbers and taxonomic diversity at that time. Isotopic studies of brachiopod shells from Sweden (Middleton *et al.*, 1988; Marshall and Middleton, 1990) using ^{12}C and ^{13}C indicate that a significant sequestering of ^{12}C in sediment took place at about the Rawtheyan-Hirnantian Stage boundary. Hirnantian brachiopod shells are enriched in ^{13}C. Wang Xiaofeng and Chai Zhifang (1989) described ^{13}C enrichment in the Hirnantian in their study of dark shales in the Ordovician-Silurian boundary interval in south China. Faunal and geochemical studies appear to be consistent in indicating a marked biomass change near the Rawtheyan-Hirnantian boundary. Slowed rates of origination characterized most organismal stocks during the Hirnantian Stage and *Glytograptus persculptus* zone. Extinction rates slowed in post-Rawtheyan-Hirnantian Stage boundary interval time, but slow origination rates during the Hirnantian into earliest Silurian resulted in marked faunal changes between the Late Ordovician and Early Silurian. Recovery and reradiation were slow until sea level rose significantly such that many shelf sea habitats not only reopened but became stable during the early part of the Silurian. Much of the Hirnantian and subsequent Early Silurian Rhuddanian was typified by environmental instabilities resulting from glaciation followed by relatively rapid global warming. Each major organismal stock responded to these environmental changes somewhat specifically, depending on its tolerances for the environmental changes.

TABLE 2.3 Significant Physical Environmental Changes, Organismal Mass Extinctions, and Clade Turnovers in the Latest Ordovician

Stage	Zone	Event
Rhuddanian	*acuminatus*	—
Hirnantian	*persculptus*	Conodont turnover Graptolite reradiation Sea-level rise Onset of glacial melting Brachiopod turnover
	extraordinarius	Glacial maximum
Rawtheyan	*pacificus*	Trilobite mass extinction Brachiopod major extinction
	complexus	

NOTE: Prominent depletions in ^{13}C (Wang Xiaofeng and Chai Zhifang, 1989; Marshall and Middleton, 1990) have been noted in samples from near the Rawtheyan-Hirnantian boundary and close to the base of the persculptus zone. Significant numbers of brachiopod extinctions may have taken place throughout the late Rawtheyan to early Rhuddanian, although the majority seemingly occurred at the levels indicated. Prominent trilobite mass mortalities appear to have taken place near the end of the Hirnantian Stage in the tropics and at the Rawtheyan-Hirnantian boundary outside the tropics. Conodont mass mortality or faunal turnover occurred at about the same time the graptolites commenced reradiation during the time of the persculptus zone. The prominent graptolite mass mortality took place close to the end of the Rawtheyan, as did that of the chitinozoa. Mass mortality among corals occurred in the tropics near the Rawtheyan-Hirnantian boundary as sea levels dropped significantly. The patterns in mass mortality and reradiation differ from organism to organism, depending on their mode of life and tolerance to change in the physical environment. As Wilde and Berry (1984) proposed, significant faunal changes took place near both the beginning and the end of glaciation. Organisms responded to major changes in ocean circulation and thermohaline density stratification at those times.

REFERENCES

Barnes, C. R., and S. M. Bergstrom (1988). Conodont biostratigraphy of the uppermost Ordovician and lowermost Silurian, in *A Global Analysis of the Ordovician-Silurian Boundary*, L. R. M. Cocks and R. B. Rickards, eds., British Museum (Natural History) Bulletin 43 (Geology Series), pp. 325-343.

Barnes, C. R., and S. H. Williams, eds. (1991). *Advances in Ordovician Geology*, Geological Survey of Canada Paper 90-9, 336 pp.

Berner, R. A. (1981). A new geochemical classification of sedimentary environments, *Journal of Sedimentary Petrology 51*, 359-365.

Berry, W. B. N., and A. J. Boucot (1973). Glacio-eustatic control of Late Ordovician-Early Silurian platform sedimentation and faunal change, *Geological Society of America Bulletin 84*, 275-284.

Berry, W. B. N., P. Wilde, and M. S. Quinby-Hunt (1987). The oceanic non-sulfidic oxygen minimum zone: A habitat for graptolites? *Geological Society of Denmark Bulletin 35*, 103-114.

Berry, W. B. N., P. Wilde, and M. S. Quinby-Hunt (1990). Late Ordovician mass mortality and subsequent Early Silurian reradiation, in *Extinction Events in Earth History*, E. G. Kauffman and O.H. Walliser, eds., Springer-Verlag, Berlin, pp. 115-123.

Beuf, S., B. Biju-Duval, O. de Chapperal, R. Rognon, O. Gariel, and A. Bennacef (1971). Les Gres du Paleozoique inferieur au Sahara—Sedimentation et discontinuities, evolution structurale d'un Craton, *Institut Francais Petrole—Science et Technique du Petrol 18*, 464 pp.

Brenchley, P. J. (1988). Environmental changes close to the Ordovician-Silurian boundary, in *A Global Analysis of the Ordovician-Silurian Boundary*, L. R. M. Cocks and R. B. Rickards, eds., British Museum (Natural History) Bulletin 43 (Geology Series), pp. 377-385.

Brenchley, P. J., and G. Newall (1984). Late Ordovician environmental changes and their effect on faunas, in *Aspects of the Ordovician System*, D. L. Bruton, ed., Palaeontological Contributions from the University of Oslo No. 295, pp. 65-79.

Brenchley, P. J., M. Romano, T. P. Young, and P. Storch (1991). Hirnantian glacio-marine diamictites—Evidence for the spread of glaciation and its effects on upper Ordovician faunas, in *Advances in Ordovician Geology*, C. R. Barnes and S. H. Williams, eds., Geological Survey of Canada Paper 90-9, 325-336.

Briggs, D. E. G., R. A. Fortey, and E. N. K. Clarkson (1988). Extinction and the fossil record of the arthropods, in *Extinction and Survival in the Fossil Record*, G. P. Larwood, ed., Clarendon Press, Oxford, pp. 171-209.

Cocks, L. R. M. (1988). Brachiopods across the Ordovician-Silurian boundary, in *A Global Analysis of the Ordovician-Silurian Boundary*, L. R. M. Cocks and R. B. Rickards, eds., British Museum (Natural History) Bulletin 43 (Geology Series), pp. 311-315.

Cocks, L. R. M., and R. B. Rickards, eds. (1988). *A Global Analysis of the Ordovician-Silurian Boundary*, British Museum (Natural History) Bulletin 43 (Geology Series), 394 pp.

Fortey, R. A. (1989). There are extinctions and extinctions: Examples from the lower Palaeozoic, *Philosophical Transactions of the Royal Society of London B325*, 327-355.

Gallardo, A. (1963). Notas sobre la densidad de la fauna bentonica en el sublittoral del norte de Chile, *Gayana Zoologica 8*, 3-15.

Grahn, Y. (1988). Chitinozoan stratigraphy in the Ashgill and Llandovery, in *A Global Analysis of the Ordovician-Silurian Boundary*, L. R. M. Cocks and R. B. Rickards, eds., British Museum (Natural History) Bulletin 43 (Geology Series), pp. 317-323.

Kastner, M. (1983). Origin of dolomite and its spatial and chronological distribution—A new insight, *American Association of Petroleum Geologists Bulletin 67*, 2156.

Koren, T. N. (1991). Evolutionary crisis of the Ashgill graptolites, in *Advances in Ordovician Geology*, C. R. Barnes and S. H. Williams, eds., Geological Survey of Canada Paper 90-9, 157-164.

Lesperance, P. J. (1988). Trilobites, in *A Global Analysis of the Ordovician-Silurian Boundary*, L. R. M. Cocks and R. B. Rickards, eds., British Museum (Natural History) Bulletin 43 (Geology Series), pp. 359-376.

Libes, S. M. (1992). *An Introduction to Marine Biogeochemistry*, John Wiley & Sons, Inc., New York, 733 pp.

Marshall, J. D., and P. D. Middleton (1990). Changes in marine isotopic composition in the Late Ordovician glaciation, *Journal of the Geological Society London 147*, 1-4.

Melchin, M. J., and C. E. Mitchell (1988). Late Ordovician mass extinction among the Graptoloidea, in *Abstracts of the Fifth International Symposium on the Ordovician System*, H. S. Williams and C. R. Barnes, eds., St. John's, Newfoundland, p. 58.

Middleton, P., J. D. Marshall, and P. J. Brenchley (1988). Isotopic evidence for oceanographic changes associated with the Late Ordovician glaciation, in *Abstracts of the Fifth International Symposium on the Ordovician System*, H. S. Williams and C. R. Barnes, eds., St. John's, Newfoundland, p. 59.

Quinby-Hunt, M. S., P. Wilde, W. B. N. Berry, and C. J. Orth (1988). The redox-related facies of black shales, *Geological Society of America Abstracts with Programs 20*, A193.

Quinby-Hunt, M. S., P. Wilde, C. J. Orth, and W. B. N. Berry (1989). Elemental geochemistry of black shales—Statistical comparison of low-calcic shales with other shales, in *Metalliferous Black Shales and Related Ore Deposits—Program and Abstracts*, R. I. Grauch and J. S. Leventhal, eds., U.S. Geological Survey Circular 1037, pp. 8-15.

Quinby-Hunt, M. S., W. B. N. Berry, and P. Wilde (1990). Chemo-facies in low-calcic black shales, *Abstracts of Papers, 13th International Sedimentological Congress*, 444.

Rickards, R. B. (1988). Graptolite faunas at the base of the Silurian, in *A Global Analysis of the Ordovician-Silurian Boundary*, L. R. M. Cocks and R. B. Rickards, eds., British Museum (Natural History) Bulletin 43 (Geology Series), pp. 345-349.

Rickards, R. B., J. E. Hutt, and W. B. N. Berry (1977). *Evolution of the Silurian and Devonian Graptoloids*, British Museum (Natural History) Bulletin 28 (Geology Series), 120 pp.

Rognon, P., B. Biju-Duval, and O. de Charpal (1972). Modeles glaciaires dans l'Ordovicien superieur saharien: Phases d'erosion et glacio-tectonique sur la bordure nord des Eglab, *Revue Geographie Physical Geologie Dynamique 14*, 507-527.

Sheehan, P. M. (1973). The relation of Late Ordovician glaciation to the Ordovician-Silurian changeover in North American brachiopod faunas, *Lethaia 6*, 147-154.

Sheehan, P. M. (1982). Brachiopod macroevolution at the Ordovician-Silurian boundary, in *Third North American Paleontological Convention Proceedings 2*, B. Mamet and M. Copeland, eds., pp. 477-481.

Sheehan, P. M., and P. J. Coorough (1990). Brachiopod zoogeography across an Ordovician-Silurian extinction event, in *Palaeozoic Palaeogeography and Biogeography*, W. S. McKerrow and C. R. Scotese, eds., The Geological Society London Memoir 12, pp. 181-187.

Vaslet, D. (1990). Upper Ordovician glacial deposits in Saudi Arabia, *Episodes 13*, 147-161.

Wang Kun, B. D. E. Chatterton, C. J. Orth, M. Attrep, Jr., and Jijin Li (1990). Geochemical analyses through the Ordovician/Silurian mass extinction boundary, Anhui Province, South China, *Geological Society of America Abstracts with Program 11*, 87.

Wang Xiaofeng, and Chai Zhifang (1989). Terminal Ordovician mass extinction and discovery of iridium anomaly—An example from the Ordovician-Silurian boundary section, eastern Yangtze Gorges area, China, *Progress of Geosciences of China 1985-1988, Vol. III*, Geological Publishing House, Beijing, pp. 11-16.

Wilde, P., and W. B. N. Berry (1984). Destabilization of the ocean density structure and its significance to marine "extinction" events, *Palaeogeography, Palaeoclimatology, Palaeoecology 48*, 143-162.

Wilde, P., W. B. N. Berry, M. S. Quinby-Hunt, C. J. Orth, L. R. Quintana, and J. S. Gilmore (1986). Iridium abundances across the Ordovician-Silurian stratotype, *Science 233*, 339-341.

Wilde, P., M. S. Quinby-Hunt, and W. B. N. Berry (1990). Vertical advection from oxic or anoxic water from the main pycnocline as a cause of rapid extinction or rapid radiation, in *Extinction Events in Earth History*, E. G. Kauffman and O. H. Walliser, eds., Springer-Verlag, Berlin, pp. 85-98.

Global Change Leading to Biodiversity Crisis in a Greenhouse World: The Cenomanian-Turonian (Cretaceous) Mass Extinction

EARLE G. KAUFFMAN
University of Colorado

ABSTRACT

The Cenomanian-Turonian (C-T) mass extinction occurred during a peak global greenhouse interval, with eustatic sea level elevated nearly 300 m above present stand; atmospheric CO_2 at least four times present levels; and global warm, more equable climates reflecting low thermal gradients from pole to equator, and from the top to the bottom of world oceans. Despite development of an oceanic anoxic event (OAE II), marine diversity was at a Cretaceous high just prior to the extinction interval; many lineages had evolved narrow adaptive ranges over millions of years under greenhouse conditions. Marine biotas were thus extinction prone. Tropical reef ecosystems experienced widespread extinction beginning near the early-middle Cenomanian boundary; major lower Cretaceous lineages of reef-building rudistid bivalves were largely extinct by middle-late Cenomanian time. Within 520,000 yr of the C-T boundary, nontropical late Cenomanian biotas experienced 45-75% species extinction, depending on the group, through a series of discrete, ecologically graded, short-term events, or steps, beginning with subtropical and warm temperate stenotopic biotas, and terminating with more broadly adapted cool temperate biotas. These extinction events were closely linked or coeval with abrupt, large-scale perturbations in the ocean-climate system, as evidenced by major fluctuations in trace elements (including Ir), stable isotopes, and organic carbon values; the rate and magnitude of these chemical and thermal perturbations progressively exceeded the adaptive ranges of various components of the marine ecosystem as the effects of late Cenomanian environmental perturbations became compounded through time. Two possible catalysts for these abrupt environmental changes are (1) expansion of the oceanic oxygen minima zone(s) to intersect both the deep ocean floor, and deeper continental shelf and epicontinental sea habitats, initiating trace element advection and chemical stirring of the oceans; and (2) oceanic impacts of meteorites and/or comets as part of the Cenomanian impact shower. Evidence is presented for both hypotheses, and a multicausal explanation for C-T mass extinction is probable. Development of an integrated real-time scale for the Cenomanian-Turonian extinction interval, blending new ^{40}Ar-^{39}Ar ages from volcanic ashes (bentonites), with 100,000- and 41,000-yr Milankovitch climate cycle deposits across the boundary, allows a precise timetable for environmental perturbation and C-T mass extinction to be developed at a resolution comparable to Quaternary studies of global change. The Cenomanian-Turonian mass extinction may thus serve as a model for the rates, patterns, causes, and consequences of a global biodiversity crisis, leading to mass extinction, in a greenhouse world.

INTRODUCTION

The modern Earth is undergoing a geologically "instantaneous" transformation, or global change, characterized by increasing concentrations of atmospheric greenhouse gases, ozone depletion, global warming, environmental deterioration, habitat destruction, and the mass extinction of species, resulting in a global biodiversity crisis (Wilson, 1988). These extraordinarily rapid global perturbations are related largely to the overpopulation of a single species, *Homo sapiens*, whose numbers may already exceed the resource-carrying capacity of the Earth. Of the estimated 30 million species of plants and animals (Wilson, 1988) that probably existed on Earth prior to man's recent population explosion, more than half live within complex, easily perturbed, tropical ecosystems (e.g., rain forests and reefs) that are currently threatened by human activity. The predicted destruction of these ecosystems may result in the loss of more than half of global biodiversity. We have thus entered an early phase of a global mass extinction, but at a rate that exceeds that for nearly all well-documented ancient mass extinction events. It is imperative that we develop integrated physical, chemical, and biological data that will help us understand the processes and consequences of global change and biodiversity decline, not just from the familiar but geologically atypical icehouse world of the Quaternary, but also from past greenhouse worlds lacking permanent polar ice and cold climates. Greenhouse worlds are characterized by higher sea level; warmer, more equable, maritime-influenced climates; expanded tropics; and a largely stenotopic global biota delicately perched on the verge of extinction. This is a world that we may soon be entering through accelerated global warming.

The focus of global change research on Quaternary history is built on three main premises: (1) there is an urgent need to understand the natural evolution of Quaternary Earth systems as a baseline for assessing the staggering impact of the human species on global environments and ecosystems during the past 9000 to 15,000 yr, but especially during the past 3000 yr (the agrarian and industrial "revolutions"); (2) the physical, chemical, and biological processes characterizing Quaternary Earth history can largely be observed, interpreted, and modeled from modern observations—the Uniformitarian approach; and (3) the preservation and resolution of Quaternary physical, chemical, and biological data relevant to understanding the dynamics of global change are unparalleled in the geologic record. These are justifiable approaches to an urgent problem. Yet, when considering the current rate of global warming associated with modern environmental changes and the expanding biodiversity crisis, it is necessary to refocus some of our global change research to understanding the greenhouse intervals that characterized so much of Earth history and to study in detail not only the processes, but also the ecological and genetic consequences, of global mass extinction. Research focused on ancient global change characteristically depends on achieving a resolution among physical, geochemical, and paleobiological data that is adequate for interpretation and predictive modeling of modern global change phenomena.

In searching for a geological test case for the study of ancient global change with relevance to the modern Earth, the Cretaceous Period emerges as one of the best candidates. In particular, the middle Cretaceous presents a unique opportunity to document and model dynamic changes in ocean-climate systems associated with a global mass extinction (Cenomanian-Turonian boundary bioevent) in a greenhouse world, and then to compare these with the environmental and ecological crisis on the modern Earth as it potentially moves from an icehouse to a greenhouse state. This chapter demonstrates that (1) resolution of middle Cretaceous physical, chemical, and biological data are sufficient to make comparisons with major trends in Quaternary ocean-climate systems, and biodiversity decline; (2) systems of chronology and regional correlation of environmental and biological changes in the Cretaceous are comparable to those used in Quaternary global change studies; (3) significant similarity in patterns of environmental change and biological response exists between Cretaceous and Quaternary extinction intervals to allow construction of predictive models from Cretaceous observations that are applicable to our understanding of the potential long-term consequences of the modern biodiversity crisis.

HIGH-RESOLUTION APPROACH TO DOCUMENTING ANCIENT ENVIRONMENTAL CHANGE

Quaternary analyses of global environmental changes, and biological responses to them, are conducted at scales of years (varves) to tens of thousands of years with nearly continuously represented data (Figure 3.1), commonly sampled at the 1- to 10-cm stratigraphic scale. Excellent preservation of original sedimentological, geochemical, and paleobiological materials allows integrated environmental and ecological analyses through small increments of time. These analyses reveal the dynamics and mechanisms of global change at several scales (e.g., Thompson, 1991; Webb, 1991). However, claims that older stratigraphic data sets are not completely enough preserved or highly enough resolved to contribute significantly to global change research and predictive modeling, are rejected. High-resolution event stratigraphic methodology (Kauffman, 1988a; Kauffman et al., 1991) provides interdisciplinary data, at 100- to >1000-yr Quaternary scales, for Phanerozoic strata.

Rock accumulation rates (RARs) for marine strata, which may preserve the most continuous and diverse record of Phanerozoic environmental changes, range from <1 cm/1000 yr (basinal fine-grained facies) to >1 m/1000 yr (e.g., in coarse-grained turbiditic, slope fan, shoreface, foreshore, and estuarine channel facies). Whereas more rapidly deposited strata allow finer stratigraphic time divisions to be sampled easily, these facies commonly reflect episodic high-energy sedimentation events separated by erosive intervals and do not preserve a long, continuous record of environmental change. More slowly but more continuously deposited basinal marine and lacustrine sequences characterized by shales, mudstones, and biogenic pelagic or hemipelagic facies provide the best Phanerozoic record of global change at scales comparable to long-term Quaternary records.

Such data do exist and have been gathered largely through the application of methods inherent in high-resolution event stratigraphy (HIRES: Kauffman, 1988a; Kauffman et al., 1991). HIRES focuses the analysis of stratigraphic sections on the centimeter-scale in search of event and cyclic stratification (see papers in Einsele et al., 1991). These events are expressed as physically unique surfaces or thin intervals; as short-term geochemical excursions from background values; as short-term evolutionary and ecological phenomena; and as depositional cycle and hemicycle boundaries, all with regional to interregional extent. Thus, stratigraphic deviations from background patterns are emphasized, and data from various disciplines are integrated into holisitic interpretations of these depositional events. Initially, these data comprise a working chronostratigraphy for regional correlations at very high levels of resolution (days to hundreds of years per event surface or thin stratigraphic interval, typically spaced at intervals hundreds to thousands of years apart; Kauffman, 1988a). The correlation potential for HIRES exceeds that of the best biozonation, geochronology, or magnetostratigraphy (Kauffman et al., 1991).

Ultimately, a diverse, high-resolution physical, chemical, biological, and cyclostratigraphic data base collected continuously over a significant interval of Phanerozoic time will enhance integrated analysis of dynamic changes in regional to global environments, and biological responses

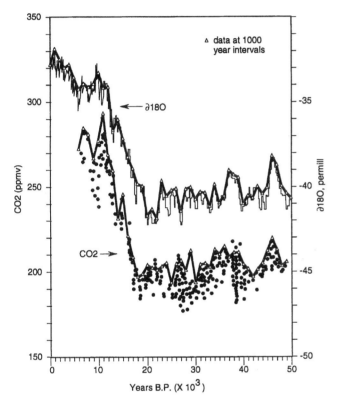

FIGURE 3.1 A comparison between the resolution of Quaternary data and time scales used in global change studies, and that possible from high-resolution Cretaceous paleoenvironmental studies in fine-grained marine facies. Quaternary data taken from Oeschger and Arquit (1991): Lower field of small black circles represent CO_2 values obtained from centimeter-scale (100- to 200-yr) intervals in ice cores at Byrd Station, Antarctica; the thin line graph above it connects $\delta^{18}O$ values from the same core and sample set. The triangles connected by a heavy dashed or solid line and superimposed on the Quaternary data set represent a typical high-resolution, centimeter-scale (1000-yr) sample interval for middle Cretaceous global change and mass extinction research, applied to the same data set. Note that all major climate trends shown by high-resolution Quaternary data are also shown by the data set representing Cretaceous sample intervals, and only the smallest fluctuations are lost between the 100-yr and 1000-yr scales.

to them, at Quaternary time scales of hundreds to tens of thousands of years. As examples, Figure 3.1 shows typical analyses of Quaternary climate change, as measured by stable isotopes and CO_2 data depicting temperature changes over the past 50,000 yr; the values for 1000-yr intervals (the typical duration of 1-2 cm of shale or limestone in Phanerozoic basinal facies) are highlighted by triangles and connected by a bold line. The thousand-year Phanerozoic level of analysis effectively replicates the significant patterns of global change in the more finely analyzed Quaternary data. Further, interregional correlation of Quaternary global change data relies heavily on the isotopic record of glaciation and sea-level fluctuations, as regulated through Milankovitch climate cyclicity. Milankovitch cyclostratigraphy is also widely recognized in ancient sedimentary sequences representing nonglacial greenhouse intervals (e.g., for the Cretaceous; Barron et al., 1985; Fischer et al., 1985; Kauffman, 1988a; Glancy et al., 1993), and extensively used for regional correlation as a part of HIRES chronostratigraphy (Hattin, 1971, 1985; Elder, 1985; Kauffman, 1988a; Kauffman et al., 1991). Quaternary and older rocks can thus be correlated at the same scales. Further, event chronostratigraphic resolution in the older stratigraphic record commonly exceeds that possible from Quaternary climate cyclostratigraphy. Whereas Quaternary scientists have the advantage of finer time scales of observation, students of older geological intervals have the advantage of longer continuous records of global change. Both groups have much to contribute to the documentation, interpretation, and prediction of global change phenomena and their short- and long-term effects on Earth ecosystems. An example of high-resolution stratigraphic analysis of an interval of significant past global change and biodiversity crisis is drawn from the Cretaceous record of North America, the Cenomanian-Turonian mass extinction interval (92 to 94 million years ago (Ma)).

THE CENOMANIAN-TURONIAN (C-T) MASS EXTINCTION—AN ANCIENT GLOBAL BIODIVERSITY CRISIS IN A CHAOTIC GREENHOUSE WORLD

An abrupt change in the global marine biota at the Cenomanian-Turonian boundary was first noted by d'Orbigny (1842-1851), who named these stages (Hancock, 1977). Raup and Sepkoski (1984, 1986) and Sepkoski (1990, 1993) statistically determined that the Cenomanian-Turonian boundary interval was a secondary mass extinction on the basis of both familial and generic-level data. They recorded a 5% loss of families and a 15% loss of genera (per-genus extinction rate 0.026), utilizing stage-level data with an average duration of 6 million years (m.y.); this broad resolution did not, however, allow rates and patterns of Cenomanian extinction to be determined. For example, a review of global Cenomanian ammonite extinction data (Collom, 1990) resolved to substage level showed several discrete extinction events among these molluscs within the Cenomanian.

A number of highly detailed regional studies of Cenomanian-Turonian extinction patterns, and associated environmental perturbations, have been conducted during the last three decades which shed light on the dynamic nature of this interval. Of particular interest are the new data from the Western Interior of North America (Figure 3.2), from western and central Europe, from North Africa, and from northern South America (Colombia, Venezuela, and Brazil). These studies do not support the idea of a single "catastrophic" boundary extinction among ecologically and genetically diverse taxa, but rather a 1.46-m.y. interval of ecologically graded extinction characterized by several discrete, short-term events with global expression ("steps" of Kauffman, 1988b), most of them in the last 520,000 yr of the Cenomanian. Each of these discrete extinction events is associated with geochemical evidence for environmental chaos in the ocean-atmosphere system— short-term, large-scale perturbations in temperature, ocean chemistry, and carbon and nutrient cycling in the sea, upon which are superimposed a somewhat exaggerated expression of Milankovitch climate cyclicity at 41,000- and 100,000-yr intervals (subsequent discussion). Collectively, these integrated, high-resolution (centimeter to meter scale) paleobiological, geochemical, and sedimentological studies provide a global perspective of the rates, patterns, causes, and consequences of the C-T mass extinction interval.

In the Western Interior Basin of North America (Figure 3.2), bed-by-bed stratigraphic and paleontologic studies by Cobban and Scott (1972) and later Cobban (1985, 1993) documented an abrupt change in ammonite assemblages across the C-T boundary; they noted the loss of typical late Cenomanian ammonites at five to six discrete levels within 4 m of the C-T boundary, spanning the biozones of *Sciponoceras gracile* (*Vascoceras diartanum* and *Euomphaloceras septemseriatum* subzones) and, above it, *Neocardioceras* spp. (*Eoumphaloceras navahopiensis*, *Neocardioceras juddi*, and *Nigericeras scotti* subzones, ascending). Cobban (1985, 1993) and Kennedy and Cobban (1991) defined the C-T boundary between the latest Cenomanian *Neocardioceras* biozone, *Nigericeras scotti* subzone, and the overlying *Watinoceras* biozone, *W. devonense* subzone. Kauffman (1975) and Kauffman et al. (1976, 1993) also noted a major change in Bivalvia at the ammonite-based C-T boundary of North America and Europe, with the disappearance of latest Cenomanian *Inoceramus pictus* and *I. tenuistriatus* lineages, overlain by the first common occurrence of early Turonian *Mytiloides*

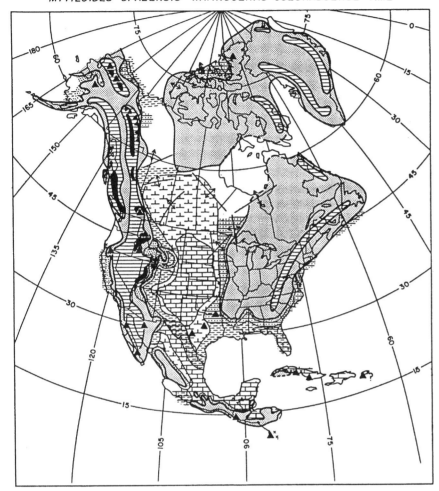

FIGURE 3.2 Paleogeographic map of the Western Interior Basin of North America during the Cenomanian-Turonian boundary interval, showing the extent of the interior epicontinental seaway (defined by bold lines enclosing sedimentary rock symbols) during maximum eustatic sea-level highstand. Note the large aerial extent of fine-grained basinal facies (limestone, marl, shale symbols), which preserve relatively complete C-T boundary sections, and from which the high-resolution sedimentologic, geochemical, and paleobiologic data were obtained for this study of Cretaceous global change associated with a mass extinction within a greenhouse world.

(*M. hattini* zone). To date, microfossils have not been successfully used to identify the ammonite-based C-T boundary. The boundary falls within the *Whiteinella archeocretacea* planktic foraminifer biozone (Caron, 1985; Eicher and Diner, 1985, 1989; Leckie, 1985) and within the upper *Lithraphidites acutum* nannoplankton biozone (Watkins, 1985; Watkins *et al.*, 1993).

Studies of the rates and patterns of marine extinction across the C-T boundary in North America began with the work of Koch (1977, 1980), who noted an overall loss of nearly 70% of known late Cenomanian molluscan species occurring in the *Sciponoceras gracile* and *Neocardioceras* biozones; the most extinction-prone molluscs were those with limited facies associations and biogeographic range. Kauffman (1984b) integrated diverse late Cenomanian molluscan data from detailed stratigraphic sections to demonstrate that the C-T extinction occurred as a series of discrete events (later called "steps"; Kauffman, 1988b). Elder (1985, 1987a,b) conducted the most detailed (10-cm scale) high-resolution stratigraphic and paleontologic analysis of the late Cenomanian macrofaunal extinctions. He recorded extinction levels as follows: 13% of 61 genera and 51% of 84 species; ammonites and inoceramid bivalves accounted for 85% of the extinction; ammonites suffered 33% extinction among 24 genera, and 74% extinction among 31 species; inoceramids suffered 92% extinction among 11 species, but no generic extinction. Harries and Kauffman (1990) and Harries (1993) extended these high-resolution studies through the early Turonian, at numerous sections across the Western Interior Basin of North America. Eicher and Diner (1985, 1989) and Leckie (1985) noted two important extinction events, or steps, among a few species of planktonic foraminifera across the C-T boundary; the late Cenomanian *Rotalipora* extinction is coincident with the

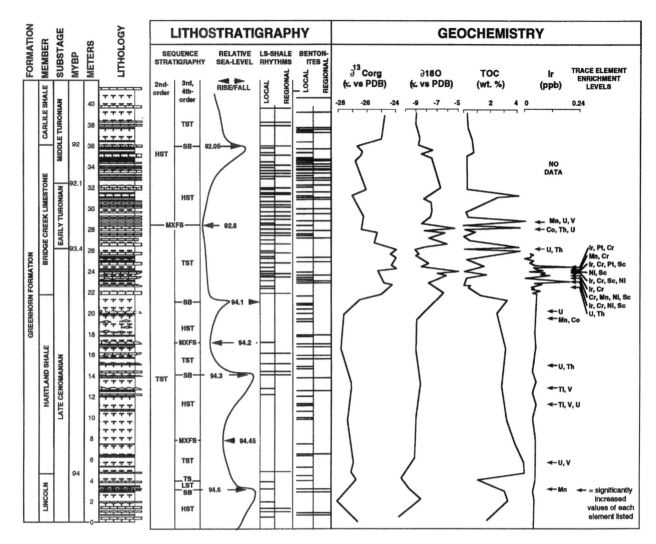

FIGURE 3.3 Compilation of high-resolution sequence stratigraphic, cyclostratigraphic, volcanic (bentonite) event-stratigraphic, and geochemical data for the Cenomanian-Turonian boundary interval at the proposed C-T boundary stratotype section (Kennedy and Cobban, 1991), along the north cuts of the Arkansas River just west of Pueblo, Colorado. Note that sea level reaches a second-order maximum flooding peak (300 m above present stand), after a series of third-order fluctuations, just above the C-T boundary. Note also the intense interval of C-T volcanism, and the rapid, large-scale geochemical perturbations of the ocean-climate system as depicted by extraordinary fluctuations in $\delta^{13}C$, $\delta^{18}O$, C_{org}, and trace element values. The broad positive global excursion of the $\delta^{13}C$ record reflects a major change in oceanic carbon cycling associated with development of oceanic anoxic event II (see Arthur et al., 1985, and references therein). Trace element enrichment levels are represented in the right-hand column by right-facing arrows and abbreviations of the primary trace elements characterizing each enrichment horizon. An Ir curve is provided to show the enrichment interval, and the two major spikes, all correlative with C-T mass extinction events (from Orth et al., 1988). The figure is plotted at the same scale as Figure 3.4, and the two may be placed together to show the entire high-resolution data set for the C-T extinction interval and adjacent background conditions.

first group of molluscan extinction events within a zone of Ir and other trace element enrichment (Figures 3.3 and 3.4) (Kauffman, 1988b, Figure 3.2), and the extinctions of the *Whiteinella* boundary fauna and the *Lithraphidites acutum* nannoplankton assemblage (Watkins, 1985; Watkins et al., 1993) occur within a lower Turonian interval that also includes macrofossil extinction step 6 (Figure 3.4) of Kauffman (1988b). These studies revealed an even more complex stepwise pattern of molluscan extinction, based on robust integrated range data, than originally noted by Kauffman (1984a); these data are now being tested for sampling biases to compensate for the Signor-Lipps (1982)

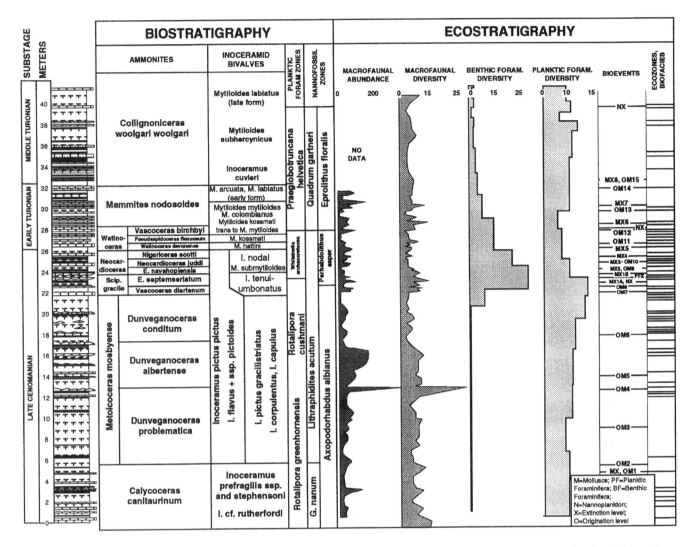

FIGURE 3.4 Compilation of ammonite, inoceramid bivalve, planktonic foraminifer, and nannofossil biozones for the C-T boundary interval (left side), as they are defined at the proposed C-T boundary stratotype just west of Pueblo, Colorado (see Figure 3.3 explanation: both figures are plotted at the same scale). To the right is a compilation of the C-T boundary ecostratigraphy at Pueblo, consisting of diversity and abundance curves, stratigraphically concentrated origination (OM-1–OM-15) and extinction events, or steps (MX, MX-1, to MX-8 for molluscs; NX for nannoplankton extinctions; PFX for the *Rotalipora* planktonic foraminifer extinction), and to the far right, a compilation of ecozone and biofacies boundaries, reflecting ecological response to abrupt changes in benthic environments. Note the focus of extinction events around the C-T boundary, mainly in the latest Cenomanian, following a long interval of origination and buildup of diversity. Note also unexplained abrupt increase in benthic foraminifer diversity, after a long interval without benthics on the seafloor, at the initiation of macrofossil and planktic foraminifer extinction.

effect, by calculating predicted ranges for taxa at 50, 75, and 95% confidence intervals, using the Koch and Morgan (1988) and Marshall (1991) statistical tests. Initial results support the stepwise extinction hypothesis (Harries, Kauffman, and Elder, in manuscript).

The global aspect of the Cenomanian-Turonian mass extinction and its stepwise pattern are demonstrated by similar high-resolution stratigraphic and paleobiologic studies in England and France, and in Colombia, South America.

Jeffries (1961, 1963) first documented the details of diverse macrofossil extinction patterns in the Plenus Marls of England; ammonite extinction patterns have more recently been studied by Kennedy (1971) and Wright and Kennedy (1981). Whereas these studies have not been directed specifically at mass extinction phenomena, they do show similarly distributed, stepwise disappearance of late Cenomanian macrofossils. Gale *et al.* (1993) have further shown that the macrofossil extinction interval occurs mainly within the early part of the $\delta^{13}C$ global excursion in the same way as it does in the Western Interior Basin of North America, and elsewhere. The most detailed study of diverse microfossil extinction patterns yet conducted for western Europe are those in Jarvis et al. (1988); these patterns can be correlated precisely to North America for cosmopolitan planktic foraminifera and nannofossils. In Colombia, Villamil and Arango (1994) have recently completed a high-resolution, integrated stratigraphic analysis for the C-T boundary interval in the Magdalena Basin region, and have shown by graphic correlation nearly identical geochemical fluctuations (stable isotopes, trace elements including Ir, organic carbon), in the same stratigraphic order as found in Europe and North America, suggesting oceanic sources for these chemical perturbations. Within this geochemical framework, two of the latest late Cenomanian molluscan extinction events have been recognized, as well as the same planktic foraminifer extinctions below (*Rotalipora* extinction) and above (*Whiteinella* extinction) the C-T boundary. Microtektite-shaped (glassy?) spheres were recently discovered at two levels in the latest Cenomanian extinction interval of Colombia by Arango and Villamil (in manuscript), one level associated with the *Rotalipora* extinction, and a higher level associated with the first major iridium spike in the sequence, about 2 m below the C-T boundary.

An apparent ecological gradation is shown by successive extinction events across the C-T mass extinction interval, suggesting varying sensitivities of different marine ecosystems to large-scale ocean-climate perturbations, and the cumulative effect of multiple perturbations through time. Johnson and Kauffman (1990) showed that extinction of tropical rudistid-dominated reefs in the Caribbean Province was initiated in the latest early and middle Cenomanian, and was largely completed by the time the extinction process began in the North Temperate Realm (e.g., western Europe and the Western Interior Basin of North America), some 520,000 yr below the C-T boundary (Kauffman, 1984a, 1988b; Elder, 1985, 1987a,b) (Figures 3.4 to 3.6). Philip (1991) observed a similar pattern of early extinction for many Cenomanian rudists in southern Europe, but with the terminal rudistid extinction apparently closer to the C-T boundary than in the Caribbean. Near or at the end of the Cenomanian reef extinctions in the Caribbean Province, marginal tropical and subtropical molluscan taxa and warm water calcareous plankton that had rapidly immigrated into the Western Interior Basin of North America along with a warm water mass from the proto-Gulf of Mexico and Caribbean basins (Kauffman, 1984b), largely became extinct 450,000 to 520,000 yr below the C-T boundary (late late Cenomanian). More temperate and cosmopolitan taxa disappeared in the latest Cenomanian and on into the basal Turonian in both North America and Europe.

The C-T mass extinction occurs in association with an extraordinary interval of rapid global environmental change, and is further unusual in that it is associated with elevated temperatures (Hay, 1988) and the highest sea-level stand of the late Phanerozoic, reaching nearly 300 m above present stand (Pitman, 1978; McDonough and Cross, 1991) (see Figure 3.2 for extent of North American interior seaway). These are conditions usually considered favorable to the diversification of life. The extinction occurs at the peak of the last great greenhouse interval, with atmospheric CO_2 at least four times present levels (Berner, 1991, 1994), and mean global sea surface temperatures 2°C (tropics) to 17°C (poles) higher than today (Hay, 1988, Figure 1). Thermal gradients were thus relatively low from pole to equator, and from surface to bottom waters of the global ocean system (Hay, 1988, Figure 1). This, in turn, promoted sluggish ocean circulation not dominantly driven by polar water masses as today, but rather by warm, oxygen-poor, hypersaline water masses (Hay, 1988, Figure 5) derived from subtropical evaporite belts (Barron,

---

FIGURE 3.5 (Opposite) Diagram showing the various methods of establishing a real-time scale for the study of the C-T mass extinction and its causes. The proposed C-T boundary stratotype section (Kennedy and Cobban, 1991) of the Bridge Creek Limestone Member, Greenhorn Formation, at Pueblo, Colorado, is shown on the left, and standard Western Interior ammonite biozones are shown in the next column. Radiometric ages (with asterisks) are from Obradovich (1993) and based on new single crystal ^{40}Ar-^{39}Ar age analyses. The calculated time scale based on assigning equal time durations to ammonite biozones between dated levels (third column) yields a duration for the Bridge Creek Limestone Member at Pueblo of 1.9 m.y. In the fourth and fifth columns (A,B), a time scale is calculated by assigning 100,000-yr intervals to prominent limestone beds (first and sixth columns) interpreted to represent the 100,000-yr Milankovitch orbital eccentricity cycle. The fourth column (A) starts with a radiometrically dated cycle in the middle of the series, and calibrates ages both up and down the column; this is the best scale, and most closely matches other radiometric ages and the Cenomanian biozone-based scale. The fifth column (B) starts at the top of the series and counts 100,000-yr intervals downward through the major Milankovitch cycles; this method is less accurate than A. By using this scale (fourth column, A), the total number of Milankovitch bedding rhythms (n = 46) is divided into the duration of the Bridge Creek Limestone Member (1.9 m.y.), yielding an average duration of 41,300 yr per bedding rhythm—the Milankovitch axial obliquity cycle. The same scale is used to document the average duration of regional event-bounded stratigraphic intervals (17,000 yr), the highest resolution of regional chronostratigraphic correlation across the C-T mass extinction interval.

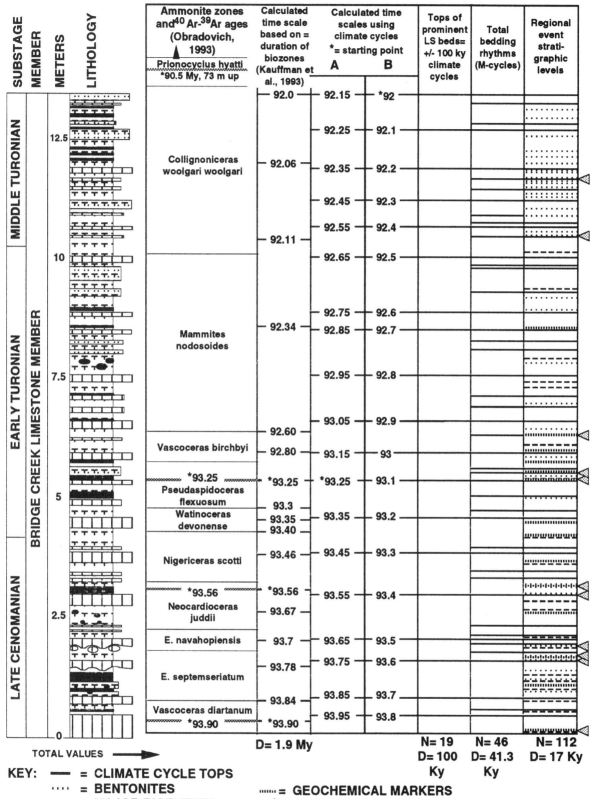

		A TIME TABLE FOR C-T EXTINCTION									
	KY	SL	VOLC	∂13C	∂18O	TOC	TR. EL.	DIV	ABUN	ORIG	EXT
RECOVERY INTERVAL	370		ACT	>	>	<			>M	OM15 OM14	MX8
	110		ACT	<	<	>		BF<			
EXTINCTION INTERVAL T / C	230			<	>	<		BF< M>	≥M	OM13	MX7 NX
	150	MFS		< >	≥ >		>Mn,U,V >Co,U,Th	PF> M>	>M		MX6
	80		ACT	<	> <			M>	<M	OM12	
	140			<	≥ >	> <	>U,Th	BF<	>M	OM11	MX5 NX
	170			>	<	<		PF<			MX4
	20			>	>	>	>Ir,Pt, Cr	BF<	<M	OM10	MX3
	20			>	< >	<	>Mn,Cr	BF>		OM9	NX MX2
	40		ACT	<	<	>	>Ir,Cr,Pt >Ni,Sc >Ir,Cr	BF> M>	>M		MX1b
	80		ACT		>	<	>Cr,Mn, Ni,Sc	BF> PF<			
	50		ACT				>Ir,Cr, Ni,Sc	BF> PF<	>M <M		PFX MX1a
PRE-EXTINCTION INTERVAL	140			>	<	<	>U,Th	PF> BF<	>M	OM8	
	220	SB	ACT	>	>	>		1st BF>		OM7	
	90		ACT				>U >Mn,Co	PF>	<M	OM6	
	90	SB	ACT				>U,Th	M>	>M		
	30									OM5	
	50						>Tl,V >Tl,V,U	M>	>M	OM4	
	120	MFS	ACT			>				OM3	
	60			>	>		>U,Th		>M	OM2	
	50			>	<	<	>Mn			OM1	MX0
	150	SB		<	>	>		M>	>M		

FIGURE 3.6 Summary diagram: a timetable for C-T mass extinction, integrating all of the high-resolution sedimentologic, geochemical, and paleobiological data (see Appendix) collected at the Pueblo, Colorado, boundary section. The data extend through the pre-extinction, mass extinction, and survival/recovery intervals. Divisions of the data into time packages, utilizing the time scale in column A of Figure 3.5, are based on unique stratigraphic combinations of physical, chemical, and biological events (from Figures 3.3 and 3.4). Right-facing arrows denote an abrupt, major increase in a value, and left-facing arrows a major decrease. Key: SL = sea level, with standard sequence stratigraphic abbreviations; VOLC = volcanic activity (ACT) or relative quiescence, as measured by the number and thickness of volcanic ashes per unit time; BF = benthic foraminifers; PF = planktonic foraminifers; M = molluscan taxa; OM = stratigraphically focused molluscan origination level; MX = focused molluscan extinction level, or step; NX = nannoplankton extinction event; PFX = planktonic foraminifer extinction level (*Rotalipora* spp.).

1993). Collectively, these conditions resulted in expansion of the oceanic oxygen minima zones, in some cases extending to the ocean floors, and the development of an oceanic anoxic event (OAE II of Schlanger and Jenkyns, 1976; Arthur et al., 1985). A globally documented positive shift in $\delta^{13}C$ values (Scholle and Arthur, 1980; Pratt, 1985; see Figure 3.3) over the 880,000-yr span of OAE II denotes a shift in global oceanic carbon cycling and increased sequestering of organic carbon in ocean basins and epicontinental seas. A strong but dynamically fluctuating increase in total organic carbon values and petroleum source rock development resulted from these paleoceanographic events. In many parts of the world, rapid, large-scale fluctuations in marine $\delta^{18}O$ also characterize this interval (e.g., Arthur et al., 1985; Pratt, 1985), denoting extraordinary temperature and/or salinity fluctuations, although the extent of these primary signals is difficult to calculate because of widespread diagenetic overprints in pelagic and hemipelagic sediments spanning the C-T boundary interval. A series of closely spaced, short-term, compositionally distinct, trace element enrichment events is associated, in North and South American, European, and North African data, with the early phases of large-scale, stable isotope and organic carbon fluctuations early in OAE II. These trace element spikes, in similar stratigraphic order and including at least two major iridium excursions (Orth et al., 1988, 1990, 1993) (see Figure 3.3), possibly represent a complex trace element advection sequence (sensu Berry et al., 1990; Wilde et al., 1990) associated with benthic touchdown of the Cenomanian OAE, and remobilization of sequestered trace elements from the seafloor. However, an extraterrestrial source for the iridium cannot be ruled out (Orth et al., 1988, 1990, 1993, and subsequent discussion). Collectively, these geochemical data suggest a major change in latest Cenomanian-earliest Turonian ocean-climate systems, from one of stability to one of great instability characterized by high-frequency, large-scale, ocean-climate perturbations acting on an extinction-prone, predominantly stenotopic global marine biota. These perturbations may have been the primary causes for the series of late Cenomanian extinction events seen worldwide, as they progressively exceeded the adaptive ranges of, first, tropical and, eventually, temperate taxa and ecosystems.

For example, using compilations (Kauffman, 1984a, 1988b; Elder, 1985, 1987a) of molluscan range data, and integrating these with stable isotope data of Arthur et al. (1985) and Pratt (1985), and trace element data of Zelt (1985) and Orth et al. (1987, 1990, 1993) for the Western Interior Basin of North America, Kauffman (1988b) noted a close correlation of the eight C-T extinction events, or steps, of Elder (1985, 1987a), and ocean-climate perturbations expressed as short-term excursions of the stable isotope and trace element record above Cenomanian background levels. In particular, the first five mass extinction steps are coincident with iridium and other noble metal trace element excursions, suggesting a cause-effect relationship. Further, the entire C-T mass extinction record in the Western Interior of North America and elsewhere occurs within the 1.2-m.y.-long stratigraphic interval of extraordinarily rapid, large-scale, fluctuations in $\delta^{13}C$, $\delta^{18}O$, C_{org}, $CaCO_3$, and trace elements. The extinction occurs against a background of exceptionally well defined 41,000- and 100,000-yr Milankovitch climate cyclicity and increasing oxygen restriction in the world's oceans (OAE II) with expansion and possibly benthic touchdown of the oceanic oxygen minimum zone(s). Yet the question remains: What initiated these dynamic changes on such a short time scale?

Whereas Orth et al. (1987, 1990, 1993) have cautiously suggested that the enriched late Cenomanian trace element suite, including Ir, was probably derived from oceanic sources (e.g., mantle outgassing during rapid Cenomanian-Turonian plate rearrangements, and/or from the last phases of outgassing of the Pacific superplume; Larson, 1991a,b), extraterrestrial sources cannot be ruled out. Whereas impacting is a stochastic process, and predictably spatially random (Grieve, 1982), it is not always temporally random. For example, Grieve (1982) has noted four large middle Cretaceous terrestrial impact craters with age error bars overlapping the Cenomanian-Turonian boundary: these are the Deep Bay Crater (100 ± 50 Ma), the Logoisk Crater (100 ± 20 Ma), the Steen River Crater (95 ± 7 Ma), and the West Hawk Crater (100 ± 50 Ma). Two additional craters have late Albian to Cenomanian ages of 100 and 100 ± 5 Ma. Alvarez and Muller (1984) and Strothers and Rampino (1990) have both noted that these craters comprise a statistical impact cluster, one of several with a proposed periodicity of 30 to 34 Ma. The late Albian-early Turonian (but predominantly Cenomanian) cluster of crater ages may therefore represent an impact storm, or shower (sensu Hut et al., 1987). Kauffman (1988b) pointed out that not only was the terrestrial impact record conservative for any temporal cluster of craters because of loss of record by subduction; weathering and erosion; the extent of younger sedimentary, vegetative, and ice cover on land; as well as the fact that many areas are still poorly explored (Grieve, 1982), but also that most extraterrestrial objects hitting Earth would fall into the sea. During greenhouse intervals (Fischer and Arthur, 1977) with elevated global sea level, as in the Cenomanian-Turonian boundary interval, up to 80% or more of the Earth was covered by water. Thus, for any single terrestrial impact recorded at these times, an additional four would be predicted to have landed in the sea, and the timespan of the impact shower would be expanded over that of the terrestrial impacting record by at least a million years. These predictions are conservative because of the loss of the shallow water and terrestrial impact record through subduction, weathering, erosion, and younger sedimentary cover. However, if they are correct, the predominantly late Albian-Cenomanian impact shower would have extended across the C-T boundary. A series of meteorites or comets impacting in the sea over a relatively short period would probably result in massive evacuation of water (as steam) and debris along the impact path, major shock waves with their resultant marine tectonic effects, compression and heating of the surrounding oceanic water mass, giant tsunamis, and ultimately rapid mixing of oceanic water masses (Melosh, 1982). Oceanic feedback from such a geologically instantaneous perturbation might produce overturn of oceanic water masses, chemical advection events, and dramatic changes in the thermal and chemical character of these water masses. These changes would be preserved in the sedimentary record as dramatic excursions in the stable isotope, trace element, and C_{org} record beyond background levels (Kauffman, 1988b). Such oceanic perturbations would be expected to result in a prolonged and complex series of feedback processes (loops) as the ocean system sought equilibrium after each impact event. Could the dramatic chemical and thermal changes in the oceans across the Cenomanian-Turonian extinction interval have been initiated by oceanic impact, as would be predicted by the known impacting record? Emerging new data support this hypothesis. Latest Cenomanian bulk sediment samples collected by Michael Rampino and the author north of Boulder, Colorado, have yielded a few shocked quartz grains, some with multiple shock lamellae (Rampino et al., 1993; M. Rampino, personal communication, August, 1993), within an interval also characterized elsewhere by trace element (including iridium) enrichment (Figure 3.3). Further, Tomas Villamil and Claudia Arango (University of Colorado) have collected numerous microtektite-shaped microspheres, possibly glass, from at least two latest Cenomanian horizons in the Villeta Group of central Colombia, one associated with the *Rotalipora* foraminifer extinction, the other with a major iridium spike and molluscan extinction event. The composition of these grains is currently being analyzed; the results will bear heavily on the hypothesis of impacting as a catalyst for dramatic C-T ocean-climate perturbations and resultant steps of mass extinction.

The high-resolution biological, geochemical, and sedimentological data base developed internationally for the C-T mass extinction interval is one of the most comprehensive data sets available for the study of an ancient biodiversity crisis. These global data provide a wealth of information that can be correlated precisely, at a high level of resolution, among different continental margins, epicontinental seas, and ocean basins through lineage and assemblage biostratigraphy, event and cycle chronostratigraphy, and on a broader scale, geochronology and sea-level history. We are now in a position to study and model the rates, patterns, causal mechanisms, and ecological and evolutionary aftermath of an ancient (C-T) mass extinction in a greenhouse world, at a very fine level of temporal resolution. This will, in turn, help us to understand the possible consequences of the modern biodiversity crisis on Earth.

A CASE HISTORY: THE PUEBLO, COLORADO, C-T BOUNDARY SECTION

In order to demonstrate the application of high-resolution stratigraphic methodology to the study of Cretaceous global change, and its relationship to a major biodiversity crisis, data are presented (Figures 3.3 to 3.5) and interpreted (Figure 3.6) for the late Cenomanian-early Turonian mass extinction interval as preserved along the north cut face of the Arkansas River where it breaches the Rock Canyon Anticline just west of Pueblo, Colorado. This locality is the standard reference section for the Greenhorn cyclothem (Scott, 1964, 1969; Cobban and Scott, 1972; papers in Pratt, 1985) and is the proposed Cenomanian-Turonian boundary stratotype section (Kennedy and Cobban, 1991). It has further been a focus for paleobiological studies of the C-T mass extinction in North America (Koch, 1977, 1980; Kauffman 1984a, 1988b; Eicher and Diner, 1985, 1989; Elder, 1985, 1987a,b; Leckie, 1985; Watkins, 1985; Watkins et al., 1993), and its evolutionary aftermath (Harries and Kauffman, 1990; Harries, 1993). The most detailed geochemical framework for the C-T boundary interval has also been developed at the Pueblo section, including stable isotope and C_{org} analyses (Arthur et al., 1985; Pratt, 1985), and trace element analyses (Zelt, 1985; Orth et al., 1988, 1990, 1993), summarized in Figures 3.3 and 3.4. Data from the Pueblo section have played a major role in the development of the highly refined Cretaceous biostratigraphy for the basin (Cobban and Scott, 1972; Kauffman et al., 1976, 1985, 1993; Cobban, 1985, 1993; Eicher and Diner, 1985; Leckie, 1985; Watkins, 1985; Caldwell et al., 1993; Watkins et al., 1993; and references therein). Pueblo is also a key section for the development of a regional system of high-resolution event chronostratigraphy (Elder and Kirkland, 1985; Kauffman, 1988a; Kauffman et al., 1991) and cyclostratigraphy (Hattin, 1971, 1985; Barron et al., 1985; Elder, 1985; Fischer et al., 1985; Kauffman, 1988a) for the C-T boundary interval in the Western Interior Basin of North America. Data resolution from this Cretaceous section compares favorably with Quaternary data for the documentation and interpretation of ancient global change as it relates to a major biodiversity crisis.

Geologically, the Pueblo C-T boundary section lies in the eastern portion of the Axial Basin (Cretaceous tectonostratigraphic zones of Kauffman, 1984b, 1988a) within the Western Interior foreland basin (Figure 3.2). The C-T boundary interval was coeval with active volcanism, thrusting and subsidence in the western part of the foreland basin (Kauffman, 1984b, 1985; Kauffman and Caldwell, 1993), broadly reflecting active middle Cretaceous plate rearrangement, late phases of development of the Pacific superplume (Larson, 1991a,b), active migration of the Caribbean Plate into the proto-Caribbean Basin (Pindell and Barrett, 1990), and numerous large-scale eruptive volcanic centers in both northwestern (Idaho, Montana) and southwestern (Arizona, New Mexico, Texas) localities. These explosive volcanic episodes produced thick, regionally distributed, marine ash or bentonite beds (Kauffman, 1985, Fig. 4). None of these volcanic ashes are associated with iridium enrichment or extinction events (Orth et al., 1987, 1990, 1993).

At Pueblo, the C-T boundary section spans the uppermost few meters of the Hartland Shale Member, and the entire Bridge Creek Limestone Member of the Greenhorn Formation (Figure 3.3) (Scott, 1964, 1969; Cobban and Scott, 1972; Elder, 1985, 1987; Elder and Kirkland, 1985; Sageman, 1985, 1992; Harries and Kauffman, 1990; Harries, 1993). This marks the latest transgressive systems tract, the maximum flooding interval (eustatic sea level 300 m above present stand), and the earliest highstand systems tract of the Greenhorn second-order sequence, or cycle (Kauffman, 1984b, 1985; Kauffman and Caldwell, 1993). The interval further shows evidence for three third-order relative sea-level fluctuations (Figure 3.3) with regional expression. Paleobathymetry at the site during sea-level highstand has been estimated between 150-300 m and >500 m (Kauffman, 1984b, 1985; and Eicher, 1969, respectively), depending on the evidence used; foraminifers give a deeper signal than do molluscs and sedimentary features.

The Western Interior epicontinental sea was episodically stratified during C-T time, with the upper portion of an expanded oceanic oxygen minimum zone (OMZ) associated with OAE II forming low-oxygen bottom waters stratified beneath a shallower, warm-temperate to subtropical oxygenated water mass; both water masses immigrated into the Western Interior Seaway from the proto-Gulf of Mexico-Caribbean region (Kauffman, 1984b, 1985; Kauffman and Caldwell, 1993), displacing cooler northern water masses poleward. The low-oxygen benthic water mass (expanded OMZ) was neither stable nor stagnant, as indicated by dynamic changes in continuously represented benthic paleocommunities (Kauffman and Sageman, 1990; Sageman et al., 1991, 1994), as well as the frequency of short-term, large-scale geochemical fluctuations (Arthur et al., 1985; Pratt, 1985; Zelt, 1985; Orth et al., 1988, 1990, 1993). Well-defined 41,000- and 100,000-yr Milankovitch climate cycles are represented by shale-limestone, shale-calcarenite, and marl-chalk bedding rhythms throughout the study interval (Arthur et al., 1985; Barron et al., 1985; Fischer et al., 1985; Pratt, 1985; Kauffman, 1988a; Figure 3.5 herein). The shales and marls represent wetter and possibly somewhat cooler climatic phases with high freshwater runoff and clay delivery to the interior basin; the limestones and calcarenites represent drier and possibly warmer climate phases, more normal marine surface conditions, and a dominance of biogenic, hemipelagic sedimentation with great reduction in clay delivery to the basin. The Milankovitch climate cycles represent a predictable aspect of Cretaceous global change, but are especially enhanced in lithological, geochemical, and biological expression during the middle Cretaceous (Barron et al., 1985; Kauffman, 1988a).

Both planktic and benthic biotas associated with the Cenomanian-Turonian boundary interval at Pueblo represent warm, normal to near-normal marine environments (excepting intervals of benthic oxygen restriction). These biotas paleobiogeographically belong to the warm-temperate to subtropical Southern Interior Subprovince and its northern biogeographic ecotone with the Central Interior Subprovince—the principal endemic center for molluscan evolution in the Western Interior Basin (Kauffman, 1984b).

Against this dynamic environmental backdrop of global sea-level fluctuations, frequent changes in stratification, temperature and water chemistry, active regional tectonism, basin subsidence and volcanism, and more predictable changes in climate related to Milankovitch cyclicity, the global Cenomanian biota became progressively more stressed and underwent accelerated extinction as a series of ecologically graded (tropical to temperate) steps or events during the last million years of the Cenomanian (Kauffman, 1984a, 1988b; Elder, 1985, 1987a,b; Johnson and Kauffman, 1990; Figures 3.4 and 3.6 herein). The final 520,000 yr of this extinction interval are especially well preserved in temperate facies of the Western Interior Cretaceous Basin. Each extinction event was tied to a major short-term shift in ocean temperature and/or chemistry, indicated by the geochemical profiles at Pueblo (Figures 3.3 and 3.6). Each event represents a breached biological threshold for a particular suite of ecosystems, starting with tropical Caribbean Province reef-associated communities (Johnson and Kauffman, 1990). The high-resolution data base on global environmental change, as exemplified by the Pueblo, Colorado section, provides important insights into the rates, patterns, causes, and consequences of the global C-T biotic crisis at scales comparable to those used for long-term sensing of Quaternary global change. This, in turn, allows critical comparisons between Quaternary icehouse and Cretaceous greenhouse worlds.

ESTABLISHING A CHRONOLOGY FOR ENVIRONMENTAL DECLINE AND MASS EXTINCTION ACROSS THE C-T BOUNDARY

To compare the rates, patterns, and magnitude of environmental changes and biological stress leading to global late Cenomanian-early Turonian mass extinction, with the Quaternary record of global change, a highly detailed real-time scale must be established for the C-T interval. Integration of this time scale with physical, chemical, and biological event surfaces/beds, and with short-term cyclostratigraphic units (e.g., Milankovitch climate cycle deposits) of regional extent, has resulted in the construction of a high-resolution chronostratigraphic system (Kauffman, 1988a; Kauffman et al., 1991), which allows precise regional correlation and interpretation, individually or collectively, of C-T boundary mass extinction events. In the C-T boundary chronology developed for the Western Interior Basin of North America, biostratigraphic zonation has been resolved to 0.18-0.33 m.y. per biozone; Milankovitch cyclostratigraphy to 41,000 yr per cycle, and event chronostratigraphic resolution to 17,000 yr per event-bounded interval. There are few systems of chronology and correlation that match this level of resolution, which has similarly been developed in western Europe and northern South America.

Figures 3.3 and 3.4 show the physical, chemical, biological, and cyclostratigraphic details of the late Cenomanian-early Turonian section at the Pueblo, Colorado standard reference section and proposed boundary stratotype (Kennedy and Cobban, 1991). The Bridge Creek Limestone spans the C-T boundary (Figure 3.5) and contains 24 bentonite beds and 13 additional persistent limonite beds, probably representing altered bentonites. These bentonites record the late phases of one of the most active intervals of volcanism in Cretaceous history (Kauffman, 1985, Figure 4). Thirteen of these bentonites (PBC4, 5, 11, 17, 19, 20, 30, 32, 44-46, 48, and 50, in Kauffman et al., 1985, p. FRS-11) are much thicker and more highly persistent than the others, and form regional chronostratigraphic marker beds (Hattin, 1971, 1985; Elder, 1985; Kauffman, 1988a). These have yielded materials for radiometric dating. Utilizing single sanidine crystal ^{40}Ar-^{39}Ar dating techniques, Obradovich (1993) has established high-resolution geochronologic tie points to ammonite biozones of Cobban (1985, 1993; Figure 3.4 herein) that allow a calculated time scale to be constructed for the entire interval, if equal time durations are assumed for all intervening ammonite biozones between dated Cenomanian-Turonian bentonite levels (Figure 3.5, second column). This commonly used calibration technique assumes constant evolutionary rates within ammonite lineages, however, which is improbable. Obradovich (1993) has dated the lowest ammonite zone in the Bridge Creek Limestone Member (*Vascoceras diartanum*), bed 63, as having a mean value of 93.9 Ma, from a bentonite that does not occur in the condensed section at Pueblo (Figure 3.5). A second age of 93.56 Ma has been obtained from the latest Cenomanian *Neocardioceras juddi* zone in a bed equivalent to the "*Neocardioceras* bentonite" of Elder and Kirkland (1985) and Elder (1985, 1987a), unit 80, PBC-11 of the Pueblo reference section (Kauffman *et al.*, 1985, p. FRS-11). A third mean age of 93.25 Ma was obtained from the early Turonian *Pseudaspidoceras flexuosum* zone in a bed equivalent to bentonite marker bed 96 (PBC20; Kauffman *et al.*, 1985, p. FRS-11). The top of the Bridge Creek Limestone Member (lowest part of the *Collignoniceras woollgari* biozone) has been given an interpolated age of 92.05 Ma based on assigning equal age ranges to ammonite biozones that lie between well-dated levels below and above this biozone (see Figure 3.5).

Based on this set of new ^{40}Ar-^{39}Ar ages, and the calculated time scale that can be constructed between them, if equal age ranges are assumed for intervening ammonite biozones, a duration of 1.95 m.y. is calculated for the Bridge Creek Limestone Member at Pueblo. Here, the member spans the entire set of C-T mass extinction steps (Kauffman, 1988b; Figures 3.4 and 3.6 herein), the complete survival interval, and the recovery interval (sensu Harries and Kauffman, 1990) up to basic recovery of the marine ecosystem in the early to early-middle Turonian. The average durations for ammonite subzones and zones, based on the exclusive ranges of biostratigraphically tested ammonite species, equals 190,000 yr for the C-T interval. This is calculated by dividing the number of ammonite biozones into the time duration between the mean values of bounding radiometric ages.

Because of the dangers in the assumption that ammonite species evolve at relatively constant rates, even through this thin interval, an independent, more highly refined dating system was constructed to test this time scale by using shale (or marl)-limestone (or chalk, calcarenite) bedding rhythms reflecting Milankovitch climate cycles. The scale of this independent chronology is adequate to evaluate the relative importance and interaction of individual physical, chemical, and biological events associated with the C-T mass extinction, survival, and recovery intervals.

At Pueblo, various authors have measured closely spaced sections of the Bridge Creek Limestone Member in great detail, and have reported between 41 and 50 shale (or marl)-limestone (or chalk, calcarenite) bedding couplets. Most sections yield between 44 and 48 bedding couplets that probably reflect Milankovitch cyclicity (Barron *et al.*, 1985; Fischer *et al.*, 1985; Kauffman, 1988a) in the form of climate-regulated dilution cycles (Pratt, 1985) and/or productivity cycles (Eicher and Diner, 1985, 1989).

At the proposed C-T boundary stratotype at Pueblo, Elder and Kirkland (1985) and subsequent work by the author have confirmed 46 Milankovitch-type bedding cycles of all scales in the Bridge Creek Limestone Member, 19 of which are capped by relatively thicker, more resistant limestone or calcarenite beds with extensive basinal dispersion (Figure 3.5; Hattin, 1971, 1985; Elder and Kirkland, 1985; Kauffman, 1988a). These thicker and more pervasive limestone units probably reflect the 100,000-yr Milankovitch orbital eccentricity cycle (Barron *et al.*, 1985; Fischer *et al.*, 1985; Kauffman *et al.*, 1987; Kauffman, 1988a) (Figure 3.5). There are an average of 2.5 smaller bedding cycles between these more prominent limestone beds, suggesting that they may represent the 41,000-yr Milankovitch axial eccentricity cycle (Figure 3.5). These interpretations are subsequently tested and confirmed.

Given the 1.95-m.y. calculated duration for the Bridge Creek Limestone, C-T boundary section at Pueblo, the 19 thicker, more regionally persistent limestone beds have an average duration of 100,000 yr, and logically represent the Milankovitch orbital eccentricity cycle (Figure 3.5). Further, the average duration for the 46 bedding cycles of all scales is 41,300 yr; these are considered representative of the 41,000-yr Milankovitch axial eccentricity cycles (Figure 3.5). This is the optimal level of age resolution for interpreting regional environmental change and the patterns of C-T mass extinction.

These Milankovitch cycle determinations provide an independent means of testing, and recalibrating, the interpolated time scale, based on assigning equal durations to ammonite biozones between radiometrically dated levels. Two methods of calibrating a new C-T boundary time scale, based on integrating radiometric ages with 100,000-yr Milankovitch climate cycles, are presented in Figure 3.5, columns A and B. The first, and most commonly practiced method, starts at a calibrated or radiometric age at the top (or base) of the long Milankovitch cycle sequence, and assigns progressively greater (or lesser) 100,000-yr intervals to the top of each bedding rhythm representing the orbital eccentricity cycle (e.g., Figure 3.5, column B). When this method is applied to the Bridge Creek Limestone Member at Pueblo, the calculated values for the 100,000-yr Milankovitch bedding rhythms deviate from the mean values of ^{40}Ar-^{39}Ar ages within the member by 100,000 to 160,000 yr, and vary from the time scale calculated on the assignment of equal durations to successive ammonite biozones by 150,000 to 360,000 yr.

A second, and preferred, calculated time scale was constructed by utilizing a mean ^{40}Ar-^{39}Ar age from the middle of the Bridge Creek Limestone Member as a starting point (93.25 Ma for the bentonite in the middle of the early Turonian *Pseudaspidoceras flexuosum* biozone; Figure 3.5, column A), and assigning progressively younger 100,000-yr durations for orbital eccentricity cycles up-section, and progressively older 100,000-yr durations for the same cycles down-section. In this method, the ages calculated from the Milankovitch cycle scale deviate by only 10,000 to 50,000 yr from other ^{40}Ar-^{39}Ar ages within the sequence, and only 10,000 to 50,000 yr from ages calculated by assigning equal range durations to late Cenomanian ammonite biozones, which suggests a relatively constant evolutionary rate. However, the same scale deviates by 440,000 to 510,000 yr from ages calculated by the same method for early Turonian ammonite biozones. This suggests that postextinction ammonite evolution takes place at much more variable rates than found among pre-extinction ammonite lineages. In both Milankovitch cycle-based time scales, therefore, the assumption that ammonite biozones have approximately equal durations, and relatively stable evolutionary rates through the C-T mass extinction-recovery interval is not supported. The calibrated time scale based on integrating new ^{40}Ar-^{39}Ar ages with the 100,000-yr Milankovitch cycle history, with calculations starting from a radiometric age in the middle of the cyclic depositional sequence, is thus the most accurate. This scale is used below to calibrate the physical, chemical, and biological components of the Cenomanian-Turonian mass extinction-recovery interval.

Even finer resolution for dating and correlation can be obtained by utilizing physical and chemical event deposits/horizons within the same section. Most of the 112 event beds/horizons in the Bridge Creek Limestone are bentonites, trace element, stable isotope, and organic carbon enrichment levels (spikes), and sea-level or climate cycle boundaries with regional to interregional dispersion. Within the time scales developed for this interval from radiometric and climate cycle data, the event-bounded intervals of the Bridge Creek Limestone Member, spanning the C-T boundary, have an average duration of 17,000 yr.

This highly refined C-T chronology for the Western Interior Cretaceous Basin, integrating radiometric, event/cycle chronostratigraphic, and biostratigraphic data, closely parallels common time scales utilized in Quaternary studies of global change (Figure 3.1), ecosystem destruction, and biodiversity loss (e.g., see papers in Bradley, 1991).

A TIMETABLE FOR CENOMANIAN-TURONIAN MASS EXTINCTION

The integration of high-resolution stratigraphic, geochemical, and paleobiological data depicting late Cenomanian-early Turonian global change, with a real time scale combining ^{40}Ar-^{39}Ar volcanic ash ages with Milankovitch climate cycle deposits of known time duration, makes it possible to evaluate the rates, patterns, and timing of the C-T mass extinction and associated environmental perturbations within a greenhouse world (see Appendix).

Figures 3.3 and 3.4 show the integrated high-resolution data for the study of the C-T mass extinction at Pueblo, Colorado. Figure 3.5 shows the derivation of the real time scale from this data base. Together, these allow derivation of the timetable for environmental perturbations and mass extinction in the C-T boundary interval at Pueblo (shown in the Appendix). From oldest (1) to youngest (22), these divisions are based collectively on physical, chemical, and biological characteristics that are unique to each interval. Figure 3.6 summarizes these relationships through time. These data can be used to model the patterns of extinction, survival, and recovery in a greenhouse world, and to enhance the development of predictive models for the modern global crisis.

INTERPRETATIONS AND CONCLUSIONS

The Cenomanian-Turonian mass extinction is among the best-studied biotic crises in Earth history, and is unusual in that it occurred during one of the greatest eustatic highstands of the Phanerozoic, in a relatively equable, warm, greenhouse world usually considered favorable for the diversification of life. High sea level and widespread development of epicontinental seas in this interval have ensured the preservation of complete C-T boundary sections worldwide. In most sections, there is little or no sedimentologic expression of the extinction interval, which is preserved primarily within relatively monotonous, fine-grained neritic, hemipelagic, and pelagic facies characterized by well-defined Milankovitch bedding cycles. However, integrated geochemical and paleobiological data, obtained through very high resolution (1 to 10 cm) stratigraphic sampling, reveal dynamic environmental changes in ocean-climate systems during this global mass extinction interval.

The C-T boundary interval initiates with a series of tropical extinctions among rudistid bivalves and other reef-associated taxa, beginning at the end of the early Cenomanian (95.9 Ma) and culminating 0.5-1.0 m.y. below the C-T boundary (about 93.9 to 94.4 Ma); dating of this final tropical extinction event and its correlation with well-studied temperate records, however, is imprecise. Johnson and Kauffman (1990) recorded a loss of three rudistid genera (0.1 per taxon extinction rate), and 32 species (0.2 per taxon extinction rate), which comprise 43% of the lower Cenomanian rudistid species, 26% of the middle Cenomanian species, and 47% of the upper Cenomanian rudistid species in the Caribbean Province. Tropical extinction patterns are similar in various well-studied Italian and French sections, where no reefs are known within 3 to 10 m of the C-T boundary.

In the Western Interior of North America, where the best integrated physical, geochemical, and temperate to subtropical biological record of the C-T extinction interval has been compiled, an integrated dating matrix of ^{40}Ar-^{39}Ar radiometric ages, and bedding rhythms, representing the 41,000-yr and 100,000-yr Milankovitch climate cycles, has allowed development of a relatively precise timetable for Cenomanian-Turonian boundary extinctions and their probable causal mechanisms. The Pueblo, Colorado, reference section is utilized to illustrate this chronology and its application to interpreting rates, patterns, and probable causes of C-T mass extinction.

The first geochemical perturbations associated with the C-T extinction interval in this setting occur 880,000 yr below the C-T boundary; the first molluscan extinction event (step) occurs 520,000 yr below the C-T boundary, coeval with initiation of a series of 10 successive trace element enrichment horizons, probably representing an oceanic advection sequence. The first five of these enrichment intervals have Ir peaks that are two to five times greater than background levels, and are precisely coeval with the first five C-T extinction steps; two of these levels have recently yielded possible microtektites and rare shocked quartz grains in Colombia and Colorado, respectively. Similar trace element enrichment patterns occur at many sections in North America, western Europe, Colombia, and North Africa. The first C-T extinction events at Pueblo also mark the first extraordinary peaks of stable isotope and organic carbon fluctuations, relative to background values. An overall enrichment of C_{org} across the C-T boundary records development of an oceanic anoxic event (OAE II). Ocean-climate systems are therefore highly perturbed, with the rates and magnitude of change in temperature and ocean chemistry well above Cretaceous background levels. In the North American Interior, extinction among subtropical to temperate lineages spans 1.46 m.y. from the latest Cenomanian through the middle-early Turonian, and appears punctuated or stepwise in nature based on regionally composited, robust range-zone plots of both macro- and microfossil species. The major part of the extinction interval spans 520,00 yr of the latest Cenomanian and comprises five steps of macrofaunal (mainly molluscan) extinction and two microfossil extinction levels. Extinction events occur successively, therefore, about every 100,000 yr. Three smaller macrofossil and two microfossil extinction events occur 230,000 to 940,000 yr above the C-T boundary within the survival and early recovery intervals.

Given that Cenomanian marine organisms over much of the world had evolved over millions of years into a warm, more equable greenhouse world characterizing Cretaceous eustatic sea-level rise, it is probable that most lineages had relatively narrow adaptive ranges with sharp ecological thresholds, especially in terms of thermal range and water chemistry, and that they were thus extinction prone. These biotas would have been easily stressed by even small-scale environmental perturbations of the water column. The very close correlation of mass extinction steps with short-term trace element enrichment events (advection events?), and with rapid, large-scale fluctuations in the temperature, salinity, and oxygen of the water column, suggests that the extraordinary rate and magnitude of these environmental shifts directly caused the series of ecologically graded C-T extinction events, or steps, as they progressively breached the narrow adaptive ranges of Cenomanian-Turonian species, and ecosystems, adapted to an equable greenhouse world. Predictably, extinction initiated among tropical reef ecosystems, and terminated with more temperate-adapted and cosmopolitan groups.

Yet the cause of these marine environmental perturbations remains an enigma; several viable hypotheses exist. Hallam (1984) and others have suggested that many mass extinctions, including the C-T boundary event, resulted from oxygen depletion in oceans and relatively deep epicontinental seas associated with expanding oxygen minima zones and the establishment of oceanic anoxic events. This mechanism would be especially effective during global greenhouse conditions, when ocean circulation was sluggish and may have been driven largely by low-oxygen warm saline bottom waters derived from marginal tropical evaporite belts, and when productivity was high. Widespread dysoxic to anoxic conditions would ecologically stress global normal marine biotas and could have contributed to C-T mass extinction. However, this hypothesis is rejected as the primary cause for C-T extinction because (1) widespread oxygen depletion in the oceans characterized most of the time interval between the Aptian Mass Extinction and the C-T boundary interval, including the long Albian OAE Ic, without causing mass extinction. If anything, these were longer and more severe intervals of oxygen depletion than the one that coincided with C-T extinction. Instead, C-T extinction comes at the end of the dysoxic to anoxic interval (Hay, 1988, Figure 3.3), during a set of dynamic geochemical changes leading to re-oxygenation of deeper portions of the oceans. C-T extinction events were more likely related to the rate and magnitude of these changes than to oxygen depletion itself. (2) Because of the long middle Cretaceous interval of oceanic oxygen depletion associated with global greenhouse warming and elevated sea level, a large proportion of deeper water marine biotas became adapted to low-oxygen conditions and are commonly associated with organic-rich black shales (e.g., many Inoceramidae and other bivalves; Kauffman and Sageman, 1990; Sageman *et al.*, 1991, 1994). Yet many of these same lineages became extinct in the C-T boundary interval as conditions shifted rapidly toward more oxygenated oceans in the basal Turonian. If anything, a case for oxygen poisoning might be made for the extinction of some low-oxygen adapted taxa.

A second explanation for C-T marine mass extinction might lie with trace element poisoning associated with advection events from the deep ocean and from epicontinental basins, as suggested for Paleozoic mass extinctions by Berry *et al.* (1990) and Wilde *et al.* (1990). Potentially toxic trace elements were greatly enriched in the oceans and deeper portions of epicontinental seas during accelerated outgassing associated with large-scale middle Cretaceous plate reorganization and development of the Pacific superplume (Larson, 1991a,b). These trace elements could have been concentrated as oxides and carbonates in basinal Cretaceous sediments in the presence of at least moderate amounts of benthic oxygen, especially during high sea level and offshore sediment starvation; trace elements could also have been sequestered in solution in low-oxygen to anoxic water masses (OMZs). Expansion and benthic touchdown of the oceanic oxygen minimum zone during the Cenomanian-Turonian OAE would have remobilized sequestered trace elements from the seafloor and caused progressive advection of potentially toxic chemicals through the water column, with profound effects on the global marine biota. If the base of the food chain was affected by these advection events, it would have caused a far-reaching set of negative ecological feedback loops within the trophic web. The precise correlation of the first five C-T extinction steps, and two of the succeeding steps, with trace element enrichment horizons in epicontinental and continental shelf deposits of four continents, strongly supports the hypothesis of advection and trace element poisoning as a partial cause of C-T mass extinction. The relatively shallow water trace element enrichment layers found to be associated with C-T extinction events in high-resolution stratigraphic analysis, probably represent depositional sites of advecting trace elements from deeper ocean sources, situated at and above the oceanic redoxcline between the OMZ and the mixing zone.

A third hypothesis for C-T mass extinction, which draws on events described in the two previous hypotheses, focuses on the extraordinary rates and magnitudes of ocean-climate changes associated with the 1.46-m.y.-long C-T boundary interval, and their effects on a global biota narrowly adapted for the most part to the equable greenhouse environments that had been developing throughout the early and middle Cretaceous. Data from Pueblo, Colorado (Figures 3.3 and 3.4) are characteristic of many global boundary sections and show a series of exceptionally rapid, large-scale shifts in organic carbon values (representing rapid shifts in benthic oxygen levels); in $\delta^{13}C$ values (representing changes in carbon cycling) within the global positive $\delta^{13}C$ excursion; in $\delta^{18}O$ values (possibly representing rapid salinity and/or temperature changes); and in trace element values (probably representing one or more oceanic advection sequences). Virtually all late Cenomanian extinction events, and some lesser ones in the early Turonian, are correlative with one or more of these rapid, large-scale geochemical fluctuations. Whereas these geochemical signals can be strongly modified by diagenetic processes, the fact that similar changes occur in virtually all well-studied global C-T boundary sections suggests that they represent a primary ocean-climate signal. Major changes in ocean chemistry and temperature around the C-T boundary could well have been the primary cause of extinction events as they progressively exceeded the narrow adaptive ranges of many stenotopic Cenomanian-Cretaceous lineages. Of special interest in this theory is the general correlation of many apparently rapid environmental fluctuations with bedding rhythms representing 41,000- and 100,000-yr Milankovitch climate cycles (Barron et al., 1985; Fischer et al., 1985; Kauffman, 1988a; Glancy et al., 1993; Figures 3.3 and 3.5 herein). On the one hand this may suggest diagenetic modification of a primary signal in carbonate-rich versus carbonate-poorer facies of the C-T boundary interval, but to the degree that it represents a primary signal, it suggests that the Milankovitch climate cyclicity may have acted as an independent catalyst that drove an environmentally perched, greenhouse ocean-climate system to even greater levels of change, at rates dictated by the climate cycles themselves.

Finally, the possibility of extraterrestrial influences on the C-T mass extinction cannot be ruled out. The precise correlation of the first five late Cenomanian extinction events, or steps, with Ir enrichment of two to four or five times background levels leaves open the possibility of extraterrestrial sources for the iridium. Orth et al. (1989, 1990, 1993) have been cautious in suggesting extraterrestrial origins, pointing out instead an apparent similarity of the overall C-T trace element suite to those originating from deep mantle outgassing, and the fact that the C-T boundary interval was also a time of major plate rearrangement and superplume development (Larson, 1991a,b). On the other hand, four temporally clustered late Albian to late Cenomanian terrestrial impact craters are known with age error bars that overlap the C-T boundary (Grieve, 1982), suggesting an impact storm, or shower (sensu Hut et al., 1987); this is a conservative estimate of terrestrial impacts, when considering the amount of Cenomanian surface that has been subducted or is now covered by younger sediments/strata, vegetation, ice, and especially water. Because impacting is predictably spatially random (Grieve, 1982), and the late Cenomanian world was 80 to 82% covered by water near eustatic highstand, it is likely that at least four of five potential impactors would have fallen in the world seas and oceans during this impact shower. This statistically projects at least 20 impacts during the Cenomanian and early Turonian interval, the majority of which would be aquatic. Oceanic impacting would cause repeated, short-term stirring events. This hypothesis further predicts an extended duration for the Cenomanian impact shower, with high probability that it would overlap the C-T mass extinction interval. Recent discoveries of possible microtektites (Colombia) and multilamellate shocked quartz grains (Colorado) at two levels within the latest Cenomanian extinction interval (discussed in text) support this hypothesis. These new finds suggest at least a partial extraterrestrial origin for late Cenomanian iridium, and for dramatic changes in ocean chemistry associated with some of the major early extinction events in the Cenomanian-Turonian interval. If the shocked quartz, at least, represents a younger Cenomanian terrestrial impact than previously known, for which a crater has not yet been found, a minimal number of five Cenomanian terrestrial impacts is now known, and these predict at least 20 additional aquatic impacts during the Cenomanian impact shower.

Is there a linkage among these varied data, their interpretations, and the causal mechanisms of mass extinction, especially the abruptness with which large-scale geochemical fluctuations and correlative extinction events initiate and perpetuate in the late Cenomanian and into the early Turonian? There is a high probability that such a relationship exists, that the C-T mass extinction was multicausal, with meteor/comet impacts into Cenomanian oceans—already highly stratified, largely oxygen-depleted, and rich in sequestered trace elements—causing rapid stirring, overturn, and advection events, and dramatically changing the thermal and chemical regimes of the water column up into the mixing zone. Each successive oceanic impact would set in motion a series of dynamic feedback processes expressed as rapid, large-scale, geochemical and thermal fluctuations such as those associated with individual mass extinction events, or steps, throughout the C-T boundary interval. These ocean-wide perturbations, acting on a largely stenotopic, extinction-prone marine biota, were apparently the direct causes of the ecologically graded C-T extinction events as they progressively exceeded the adaptive ranges of, first, lineages within tropical ecosystems, secondly subtropical to warm temperate lineages, and eventually more temperate lineages. Between impacting events, oceanic feedback processes, perhaps driven independently in part by Milankovitch climate cycles, might continue for perhaps thousands of years, seeking equilibrium. Yet each successive oceanic impact within the Cenomanian shower would reset the perturbation clock, driving extinction for hundreds of thousands of years and, in the end, cumulatively affecting more extinction resistant, temperate ecosystems.

This hypothesis best fits the available high-resolution data base for the C-T boundary interval, one of the best-studied mass extinction intervals in the world. However, it is a hypothesis in need of extensive testing, especially in the recognition of impacting events in the deep ocean through sedimentological, paleobiological, and geochemical sensing.

Ongoing research on the C-T mass extinction interval thus provides us with an extensive integrated data base, blending sedimentologic, geochemical, and paleobiological data at a level of stratigraphic resolution that equals that of broader-scale (100- to 1000-yr observational scale) Quaternary studies of global change. This study reveals the complex dynamics of change within ocean-climate systems, and biological response to them, at peak development of a greenhouse world—a world toward which we may be heading rapidly as a result of modern global warming and ozone depletion. Of greatest importance is the ability, in older Cretaceous rocks, to develop a timetable for extinction, survival, and recovery spanning one of the major global mass extinctions. From these data, we can develop predictive models, within a high-resolution temporal framework, for other mass extinctions, including the one currently in progress as a result of overpopulation, habitat and ecosystem destruction, and resource depletion at the hands of the human species. It should be of concern to us that modern rates of biodiversity loss exceed the extinction rates of well-studied Cretaceous mass extinctions, including those clearly associated with large bolide impacts. Even more frightening is the prediction from the fossil record that the complex tropical ecosystems, reefs, and rain forests of today, which contain more than half of the global biodiversity, are commonly the first to disappear in ancient mass extinctions and the last to ecologically recover, some 2 to 10 m.y. later (Kauffman and Fagerstrom, 1993). These tropical ecosystems are the harbingers of the mass extinction process and long-term biological crisis on Earth. They alone, once gone, account for enough global biodiversity loss to reach mass extinction levels (>50% of global species). At the rate of global destruction today, tropical rain forests and to a lesser degree tropical reefs will be largely decimated by the year 2500, and certainly by A.D. 3000. Simberloff (1984) estimated that only 2.5% of the neotropical rainforests would remain by 2050, with a resultant loss of more than 50% of the biodiversity of the Western Hemisphere, at current rates of destruction. This may be conservative. Even so, if research on ancient mass extinctions can act as a model for the future of the modern biodiversity crisis, it may be millions of years, without human perturbation, before similar ecosystems will become reestablished in the tropics.

Research on ancient mass extinction, survival, and recovery (radiation) intervals can play an important role in understanding the consequences of modern global change phenomena, and in making long-term predictions concerning the present biodiversity crisis—which is already spiraling toward mass extinction levels.

REFERENCES

Alvarez, W., and R. A. Muller (1984). Evidence from crater ages for periodic impacts in Earth, *Nature 308*, 718-720.

Arthur, M. A., W. E. Dean, R. M. Pollastro, G. E. Claypool, and P. A. Scholle (1985). Comparative geochemical and mineralogical studies of two cyclic transgressive pelagic limestone units, Cretaceous Western Interior Basin, U.S., in *Fine-Grained Deposits and Biofacies of the Cretaceous Western Interior Seaway: Evidence of Cyclic Sedimentary Processes*, L. M. Pratt, E. G. Kauffman, and F. B. Zelt, eds., Society of Economic Paleontologists and Mineralogists, 2nd Annual Midyear Meeting, Golden, Colo., Field Trip Guidebook 4, pp. 16-27.

Barron, E. J. (1993). *Atlas of Cretaceous Climate Model Results*, Earth System Science Center, Pennsylvania State University, pp. 1-36.

Barron, E. J., M. A. Arthur, and E. G. Kauffman (1985). Cretaceous rhythmic bedding sequences—A plausible link between orbital variations and climate, *Earth and Planetary Science Letters 72*, 327-340.

Berner, R. A. (1991). A model for atmospheric CO_2 over Phanerozoic time, *American Journal of Science 291*, 339-376.

Berner, R. A. (1994). Geocarb II: A revised model for atmospheric CO_2 over Phanerozoic time, *American Journal of Science 294*, 56-91.

Berry, W. B. N., P. Wilde, and M. S. Quinby-Hunt (1990). Late Ordovician mass mortality and subsequent Early Silurian re-radiation, in *Extinction Events in Earth History*, E. G. Kauffman and O. H. Walliser, eds., Springer-Verlag, Berlin, pp. 115-124.

Bradley, R. S., ed. (1991). *Global Changes of the Past*, Office for Interdisciplinary Earth Studies, University Consortium for Atmospheric Research, Boulder, Colo., 514 pp.

Caldwell, W. G. E., R. Diner, D. L. Eicher, S. P. Fowler, B. R. North, C. R. Stelck, and L. v.H. Wilhelm (1993). Foraminiferal biostratigraphy of Cretaceous marine cyclothems, in *Evolution of the Western Interior Basin*, W. G. E. Caldwell and E. G. Kauffman, eds., Geological Association of Canada Special Paper 39, pp. 477-520.

Caron, M. (1985). Cretaceous planktic foraminifera, in *Plankton Stratigraphy*, H. M. Bolli, J. B. Saunders, and K. Perch-Nielsen, eds., Cambridge University Press, Cambridge, pp. 17-86.

Cobban, W. A. (1985). Ammonite record from Bridge Creek Member of Greenhorn Limestone at Pueblo Reservoir State Recreation Area, Colorado, in *Fine-Grained Deposits and Biofacies of the Cretaceous Western Interior Seaway: Evidence of Cyclic Sedimentary Processes*, L. M. Pratt, E. G. Kauffman, and F. B. Zelt, eds., Society of Economic Paleontologists and Mineralogists, 2nd Annual Midyear Meeting, Golden, Colo., Field Trip Guidebook 4, pp. 135-138.

Cobban, W. A. (1993). Diversity and distribution of Late Cretaceous ammonites, Western Interior, United States, in *Evolution of the Western Interior Basin*, W. G. E. Caldwell and E. G. Kauffman, eds., Geological Association of Canada Special Paper 39, pp. 435-451.

Cobban, W. A., and G. R. Scott (1972). *Stratigraphy and Ammonite Fauna of the Graneros Shale and Greenhorn Limestone near Pueblo, Colorado*, U.S. Geological Survey Professional Paper 645, 108 pp.

Collom, C. J. (1990). The taxonomic analysis of mass extinction intervals: An approach to problems of resolution as shown by Cretaceous ammonite genera (global) and species (Western Interior of the United States), in *Extinction Events in Earth History*, E. G. Kauffman and O. H. Walliser, eds., Springer-Verlag, Berlin, pp. 265-276.

Eicher, D. L. (1969). Paleobathymetry of Cretaceous Greenhorn Sea in eastern Colorado, *American Association of Petroleum Geologists Bulletin 53*(5), 1075-1090.

Eicher, D. L., and R. Diner (1985). Foraminifera as indicators of water mass in the Cretaceous Greenhorn Sea, Western Interior, in *Fine-Grained Deposits and Biofacies of the Cretaceous Western Interior Seaway: Evidence of Cyclic Sedimentary Processes*, L. M. Pratt, E. G. Kauffman, and F. B. Zelt, eds., Society of Economic Paleontologists and Mineralogists, 2nd Annual Midyear Meeting, Golden, Colo., Field Trip Guidebook 4, pp. 60-71.

Eicher, D. L., and R. Diner (1989). Origin of the Cretaceous Bridge Creek Cycles in the Western Interior, United States, *Palaeogeography, Palaeoecology, Palaeoclimatology 74*, 127-146.

Einsele, G., W. Ricken, and A. Seilacher, eds. (1991). *Cycles and Events in Stratigraphy*, Springer-Verlag, Berlin, 955 pp.

Elder, W. P. (1985). Biotic patterns across the Cenomanian-Turonian extinction boundary, Pueblo, Colorado, in *Fine-Grained Deposits and Biofacies of the Cretaceous Western Interior Seaway: Evidence of Cyclic Sedimentary Processes*, L. M. Pratt, E. G. Kauffman, and F. B. Zelt, eds., Society of Economic Paleontologists and Mineralogists, 2nd Annual Midyear Meeting, Golden, Colo., Field Trip Guidebook 4, pp. 157-169.

Elder, W. P. (1987a). Cenomanian-Turonian (Cretaceous) Stage Boundary Extinctions in the Western Interior of the United States, Unpublished Ph.D. thesis, University of Colorado, Boulder, 660 pp.

Elder, W. P. (1987b). The paleoecology of the Cenomanian-Turonian (Cretaceous) stage boundary extinction at Black Mesa, Arizona, *Palaios 2*, 24-40.

Elder, W. P., and J. I. Kirkland (1985). Stratigraphy and depositional environments of the Bridge Creek Limestone Member of the Greenhorn Limestone at Rock Canyon Anticline near Pueblo, Colorado, in *Fine-Grained Deposits and Biofacies of the Cretaceous Western Interior Seaway: Evidence of Cyclic Sedimentary Processes*, L. M. Pratt, E. G. Kauffman, and F. B. Zelt, eds., Society of Economic Paleontologists and Mineralogists, 2nd Annual Midyear Meeting, Golden, Colo., Field Trip Guidebook 4, pp. 122-134.

Fischer, A. G., and M. A. Arthur (1977). Secular variations in the pelagic realm, in *Deep-Water Carbonate Environments*, H. E. Cook and P. Enos, eds., Society of Economic Paleontologists and Mineralogists Special Publication 25, pp. 19-50.

Fischer, A. G., T. Herbert, and I. Premoli Silva (1985). Carbonate bedding cycles in Cretaceous pelagic and hemipelagic sequences, in *Fine-Grained Deposits and Biofacies of the Cretaceous Western Interior Seaway: Evidence of Cyclic Sedimentary Processes*, L. M. Pratt, E. G. Kauffman, and F. B. Zelt, eds., Society of Economic Paleontologists and Mineralogists, 2nd Annual Midyear Meeting, Golden, Colo., Field Trip Guidebook 4, pp. 1-10.

Gale, A. S., H. C. Jenkyns, W. J. Kennedy, and R. M. Cornfield (1993). Chemostratigraphy versus biostratigraphy: Data from around the Cenomanian-Turonian boundary, *Journal of Geological Society of London* 150, 29-32.

Glancy, T. J., Jr., M. A. Arthur, E. J. Barron, and E. G. Kauffman (1993). A paleoclimate model for the North American Cretaceous (Cenomanian-Turonian) epicontinental sea, in *Evolution of the Western Interior Basin*, W. G. E. Caldwell and E. G. Kauffman, eds., Geological Association of Canada Special Paper 39, pp. 219-241.

Grieve, R. A. F. (1982). The record of impact on Earth: Implications for a major Cretaceous/Tertiary boundary impact, in *Geological Implications of Impacts of Large Asteroids and Comets on the Earth*, L. T. Silver and P. H. Schultz, eds., Geological Society of America Special Paper 190, pp. 25-38.

Hallam, A. (1984). The causes of mass extinction, *Nature* 308, 686-687.

Hancock, J. M. (1977). The historical development of biostratigraphic correlation, in *Concepts and Methods of Biostratigraphy*, E. G. Kauffman and J. E. Hazel, eds., Dowden, Hutchinson, and Ross, Stroudsburg, Pa., pp. 3-22.

Harries, P. J. (1993). Patterns of Repopulation following the Cenomanian-Turonian (Upper Cretaceous) Mass Extinction, Unpublished Ph.D. thesis, University of Colorado, Boulder, 356 pp.

Harries, P. J., and E. G. Kauffman (1990). Patterns of survival and recovery following the Cenomanian-Turonian (Late Cretaceous) mass extinction in the Western Interior Basin, United States, in *Extinction Events in Earth History*, E. G. Kauffman and O. H. Walliser, eds., Springer-Verlag, Berlin, pp. 277-298.

Hattin, D. E. (1971). Widespread, synchronously deposited, burrow-mottled limestone beds in the Greenhorn Limestone (upper Cretaceous) of Kansas and southeastern Colorado, *American Association of Petroleum Geologists Bulletin* 55, 412-431.

Hattin, D. E. (1985). Distribution and significance of widespread, time-parallel pelagic limestone beds in Greenhorn Limestone (upper Cretaceous) of the central Great Plains and southern Rocky Mountains, in *Fine-Grained Deposits and Biofacies of the Cretaceous Western Interior Seaway: Evidence of Cyclic Sedimentary Processes*, L. M. Pratt, E. G. Kauffman, and F. B. Zelt, eds., Society of Economic Paleontologists and Mineralogists, 2nd Annual Midyear Meeting, Golden, Colo., Field Trip Guidebook 4, pp. 28-37.

Hay, W. W. (1988). Paleoceanography: A review for the GSA centennial, *Geological Society of America Bulletin* 100, 1934-1956.

Hut, P., W. Alvarez, W. P. Elder, T. A. Hansen, E. G. Kauffman, G. Keller, E. M. Shoemaker, and P. R. Weissman (1987). Comet showers as a possible cause of stepwise extinctions, *Nature* 329, 118-126.

Jarvis, I., G. A. Carson, M. K. E. Cooper, M. B. Hart, P. N. Leary, B. A. Tocher, D. Horne, and A. Rosenfeld (1988). Microfossil assemblages and the Cenomanian-Turonian (Late Cretaceous) oceanic anoxic event, *Cretaceous Research* 9, 3-103.

Jeffries, R. P. S. (1961). The paleoecology of the *Actinocamax plenus* subzone (lower Turonian) in the Anglo-Paris Basin, *Palaeontology* 4, 609-647.

Jeffries, R. P. S. (1963). The stratigraphy of the *Actinocamax plenus* subzone (Turonian) in the Anglo-Paris Basin, *Geological Association of London* 74, 1-33.

Johnson, C. C., and E. G. Kauffman (1990). Originations, radiations, and extinctions of Cretaceous rudistid bivalve species in the Caribbean Province, in *Extinction Events in Earth History*, E. G. Kauffman and O. H. Walliser, eds., Springer-Verlag, Berlin, pp. 305-324.

Kauffman, E. G. (1975). Dispersal and biostratigraphic potential of Cretaceous benthonic Bivalvia in the Western Interior, in *The Cretaceous System in the Western Interior of North America*, W. G. E. Caldwell, ed., Geological Association of Canada Special Paper 13, pp. 163-194.

Kauffman, E. G. (1984a). The fabric of Cretaceous marine extinctions, in *Catastrophes and Earth History—The New Uniformitarianism*, W. A. Berggren and J. Van Couvering, eds., Princeton University Press, Princeton, N.J., pp. 151-246.

Kauffman, E. G. (1984b). Paleobiogeography and evolutionary response dynamic in the Cretaceous Western Interior Seaway of North America, in *Jurassic-Cretaceous Biochronology and Paleogeography of North America*, G. E. G. Westermann, ed., Geological Association of Canada Special Paper 13, pp. 273-306.

Kauffman, E. G. (1985). Cretaceous evolution of the Western Interior Basin of the United States, in *Fine-Grained Deposits and Biofacies of the Cretaceous Western Interior Seaway: Evidence of Cyclic Sedimentary Processes*, L. M. Pratt, E. G. Kauffman, and F. B. Zelt, eds., Society of Economic Paleontologists and Mineralogists, 2nd Annual Midyear Meeting, Golden, Colo., Field Trip Guidebook 4, pp. iv-xiii.

Kauffman, E. G. (1988a). Concepts and methods of high-resolution event stratigraphy, *Annual Review of Earth and Planetary Sciences* 16, 605-654.

Kauffman, E. G. (1988b). The dynamics of marine stepwise mass extinction, in *Paleontology and Evolution, Extinction Events*, M. Lamolda, E. G. Kauffman, and O. H. Walliser, eds., *Review of Espanola Paleontology*, n. extraordinario, pp. 57-71.

Kauffman, E. G., and W. G. E. Caldwell (1993). The Western Interior Basin in space and time, in *Evolution of the Western Interior Basin*, W. G. E. Caldwell and E. G. Kauffman, eds., Geological Association of Canada Special Paper 39, pp. 1-30.

Kauffman, E. G., and J. A. Fagerstrom (1993). The Phanerozoic evolution of reef diversity, in *Species Diversity in Ecological Communities*, R. Ricklefs and D. Schluter, eds., University of Chicago Press, Chicago, pp. 315-329.

Kauffman, E. G., and B. B. Sageman (1990). Biological sensing of benthic environments in dark shales and related oxygen-restricted facies, in *Cretaceous Resources, Events, and Rhythms: Background and Plans for Research*, R. N. Ginsburg and B. Beaudoin, eds., Kluwer Academic Publishers, Dordrecht, pp. 121-138.

Kauffman, E. G., W. A. Cobban, and D. E. Eicher (1976). Albian through lower Coniacian strata, biostratigraphy, and principal events, Western Interior, United States, in *Ann. Mus. d'Hist. Nat. Nice T. IV*, G. Thomel and R. Reyment, eds., pp. XXIII.1-XXIII.52.

Kauffman, E. G., L. M. Pratt, *et al.* (1985). A field guide to the stratigraphy, geochemistry, and depositional environments of the Kiowa-Skull Creek, Greenhorn, and Niobrara marine cycles in the Pueblo, Canon City area, Colorado, in *Fine-Grained Deposits and Biofacies of the Cretaceous Western Interior Seaway: Evidence of Cyclic Sedimentary Processes*, L. M. Pratt, E. G. Kauffman, and F. B. Zelt, eds., Society of Economic Paleontologists and Mineralogists, 2nd Annual Midyear Meeting, Golden, Colo., Field Trip Guidebook 4, pp. FRS-1 - FRS-26.

Kauffman, E. G., B. B. Sageman, W. P. Elder, and E. R. Gustason (1987). High-resolution event stratigraphy, Greenhorn Cyclothem (Cretaceous: Cenomanian-Turonian), Western Interior Basin of Colorado and Utah, Geological Society of America, Rocky Mountain Section Meeting, Field Guide, Boulder, Colo., 198 pp.

Kauffman, E. G., W. P. Elder, and B. B. Sageman (1991). High-resolution correlation: A new tool in chronostratigraphy, in *Cycles and Events In Stratigraphy*, G. Einsele, W. Ricken, and A. Seilacher, eds., Springer-Verlag, Berlin, pp. 795-819.

Kauffman, E. G., B. B. Sageman, J. I. Kirkland, W. P. Elder, P. J. Harries, and T. Villamil (1993). Molluscan biostratigraphy of the Western Interior Cretaceous Basin, in *Evolution of the Western Interior Basin*, W. G. E. Caldwell and E. G. Kauffman, eds., Geological Association of Canada Special Paper 39, pp. 397-434.

Kennedy, W. J. (1971). *Cenomanian Ammonites of Southern England*, Palaeontology Special Paper 8, 134 pp.

Kennedy, W. J., and W. A. Cobban (1991). Stratigraphy and interregional correlation of the Cenomanian-Turonian transition in the Western Interior of the United States near Pueblo, Colorado, a potential boundary stratotype for the base of the Turonian stage, *Newsletters on Stratigraphy 24*, 1-33.

Koch, C. F. (1977). Evolutionary and ecological patterns of upper Cenomanian (Cretaceous) mollusc distribution in the Western Interior of North America, Unpublished Ph.D. thesis, George Washington University, Washington, D.C., v. 1, 72 pp.

Koch, C. F. (1980). Bivalve species duration, areal extent and population size in a Cretaceous sea, *Paleobiology 6*, 184-192.

Koch, C. F., and J. P. Morgan (1988). On the expected distribution of species ranges, *Paleobiology 14*, 126-138.

Larson, R. L. (1991a). Latest pulse of the Earth; Evidence for a mid-Cretaceous superplume, *Geology 19*, 547-550.

Larson, R. L. (1991b). Geological consequences of superplumes, *Geology 19*, 963-966.

Leckie, R. M. (1985). Foraminifera of the Cenomanian-Turonian boundary interval, Greenhorn Formation, Rock Canyon Anticline, Colorado, in *Fine-Grained Deposits and Biofacies of the Cretaceous Western Interior Seaway: Evidence of Cyclic Sedimentary Processes*, L. M. Pratt, E. G. Kauffman, and F. B. Zelt, eds., Society of Economic Paleontologists and Mineralogists, 2nd Annual Midyear Meeting, Golden, Colo., Field Trip Guidebook 4, pp. 139-150.

Marshall, C. R. (1991). Estimation of taxonomic ranges from the fossil record, in *Analytical Paleobiology*, N. L. Gilinsky and P. W. Signor, eds., Paleontological Society, Short Courses in Paleontology No. 4, pp. 19-38.

McDonough, K. J., and T. Cross (1991). Late Cretaceous sea level from a paleoshoreline, *Journal of Geophysical Research 96*, 6591-6607.

Melosh, H. J. (1982). The mechanics of large meteoroid impacts in the Earth's oceans, in *Geological Implications of Impacts of Large Asteroids and Comets on the Earth*, L. T. Silver and P. H. Schultz, eds., Geological Society of America Special Paper 190, pp. 121-128.

Obradovich, J. (1993). A Cretaceous time scale, in *Evolution of the Western Interior Basin*, W. G. E. Caldwell and E. G. Kauffman, eds., Geological Association of Canada Special Paper 39, pp. 379-396.

Oeschger, H., and A. Arquit (1991). Resolving abrupt and high-frequency global changes in the ice-core record, in *Global Changes of the Past*, R. S. Bradley, ed., Office for Interdisciplinary Earth Studies, UCAR, Boulder, Colo., pp. 175-200.

Orth, C. J., M. Attrep, Jr., X. Mao, E. G. Kauffman, R. Diner, and W. P. Elder (1988). Iridium abundance maxima in the upper Cenomanian extinction interval, *Geophysical Research Letters 15*, 346-349.

Orth, C. J., M. Attrep, Jr., and L. R. Quintana (1990). Iridium abundance patterns across bio-event horizons in the fossil record, in *Global Catastrophes in Earth History*, V. L. Sharpton and P. D. Ward, eds., Geological Society of America Special Paper 247, pp. 45-60.

Orth, C. J., M. Attrep, Jr., L. R. Quintana, W. P. Elder, E. G. Kauffman, R. Diner, and T. Villamil (1993). Elemental abundance anomalies in the late Cenomanian extinction interval: A search for the source(s), *Earth and Planetary Science Letters 117*, 189-204.

Philip, J. (1991). L'enregistrement des evenements de la limite Cenomanien/Turonien sur les plates-formes carbonatees de la Tethys, in *Colloque International sur les Evenements de la Limite Cenomanien-Turonien*, Geologie Alpine, Mem. H. S. No. 17, Grenoble, pp. 105-106.

Pindell, J. L., and S. F. Barrett (1990). Geological evolution of the Caribbean Region; A plate-tectonic perspective, in *The Geology of North America: H. The Caribbean Region*, G. Dengo and J. P. Case, eds., Geological Society of America, Boulder, Colo., pp. 405-432

Pitman, W. C., III (1978). Relation between eustasy and stratigraphic sequences of passive margins, *Geological Society of America Bulletin 89*, 1389-1403.

Pratt, L. M. (1985). Isotopic studies of organic matter and carbonate in rocks of the Greenhorn marine cycle, in *Fine-Grained Deposits and Biofacies of the Cretaceous Western Interior Seaway: Evidence of Cyclic Sedimentary Processes*, L. M. Pratt, E. G. Kauffman, and F. B. Zelt, eds., Society of Economic Paleontologists and Mineralogists, 2nd Annual Midyear Meeting, Golden, Colo., Field Trip Guidebook 4, pp. 38-48.

Rampino, M. R., R. B. Strothers, B. O'Neil, B., and B. Haggerty (1993). Asteroid impacts, mass extinction events, and flood basalt eruptions—An external driver, *Stratigraphic Record of Global Change*, Abstract Volume, 1990 SEPM Meeting, Pennsylvania State University, pp. 57-58.

Raup, D. M., and J. J. Sepkoski (1984). Periodicity of extinctions in the geologic past, *Proceedings of the National Academy of Sciences USA 81*, 801-805.

Raup, D. M., and J. J. Sepkoski (1986). Periodic extinctions of families and genera, *Science 231*, 833-836.

Sageman, B. B. (1985). High-resolution stratigraphy and paleobiology of the Hartland Shale Member: Analysis of an oxygen-deficient epicontinental sea, in *Fine-Grained Deposits and Biofacies of the Cretaceous Western Interior Seaway: Evidence of Cyclic Sedimentary Processes*, L. M. Pratt, E. G. Kauffman, and F. B. Zelt, eds., Society of Economic Paleontologists and Mineralogists, 2nd Annual Midyear Meeting, Golden, Colo., Field Trip Guidebook 4, pp. 110-121.

Sageman, B. B. (1992). High-Resolution Event Stratigraphy, Carbon Geochemistry, and Paleobiology of the Upper Cenomanian Hartland Shale Member (Cretaceous), Greenhorn Formation, Western Interior, U. S., Unpublished Ph.D. thesis, University of Colorado, Boulder, 532 pp.

Sageman, B. B., P. B. Wignall, and E. G. Kauffman (1991). Biofacies models for organic-rich facies in epicontinental seas: Tool for paleoenvironmental analysis, in *Cycles and Events in Stratigraphy*, G. Einsele, W. Ricken, and A. Seilacher, eds., Springer-Verlag, Berlin, pp. 542-564.

Sageman, B. B., E. G. Kauffman, P. H. Harries, and W. P. Elder (1994). Cenomanian-Turonian bioevents and ecostratigraphy: Contrasting scales of local, regional, and global events, in *Paleontological Event Horizons*, C. Brett and G. Baird, eds., Columbia University Press, New York (in press).

Schlanger, S. O., and H. C. Jenkyns (1976). Cretaceous oceanic anoxic events: Causes and consequences, *Geol. Mijnbouw 55*, 179-184.

Scholle, P. A., and M. A. Arthur (1980). Carbon isotope fluctuations in Cretaceous pelagic limestones: Potential stratgraphic and petroleum exploration tool, *American Association of Petroleum Geologists Bulletin 64*, 67-87.

Scott, G. R. (1964). Geology of the northwest and northeast Pueblo Quadrangles, Colorado, *U.S. Geological Survey, Miscellaneous Geological Investigations Map I-408*.

Scott, G. R. (1969). General and engineering geology of the northern part of Pueblo, Colorado, *U.S. Geological Survey Bulletin 1262*, 131 pp.

Sepkoski, J. J. (1990). The taxonomic structure of periodic extinction, in *Global Catastrophes in Earth History*, V. L. Sharpton and P. D. Ward, eds., Geological Society of America Special Paper 247, pp. 33-44.

Sepkoski, J. J., Jr. (1993). Ten years in the library: New data confirm paleontological patterns, *Paleobiology 19*(1), 43-51.

Signor, P. W., III, and J. H. Lipps (1982). Sampling biases, gradual extinction patterns, and catastrophes in the fossil record, in *Geological Implications of Impacts of Large Asteroids and Comets on the Earth*, L. T. Silver and P. H. Schultz, eds., Geological Society of America Special Paper 190, pp. 291-303.

Simberloff, D. S. (1984). Mass extinction and the destruction of moist tropical forests, *Zh. Obshch. Biol. 45*, 767-778.

Strothers, R. B., and M. R. Rampino (1990). Periodicity in flood basalts, mass extinctions, and impacts; A statistical view and a model, in *Global Catastrophes in Earth History*, V. L. Sharpton and P. D. Ward, eds., Geological Society of America Special Paper 247, pp. 9-18.

Thompson, L. G. (1991). Ice core records with emphasis on the global record of the last 2000 years, in *Global Changes of the Past*, R. S. Bradley, ed., Office for Interdisciplinary Earth Studies, UCAR, Boulder, Colo., pp. 201-224.

Villamil, T., and C. Arango (1994). High-resolution analysis of the Cenomanian-Turonian boundary in Colombia: Evidence for sea level rise, condensation, and upwelling, in *Mesozoic-Cenozoic Stratigraphy and Evolution of Northern South America: Implications for Eustasy from a Cretaceous-Eocene Passive Margin*, J. Pindell and C. Drake, eds., Geological Society of America Special Paper (in press).

Watkins, D. K. (1985). Biostratigraphy and paleoecology of calcareous nannofossils in the Greenhorn marine cycle, in *Fine-Grained Deposits and Biofacies of the Cretaceous Western Interior Seaway: Evidence of Cyclic Sedimentary Processes*, L. M. Pratt, E. G. Kauffman, and F. B. Zelt, eds., Society of Economic Paleontologists and Mineralogists, 2nd Annual Midyear Meeting, Golden, Colo., Field Trip Guidebook 4, pp. 151-156.

Watkins, D. K., T. J. Bralower, J. M. Covington, and C. G. Fischer (1993). Biostratigraphy and paleoecology of upper Cretaceous calcareous nannofossils in the Western Interior Basin, North America, in *Evolution of the Western Interior Basin*, W. G. E. Caldwell and E. G. Kauffman, eds., Geological Association of Canada Special Paper 39, pp. 521-537.

Webb, T., III (1991). The spectrum of temporal climate variability: Current estimates and the need for global and regional time series, in *Global Changes of the Past*, R. S. Bradley, ed., Office for Interdisciplinary Earth Studies, UCAR, Boulder, Colo., pp. 61-81.

Wilde, P., M. S. Quinby-Hunt, and W. B. N. Berry (1990). Vertical advection from oxic or anoxic water from the main pycnocline as a cause of rapid extinction or rapid radiations, in *Extinction Events in Earth History*, E. G. Kauffman and O. H. Walliser, eds., Springer-Verlag, Berlin, pp. 85-98.

Wilson, E. O. (1988). The current state of biological diversity, in *Biodiversity*, National Academy Press, Washington, D.C., pp. 3-20.

Wright, C. W., and W. J. Kennedy (1981). *The Ammonoidea of the Plenus Marls and the Middle Chalk*, Palaeontological Society Monograph, 148 pp.

Zelt, F. B. (1985). Paleoceanographic events and lithologic/geochemical facies of the Greenhorn marine cycle (upper Cretaceous) examined using natural gamma-ray spectrometry, in *Fine-Grained Deposits and Biofacies of the Cretaceous Western Interior Seaway: Evidence of Cyclic Sedimentary Processes*, L. M. Pratt, E. G. Kauffman, and F. B. Zelt, eds., Society of Economic Paleontologists and Mineralogists, 2nd Annual Midyear Meeting, Golden, Colo., Field Trip Guidebook 4, pp. 49-59.

APPENDIX

Rates, Patterns, and Timing of the C-T Mass Extinction and Associated Environmental Perturbations (see Figure 3.6)

LATE CENOMANIAN BACKGROUND CONDITIONS

The extinction of Caribbean reef communities reached its peak prior to the events described below for the subtropical to temperate Western Interior Province. Precise correlation between the tropics and the north temperate realm is difficult at present. Oxygen isotope values are relative to the Pedee belemnite standard (PDB).

(1) 150,000-yr duration: A moderate negative shift in $\delta^{13}C_{org}$ to lowest Cenomanian values (–27.7‰); positive shifts in $\delta^{18}O$ and total organic carbon (TOC; wt%) to average late Cenomanian values (–8‰ and 3.5–4 wt%, respectively). Several moderately elevated peaks in molluscan diversity and abundance. Organic carbon storage occurs in the benthic zone of dysoxic stratified seas, but with dynamic, episodically oxygenated bottom waters. Falling sea level culminates in a third-order sequence boundary at the top of this interval.

(2) 50,000-yr duration: A moderate positive shift in $\delta^{13}C_{org}$ to average late Cenomanian values (–26.5 to –27‰) a negative shift in $\delta^{18}O$ values to a late Cenomanian low of –10.5‰, and major decrease in TOC to 1.2 wt%. An oxygenated interval in the benthic zone during early sea-level rise leads to moderate benthic molluscan diversity and sequestering of Mn in benthic sediments. A modest extinction among low-oxygen adapted benthic molluscs and ammonites is followed by an origination and/or immigration event among benthic molluscs adapted to somewhat higher oxygen levels.

(3) 60,000-yr duration: Return of $\delta^{18}O$ values to normal late Cenomanian values; a major increase in TOC to a late Cenomanian high of 4.5 wt%, and sharp increase in U-Th values, both indicating strong benthic oxygen restriction and a stratified water column during late sea-level rise. High surface productivity leads to organic carbon storage. Despite this, a significant origination event occurs, especially among low-oxygen adapted benthic molluscs and pelagic ammonites, leading to a modest increase in molluscan diversity.

(4) 120,000-yr duration: Relatively stable oceanographic conditions associated with third-order sea-level highstand and active volcanism in the basin. Organic carbon levels remain very high, reflecting a stratified, dysoxic middle to lower water column. An origination interval among mainly benthic molluscs leads to moderate abundance and diversity values.

(5) 50,000-yr duration: Two abrupt, major trace element (Ti, V, U) enrichment events occur against otherwise stable background geochemistry during a modest third-order sea-level fall. These trace element spikes are associated with major short-term increases in molluscan abundance and diversity, and a moderate origination event, mainly among benthic bivalves. Biological trends possibly reflect elevated nutrient levels.

(6) 30,000-yr duration: Stable oceanographic background conditions are associated with late phases of a third-order relative sea-level fall and a molluscan origination event. A third-order sequence boundary, reflecting maximum relative sea-level fall, caps the sequence.

(7) 90,000-yr duration: Relatively stable oceanographic background conditions during early sea-level rise are associated with active explosive volcanism and ash falls, and with an abrupt U-Th enrichment. Significant increases in molluscan diversity and abundance may reflect nutrient enrichment.

(8) 90,000-yr duration: The final stable phase of oceanographic background conditions prior to the C-T perturbed interval is associated with active explosive volcanism; numerous ash falls: and successive, short-term, trace element enrichment events (Mn-Co followed by U). An unexplained decrease in molluscan diversity and abundance (extinction?) is followed by a modest origination event. Planktonic foraminifers show a marked diversity increase (10-11 species) after a long period of stable background levels. Lowering of relative sea level continues.

(9) 220,000-yr duration: Initiation of the geochemically perturbed interval of the C-T boundary sequence, 880,000 yr below the boundary, is associated with increased volcanism and rapid northward immigration of southern warm water masses; major positive excursions of $\delta^{13}C_{org}$ (initiating a global oceanographic event), $\delta^{18}O$, and TOC reflect a trend toward normalization of marine surface waters in the basin, but significant stratification of the underlying water column, lowering of benthic oxygen levels, and possible immigration and expansion of the oceanic oxygen-minimum zone. These events mark initiation of the global oceanic anoxic event (OAE II) that characterizes the C-T boundary interval worldwide. They are associated with rapid northward immigration of warm-temperate to subtropical molluscs and microbiotas from Gulf Coast and Caribbean sources, a marked origination event among shallow water benthic and pelagic lineages, and the first reoccur-

rence of benthic foraminifers after >700,000 yr of environmental exclusion from the benthic zone. There is a general decrease in benthic molluscan diversity.

(10) 140,000-yr duration: Oceanic environments continued to show dynamic fluctuations as southern water masses expanded their rapid northward immigration, and global sea level approached its highest Mesozoic-Cenozoic stand. The value of $\delta^{13}C_{org}$ reached its first of four positive late Cenomanian peaks (−24.5‰), $\delta^{18}O$ declined sharply, U-Th became enriched in benthic sediments, and TOC decreased rapidly to near zero after a million years at high levels, reflecting a major short-term oxygenation event in the benthic zone. Moderate increases in molluscan and planktic foraminiferal diversity, a major diversification among benthic foraminifers, and a final origination event among warm water molluscs, characterize this interval. Thus, conditions for life seemed highly favorable, with increasing benthic and pelagic diversity, just prior to the abrupt initiation of the C-T mass extinction interval during the next 50,000 yr interval.

LATE CENOMANIAN MASS EXTINCTION

High-resolution geochemical and paleobiological data suggest that, after the widespread elimination of reef ecosystems in the Caribbean Province by middle late Cenomanian time (Johnson and Kauffman, 1990), the last 520,000 yr of the Cenomanian was characterized by chaotic, short-term, large-scale perturbations in the ocean-climate system. The rate and magnitude of these changes exceeded the adaptive ranges of diverse, extinction-prone marine organisms, narrowly adapted to an equable, maritime-dominated, warm greenhouse world, near one of the highest peaks of Phanerozoic eustatic sea-level rise and global warming. These successive perturbations, most of less than 100,000-yr duration, caused steps of mass extinction that were ecologically graded, first affecting tropical to subtropical taxa, subsequently warm temperate lineages, and finally the cooler temperate and more cosmopolitan elements of marine ecosystems. These environmental perturbations, and resultant C-T extinction events, are as follows.

(11) 50,000-yr duration: Stable isotope and TOC values are similar to (10) and atypical of late Cenomanian background conditions. This interval is characterized by the first of 11 successive phases of trace element enrichment (Ir, C, Ni, Sc peaks) above background levels; these are interpreted as a dramatic series of oceanic advection events associated with benthic touchdown of the oceanic oxygen minimum zone during OAE II, remobilization of sequestered trace elements in oxygen-deficient waters, and their reprecipitation on oxygenated substrates above the redoxcline. This trace element enrichment and the numerous volcanic ash falls changed marine chemistry and nutrient levels. Nutrient enhancement may have initially caused increases in diversity and abundance of planktic and benthic foraminifers, and molluscs. The initial regional step of the C-T mass extinction (MX1A), however, eliminated several subtropical to warm-temperate molluscan lineages, and was correlative with the first trace element enrichment peak, including iridium. Contouring of trace element values at numerous localities in the basin (Orth et al., 1993) shows a proto-Caribbean source, with values diminishing northward into Canada. A second extinction event, characterized by the abrupt loss of keeled rotaliporid foraminifers (PFX), occurs at the boundary between this and the succeeding interval, with initiation of a second zone of Ir and other trace element enrichment (see 12), and possible microtektite concentrations (in Colombia).

(12) 80,000-yr duration: A second short-term trace element spike (C, Mn, Ni, Sc advection) characterizes this interval. An abrupt positive $\delta^{18}O$ excursion is coupled with further drop in TOC values and continued active volcanism. Nutrient enrichment may have led to increases in benthic foraminifer diversity and abundance, but cannot explain decline in both molluscs and planktic foraminifers, reflecting the effects of the planktic foraminifer extinction (PFX) event ("Rotalipora extinction") at the base of the interval.

(13) 40,000-yr duration: One of the most geochemically dynamic intervals of the C-T extinction, associated with high sea level and active volcanism. Two successive trace element enrichment levels, the first characterized by Ir and C, the second by Ni and Sc spikes, may record a rapid trace element advection during downward expansion of OAE II to the seafloor. Modest negative excursions of $\delta^{13}C$ and $\delta^{18}O$, and a small positive C_{org} spike, are associated with the trace element excursions, describing dynamic ocean-climate systems. Significant increases in abundance and diversity of molluscs, and planktic and benthic foraminifers, may reflect increased nutrient levels. A short-term extinction step (MX1B), during which some subtropical and warm temperate molluscs disappeared, was associated with iridium enrichment.

(14) 20,000-yr duration: Rapid, large-scale geochemical fluctuations continue as perturbations of the ocean-climate system intensify. Two short-term trace element enrichment levels, with enhanced Ir, Pt, Sc and C followed by Mn and C enrichment, record additional trace element advection from the seafloor. A second major positive $\delta^{13}C$ spike within the global $\delta^{13}C$ excursion reaches maximum Cretaceous levels in this region (−24.2‰), and is associated with sharp positive followed by rapid negative $\delta^{18}O$

excursions, and depletion of C_{org} to near zero. Benthic foraminifers show a significant diversity increase: a modest origination event (OM9) occurs among molluscs. The major effects of these geochemical perturbations are two successive, large-scale mass extinction events—a first among subtropical and warm temperate molluscan lineages (MX2), directly correlative with the Ir-rich level, and shortly thereafter (15), a major extinction among southern, warm water elements (MX3).

(15) 20,000-yr duration: As in the stratigraphically lower sections (13, 14), rapid geochemical fluctuations in trace elements, stable isotopes, and C_{org} levels describe a highly perturbed, unstable ocean-climate system. Significant positive excursions occur in $\delta^{13}C$, $\delta^{18}O$, and C_{org} values, and a final major trace element enrichment layer (advection event) containing Ir, Pt, and C is directly associated with one of the largest warm temperate molluscan extinction events of the C-T interval (MX3). This is followed rapidly by a modest molluscan radiation (OM10), but the diversity and abundance of molluscs and benthic foraminifers decrease sharply through the interval.

(16) 170,000-yr duration: Geochemical fluctuations wane, but a strong positive $\delta^{13}C$ spike marks the last regionally correlative spike within the global positive $\delta^{13}C$ interval. It is associated with sharp negative $\delta^{18}O$ and C_{org} excursions, reduction in molluscan and planktic foraminifer diversity and abundance, and a moderately strong temperate molluscan extinction step (MX4).

HIGHEST CENOMANIAN EVENTS

(17) 140,000-yr duration: Dynamic changes in the ocean-climate system return, with very rapid positive to negative $\delta^{13}C$ excursions, negative to positive $\delta^{18}O$ excursions, very high global sea level, and a moderate U, Th, and trace element enrichment level. The rapidity and large scale of these oceanographic fluctuations exceeded the adaptive ranges of a large diversity of warm temperate molluscan genera and species. This produced the largest (terminal) Cenomanian extinction event in the C-T mass extinction interval, predominantly among temperate and cosmopolitan ammonite and bivalve lineages (MX5), which disappeared at or just below the Cenomanian-Turonian boundary.

BASAL TURONIAN EVENTS

(18) 80,000-yr duration: The last extinction steps of the C-T boundary interval occur in association with population expansion among surviving clades, and early radiations of new lineages in the basal Turonian (18-20 herein). This interval is characterized by active volcanism, a sharp negative excursion in $\delta^{13}C$, and rapid negative-positive-negative excursions in $\delta^{18}O$. Marked decline in molluscan abundance and diversity is followed shortly by the first Turonian origination event (OM12). Sea level is nearly at its highest Mesozoic stand (>300 m above present stand).

(19) 150,000-yr duration: Maximum eustatic highstand (>300 m above present stand) was accompanied by sharp increase, then decrease in $\delta^{18}O$ values; a sharp decrease, then increase in TOC levels; and the final two trace element enrichment levels (Co, U, Th, followed by Mn, U, and V enrichment) of the C-T oceanic advection interval. A modest buildup in diversity and abundance among temperate molluscs was interrupted by a small extinction event (MX6).

FINAL PHASES OF THE C-T MASS EXTINCTION

(20) 230,000-yr duration: Strong negative excursions of $\delta^{13}C$ and TOC, and a strong positive excursion of $\delta^{18}O$, record the final major perturbations of the C-T boundary interval. Molluscan diversity and abundance fluctuate wildly within the interval, from low to high values. An important molluscan radiation (OM13) is sandwiched between a major nannoplankton extinction (NX) and a moderate temperate molluscan extinction step (MX7). Above this, extinction events no longer cancel out radiation events, and overall diversity builds gradually as ecosystems recover.

EARLY RECOVERY INTERVAL

(21) 110,000-yr duration: Despite active volcanism and ash fall, sharp reductions in $\delta^{13}C$ and $\delta^{18}O$ values, and a major increase in TOC, the low diversity survivor and recovery faunas of the early Turonian remain fairly stable, with exceptionally low diversity levels noted only among benthic foraminfers.

(22) 370,000-yr duration: The final phases of geochemical perturbations associated with the C-T mass extinction interval occur at this level. The $\delta^{13}C$ and $\delta^{18}O$ values show strong positive excursions, whereas TOC drops dramatically to near zero. Increases in abundance and diversity among molluscs reflect two major origination events (OM14, OM15) during broad early Turonian radiations. These are associated with the final small extinction step (MX8) of the 1.5- to 2-m.y. long C-T mass extinction interval including the paleotropics). This small extinction affects predominantly mid- to north-temperate and cosmopolitan molluscan lineages.

END OF SAMPLING INTERVAL

4
Cretaceous-Tertiary (K/T) Mass Extinction: Effect of Global Change on Calcareous Microplankton

GERTA KELLER
Princeton University

KATHARINA v. SALIS PERCH-NIELSEN
Swiss Federal Institute of Technology

ABSTRACT

The effects of the Cretaceous-Tertiary (K/T) boundary global change on calcareous nannoplankton and planktic foraminifera are most severe in low latitudes and negligible in high latitudes. In low latitudes, species extinctions are complex and prolonged beginning during the final 100,000 to 300,000 yr of the Cretaceous, accelerating across the K/T boundary, and reaching maximum negative conditions between 10,000 and 40,000 yr into the Tertiary accompanied by low primary productivity. In high latitudes, no significant species extinctions occurred at or near the K/T boundary, and all dominant species thrived well into the early Tertiary. Return to a more stable ecosystem and to increased marine productivity in low latitudes does not occur until about 250,000 to 350,000 yr after the K/T boundary, coincident with the extinction of Cretaceous survivors in high latitudes. Within this transition interval, habitats of deep- and intermediate-dwelling tropical planktic foraminiferal species are gradually and selectively eliminated in low latitudes, and by K/T boundary time only cosmopolitan surface dwellers survive. This implies the disruption of the water-mass structure, change in the thermocline, and a drop in surface productivity. Although no single cause is likely to account for these different prolonged and dramatic faunal and environmental changes between low and high latitudes, long-term oceanic instability associated with sea-level, temperature, salinity, and productivity fluctuations may account for most of the faunal changes observed in planktic foraminifera. However, other environmental changes (e.g., volcanism, bolide impact) may have accelerated the demise of the low-latitude Cretaceous fauna already on the decline.

INTRODUCTION

The Cretaceous-Tertiary boundary transition is marked by one of the most dramatic environmental changes in the Earth's history, with both cause and effect still vigorously disputed. Presently the most popular theory of the cause of this global change is a large extraterrestrial bolide impact. Supporters of this theory cite anomalously high concentrations of noble elements and shocked mineral grains in a thin boundary clay layer in the marine and terrestrial realm as sufficient evidence of a bolide impact. In addition, Alvarez et al. (1980) viewed the reported sudden extinction of all but one Cretaceous planktic foraminiferal species as evidence of the catastrophic effect of this impact on Earth's biota. Alternative theories invoke long-term Earth-derived causes such as large-scale volcanism (McLean, 1985; Officer and Drake, 1985), mantle plume activity (Loper and McCartney 1986, 1988), sea-level fluctuations, and climate changes (Hallam, 1989) as seen in the geological record spanning the K/T boundary transition, as well as the decline and extinction of many fossil groups during the late Maastrichtian (Kauffman, 1984; Sloan et al., 1986; Keller, 1888, 1989b; Keller and Barrera, 1990; Canudo et al., 1991). Although no evidence presented to date conclusively supports either extraterrestrial or Earth-derived causes, the magnitude of this global change remains undisputed. What is unclear, however, is the nature, tempo, and geographic extent of this mass extinction.

Regardless of the ultimate cause of this global change, the effect on life and especially on calcareous marine microplankton can be evaluated from the geological record. Shells of calcareous nannofossils and planktic foraminifera contribute the bulk of marine sediments and provide a record not only of species extinction and evolution, but also of population dynamics in response to environmental perturbations. A record of climate and productivity changes is retained in the carbon and oxygen isotope ratios of their shells. No other fossil group provides as much information on the changing environment across the K/T boundary transition.

Despite the rich information contained in the calcareous marine microplankton, specialists still differ considerably in their interpretation of the extinction record, with one group arguing for catastrophic extinctions of nearly the entire fauna and flora at the K/T boundary and the other group arguing for an extended extinction period. Among nannofossil experts, the difference in opinion centers on whether Cretaceous species in Tertiary sediments are reworked or survivors (Perch-Nielsen, 1979a,b; Perch-Nielsen et al., 1982; Thierstein, 1982; Jiang and Gartner, 1986). Unfortunately, the very small size of nannofossils, and their easy transportation and redeposition, make determination of in situ assemblages a nontrivial problem, and neither view can be proven beyond reasonable doubt.

Among foraminiferal experts, differences rest largely on two apparently contradictory extinction records, the deep sea versus continental shelf or neritic environments. Deep-sea records show all but one Cretaceous species extinct at the K/T boundary, apparently supporting a catastrophic impact (e.g., Smit and Romein, 1985; D'Hondt and Keller, 1991). In contrast, neritic sections show an extended period of extinctions beginning before the K/T boundary and continuing well into the Tertiary; thereby providing only weak support for a geologically instantaneous catastrophic event (Keller, 1988, 1989a,b; Canudo et al., 1991; Huber, 1991). Impact supporters reconcile these different records by interpreting Cretaceous species present above the K/T boundary as reworked (e.g., Smit, 1982, 1990; Smit and Romein, 1985).

Unlike the unresolved problem of reworked nannofossils, reworked Cretaceous foraminiferal species in Tertiary deposits can be positively identified by their physical and preservational characteristics, as well as by their $\delta^{13}C$ isotope signal (Barrera and Keller, 1990; Keller et al., 1993). Based on these criteria, up to one-third of the Cretaceous species may have survived the K/T boundary event. In contrast, the near total extinction of Cretaceous species in the deep sea appears to be an artifact of a temporally incomplete stratigraphic record as discussed below (MacLeod and Keller, 1991a,b). Recently, the study of marine K/T sections in northern and southern high latitudes has revealed no significant species extinctions and the survival of nearly all species well into the Tertiary, which indicates that the effects of the K/T mass extinction may have been limited to lower latitudes (Keller, 1993; Keller et al., 1993).

In this chapter we examine the effects of global environmental changes on calcareous marine microplankton across the K/T boundary in deep-sea and continental shelf sequences. We have chosen the temporally most complete sections known to date: El Kef in Tunisia, Caravaca in Spain, Brazos River in Texas, Deep-Sea Drilling Program (DSDP) Site 528 in the South Atlantic, and Ocean Drilling Program (ODP) Site 738C in the Indian Antarctic Ocean (Figure 4.1). These four sections span bathyal to neritic environments from high to low latitudes and thus provide a cross section of marine ecosystems. We address four critical and interrelated topics:

1. How complete are K/T boundary sequences?
2. What is the nature of species extinctions?
3. What is the effect on dominant species populations?
4. Are specific habitats selectively destroyed?

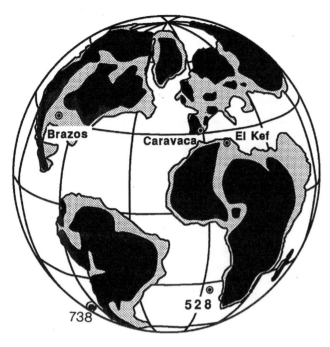

FIGURE 4.1 Locations of K/T boundary sections examined for this study plotted on a paleogeographic reconstruction of continental positions at the time of the K/T boundary (66.4 Ma). White = ocean basins; light stipple = continental platforms; black = inferred extent of terrestrial exposure.

HOW COMPLETE ARE K/T BOUNDARY SECTIONS?

To test competing causal hypotheses or to evaluate the effect of global change on marine microplankton, accurate estimates of the stratigraphic and temporal completeness of individual K/T boundary sections are needed. In current stratigraphic analysis the blueprint for judging complete and continuously deposited K/T boundary sequences seems to be the presence and relative thickness of a boundary clay, the presence of nannofossil zones *Micula prinsii* and *Neobiscutum romeinii* and planktic foraminiferal Zone *Guembelitria cretacea* (P0) followed by the *Parvularugoglobigerina eugubina* Zone (P1a) (Perch-Nielsen et al., 1982; Smit, 1982; Smit and Romein, 1985). By this definition, many K/T boundary sections have been judged as "relatively complete" even though the earliest Tertiary Zone P0 or the *N. romeinii* Zone are missing. Moreover, this method of judging chronostratigraphic completeness does not allow the recognition of the existence of intrazonal hiatuses, and may thus lead to erroneous interpretations of the nature and rate of environmental change.

Foraminiferal workers have generally examined deep-sea sections, which were believed to be more complete than shallower continental shelf sections, and found that virtually all planktic foraminiferal species disappeared simultaneously at the K/T boundary. This pattern of species extinctions is illustrated in Figure 4.2 for DSDP Site 528, which contains a relatively continuous sedimentary record with a thin laminated boundary interval (in the core catcher) (D'Hondt and Keller, 1991). Yet, as in nearly all deep-sea sections, the basal Tertiary Zone P0 and probably part of Zone P1a are missing. By judging DSDP Site 528, as well as all other low- to mid-latitude deep-sea sites that show the same sudden mass extinction and absence of the basal Tertiary Zone P0 as representing a temporally complete record, the obvious interpretation was a geologically instantaneous catastrophic event such as a bolide impact. This interpretation has even been applied to the stratigraphically more complete K/T boundary sequences at El Kef and Caravaca by Smit (1982, 1990), based on the assumption that all Cretaceous species present above the K/T boundary (except *G. cretacea*) are reworked (see Canudo et al., 1991, for a discussion of this problem).

Among nannofossil workers, Worsley (1974) speculated that major hiatuses characterized all deep-sea sections and shorter hiatuses were present in marine-shelf sections. This observation tended to be supported by the later discovery of the new uppermost Maastrichtian nannofossil Zone *Micula prinsii* and an unnamed basal Tertiary interval below the *Neobiscutum romeinii* subzone at El Kef by Perch-Nielsen (1979a). Despite this early recognition of an incomplete K/T boundary record in the deep sea by both foraminiferal and nannofossil workers, few attempts at systematic chronostratigraphic analysis have been made and only for a single or a few sections (Thierstein, 1982; Herbert and D'Hondt, 1990), largely because high-resolution stratigraphic records have not been available.

Thus, what has been missing from the K/T controversy is a comprehensive chronostratigraphic and biostratigraphic data synthesis for all K/T transitions in marine depositional settings that would allow determination of the temporal completeness of each section and thereby clarify the nature of the extinction record. Such an analysis has recently been completed by MacLeod and Keller (1991a,b), who used the graphic correlation method of Shaw (1964) to compare the distribution of biostratigraphic and lithostratigraphic datums in 15 relatively complete boundary sections. These comparisons resulted in a composite estimate of 76 latest Maastrichtian to Early Paleocene (Zones P0-P1c) biostratigraphic datums for planktic foraminiferal and calcareous nannofossil species that corrects for intersequence diachroneity and allows correlations for individual sections to be made within a common chronostratigraphic model. Based on this method, analysis of 29 K/T boundary sections suggests that short, global, intrazonal hiatuses of varying duration are present in virtually

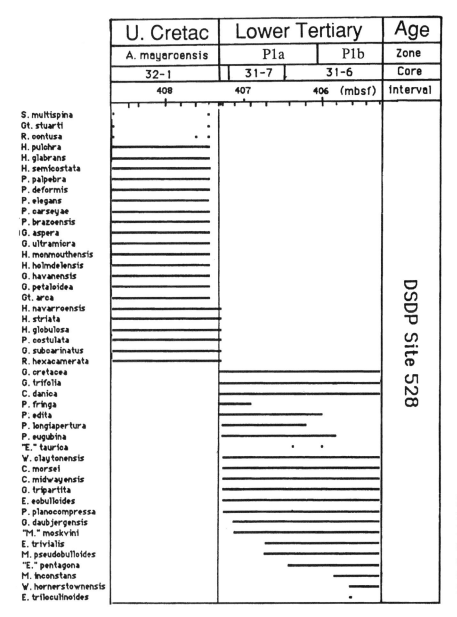

FIGURE 4.2 Stratigraphic ranges of planktic foraminifera at DSDP Site 528 (data from D'Hondt and Keller, 1991). Note that the abrupt termination of all Cretaceous species and sudden appearance of Tertiary species are the result of a hiatus that encompasses Zone P0.

all sections previously considered relatively complete. Moreover, dramatically different patterns of sediment accumulation appear to characterize neritic and deep-sea depositional environments following the K/T boundary.

Figure 4.3 illustrates this pattern for the sections discussed in this chapter along with the sea-level curve of Brinkhuis and Zachariasse (1988) based on dinoflagellate data from El Kef. Minimum estimates for hiatuses are given in black; maximum estimates are shown in stippled pattern. At DSDP Sites 528 and 577 as well as 10 other deep-sea sections examined (MacLeod and Keller, 1991a,b) an interval of nondeposition or short hiatus, which includes Zone P0 and the lower part of Zone P1a, characterizes the earliest Tertiary. This hiatus is coincident with a rapid transgression following the pre-K/T boundary maximum regression. Sediment accumulation occurs at this time, mainly in mid to outer shelf and upper bathyal environments as indicated by the presence of Zone P0, although a condensed interval or very short hiatus may be present in these sections also (Brazos, Caravaca, Agost). This sediment pattern suggests that the rapid sea-level transgression in Zone P0 trapped terrigenous sediment and organic carbon on continental shelves and temporarily deprived deep ocean basins of an inorganic sediment source and enhanced carbonate dissolution (Berger, 1970; Berger and Winterer, 1974; Loutit and Kennett, 1981; Haq et al., 1987; Donovan et al., 1988). Increased carbonate dissolution in the basal Tertiary (Zone P0) is widely recognized in

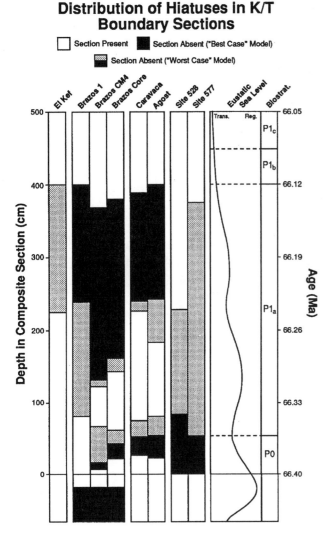

FIGURE 4.3 Extent and temporal distribution of intrazonal hiatuses plotted against the eustatic sea-level curve of Brinkhuis and Zachariasse (1988). Biostratigraphic zonation from Keller (1988) as modified by Keller and Benjamini (1991).

virtually all marine K/T boundary sequences (Keller, 1988, 1993; Canudo et al., 1991; D'Hondt and Keller, 1991; Schmitz et al., 1992). A second short hiatus is frequently present near the P1a/P1b Zone boundary (Figure 4.3), and like the P0/P1a, a hiatus may mark a sea-level lowstand. With the possible exception of El Kef, these two hiatuses appear to be present in the Spanish, Texan, and Israeli sections (Keller et al., 1990; MacLeod and Keller, 1991a,b). At El Kef an interval of reduced sedimentation (stippled pattern) may be present. A similar hiatus and dissolution pattern was observed at ODP Site 738 in the Indian Antarctic Ocean, where deposition occurred at about 1000-m depth (Keller, 1993).

It is evident from these analyses that neritic sequences such as those at El Kef and Brazos River, upper bathyal to outer neritic sequences such as those at Agost and Caravaca, and the upper slope Antarctic Ocean Site 738 contain the most detailed records of biological and environmental changes for the K/T transitions. Therefore, it is imperative that the effect of global change on marine microplankton be examined in these temporally most complete K/T boundary transitions.

SPECIES RESPONSE TO K/T DISTURBANCE

Planktic Foraminifera

Are species extinctions across the K/T boundary geologically instantaneous or successive and taxonomically selective? If species extinctions appear geologically instantaneous as illustrated for planktic foraminifera in deep-sea Site 528 (Figure 4.2), then the environmental change must have been caused by a major catastrophe, provided there is no hiatus in the sediment record. However, as discussed in the previous section, a geologically instantaneous mass extinction at the K/T boundary must now be ruled out because this pattern is a result of truncation by a hiatus (Figures 4.2 and 4.3). If species extinctions appear systematic, sequentially eliminating certain morphologies, geographically limited taxa, or habitats over an extended time interval, then long-term environmental changes such as changes in climate, sea level, rates of seafloor spreading, volcanism, and atmospheric pCO_2 are likely causes. However, an extended species extinction pattern does not necessarily rule out the possibility that an extraterrestrial impact (albeit of lesser magnitude than that proposed by the impact theory) occurred some time during this interval. However, recognizing such an impact, and separating its effects on marine microplankton from already ongoing environmental changes, are difficult, if not impossible.

The pattern of planktic foraminiferal species extinctions in low latitudes across the K/T boundary transition is best exemplified by the Tunisian El Kef section, which contains the most complete boundary sequence known to date. Figure 4.4 illustrates the sequence of species extinctions at El Kef along with illustrations of species and their sizes relative to each other. By comparing the apparent species extinction pattern in deep-sea Site 528 (Figure 4.2) with that of the mid to outer neritic El Kef section, it is immediately obvious that the patterns of extinction and evolution are dramatically different. Whereas species first and last appearances cluster at the K/T boundary at Site 528, an extended pattern beginning at least 25 cm below the boundary and ending about 4 m above it (Zone P1a) is observed at El Kef. (Note that the scale is reduced for the late Maastrichtian interval, relative to the Paleocene and

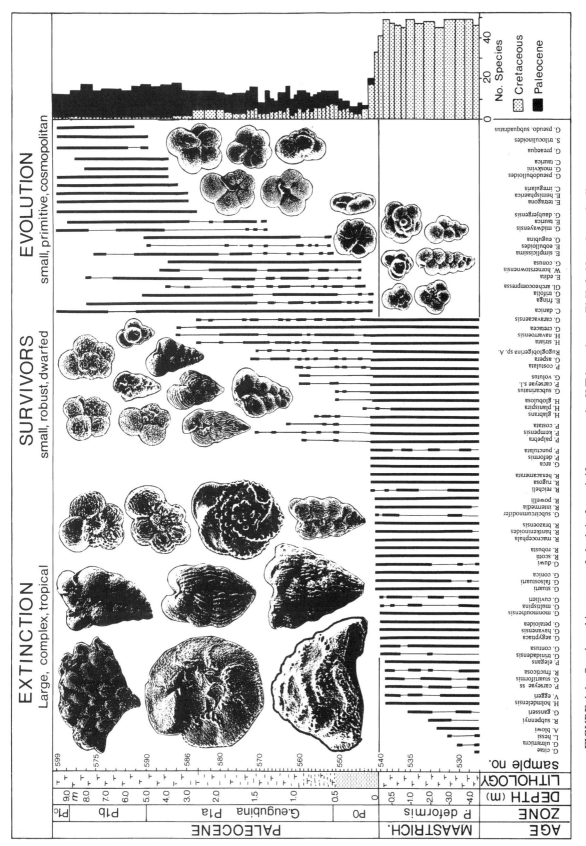

FIGURE 4.4 Stratigraphic ranges of planktic foraminifera across the K/T boundary at El Kef. Note the early disappearance of tropical, large, complex, deep, and intermediate water dwellers at or before the K/T boundary and the survival of smaller cosmopolitan surface water dwellers. Evolving Tertiary taxa are small, simple, and unornamented.

hence concentrates these extinctions near the boundary.) Fourteen species or 33% disappear at 25 cm (6 species) and 7 cm (8 species) below the K/T boundary. Twelve species extinctions (28%) coincide with the K/T boundary, and 6 species (14%) disappear 15 cm above the boundary. Of the remaining 16 Cretaceous species (38%), 8 species disappear in Zones P0 or basal Zone P1a; the remaining 8 species die out gradually in Zones P1a or basal P1b. This extended species extinction pattern is not unique to El Kef, but has also been observed at other low-latitude sections including Brazos River (Keller, 1989a,b), Agost, and Caravaca (Canudo et al., 1991). Such similar extinction patterns in regions as widely separated as Tunisia, Spain, and Texas argue against this being an artifact of sample spacing or reworking.

Moreover, species extinctions appear to be systematic, affecting tropical complex, large, and highly ornamented morphologies first (*globotruncanids*, *racemiguembelinids*, *planoglobulinids*) followed by subtropical and somewhat smaller morphologies (*pseudotextularids*, *rugoglobigerinids*; Figure 4.4). Still, smaller, less ornamented, robust cosmopolitan species with triserial, biserial, and rotalid morphologies (*guembelitrids*, *heterohelicids*, *globigerinellids*) survive the K/T boundary event and disappear gradually during the first 200,000 yr of the Tertiary. Within Tertiary sediments, these survivor species are generally dwarfed, with specimen size averaging about half the size of their Cretaceous ancestors. Their preservation is indistinguishable from the Tertiary fauna in the same samples, indicating that their origin is not likely due to redeposited Cretaceous sediments (Keller, 1988, 1989a,b). Moreover, stable isotopic signatures of the Cretaceous species *Guembelitria cretacea* and *Heterohelix globulosa*, unambiguously measure Tertiary values (Barrera and Keller, 1990; Keller et al., 1993). However, despite their ubiquitous presence in lower Danian sediments of continental shelf sections, these Cretaceous foraminifers have been routinely interpreted as reworked and consequently ignored or eliminated from faunal distribution lists (Hofker, 1960; Olsson, 1960; Berggren, 1962; Smit, 1982, 1990; Olsson and Liu, 1993). This interpretation has led to erroneously depicting a sudden extinction of all but one (*G. cretacea*) Cretaceous species exactly at the K/T boundary, as most recently illustrated by Smit (1990) and challenged by Canudo et al. (1991). Moreover, recent studies of high-latitude sequences in Denmark (Nye Klov) and the Indian Antarctic Ocean (ODP Site 738C) have shown that virtually all species survived the K/T boundary and thrived 200,000 to 300,000 yr into the Tertiary (Keller, 1993; Keller et al., 1993). This Cretaceous survivor fauna includes the same cosmopolitan taxa that survived the K/T boundary in low latitudes. In contrast to the more specialized tropical and subtropical taxa, these cosmopolitan species were probably tolerant of wide-ranging temperature, oxygen, salinity, and nutrient conditions.

Evolution of Tertiary species begins immediately after the K/T boundary with the appearance of very small, unornamented, and "primitive" morphologies (Figure 4.4). In these aspects they are similar to the surviving Cretaceous fauna. Cretaceous survivors gradually disappeared as Tertiary species diversified and somewhat larger morphotypes appeared. Species diversity, which declined near the K/T boundary from an average of 45 species to about 10 species in the earliest Tertiary Zone P0 and increased to about 15 species in Zone P1a (Figure 4.4) failed to recover until Zone P1c or about 300,000 yr after the K/T boundary. In high latitudes, this recovery coincides with the disappearance of the Cretaceous survivor taxa (Keller, 1993; Keller et al., 1993).

Calcareous Nannoplankton

Detection of the calcareous nannoplankton species extinction pattern across the K/T boundary transition is seriously hampered by the ease with which these specimens are reworked and the fact that there is usually no way to tell a reworked from an in situ nannofossil based on the state of preservation. For instance, Jiang and Gartner (1986) noted that of nearly 70 species considered to be "vanishing Cretaceous species," only 2 were not found in at least one of up to 60 samples from the 6 m of sediments representing Tertiary Zones NP1 and NP2. The relative abundance of common Cretaceous species, however, decreases gradually up-section as illustrated in Figure 4.5 (Jiang and Gartner, 1986). Recently, a quantitative study of Antarctic Ocean ODP Site 690 by Pospichal and Wise (1990) has revealed a very similar gradual decrease of Cretaceous species in the early Tertiary (Zone CP1a; Figure 4.6). A similar pattern is also observed at El Kef where a rapid decrease of Cretaceous species also occurs in nannofossil Zone NP1, but few are still present in Zones NP2 to NP3 (Figure 4.7; Perch-Nielsen et al., 1982). No quantitative data are available from DSDP Sites 577 and 528. However, at the South Atlantic Site 524, Percival (1984) also noted a gradual decline and subsequent disappearance of Cretaceous species in Zone NP1. These patterns of gradual decline of Cretaceous taxa in Tertiary sediments in sections spanning from low to high latitudes are very similar to those observed in planktic foraminifera and strongly suggest survivorship of at least some Cretaceous taxa.

As among planktic foraminifera, the vanishing Cretaceous species in lower Tertiary sediments are essentially the same group of species that are common in terminal Cretaceous sediments: *Arkhangelskiella cymbiformis*, *Cretarhabdus crenulatus*, *Eiffellithus turriseiffelii*, *Micula*

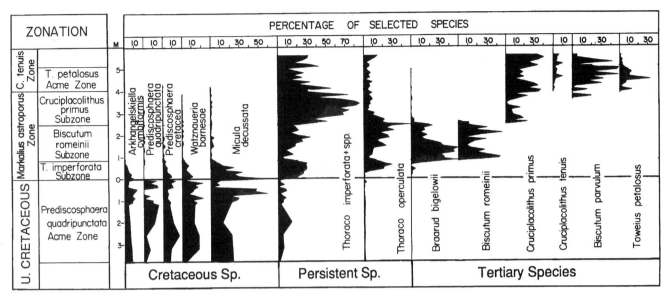

FIGURE 4.5 Abundance of dominant calcareous nannofossils (percent) across the K/T boundary in a Brazos River section (modified from Jiang and Gartner, 1986).

decussata, *Prediscosphaera cretacea*, *Nephaolithus frequens*, *Kamptnerious magnificus*, and *Watznaueria barnesae*. An exception is *Prediscosphaera quadripunctata* (*P. stoveri* of other authors), which becomes rare above the boundary. Species of *Braarudosphaera* and the calcareous dinoflagellate genus *Thoracosphaera* are the only persistent species that become considerably more common above the boundary in the *T. imperforata* subzone.

Measurements of the size distribution of several Maastrichtian species (*Arkhangelskiella cymbiformis*, *Eiffellithus turriseiffelii*, *Micula decussata*, *Prediscosphaera cretacea s. ampl.*) from samples just below and just above the boundary in El Kef indicate no significant differences in the size of the specimens, but stable isotope values of $\delta^{13}C$ and $\delta^{18}O$ from fine fraction carbonate show dramatic changes across the boundary (Perch-Nielsen *et al.*, 1982). Since the fine fraction carbonate consists primarily of Maastrichtian calcareous nannofossils (100% below the boundary and >90% above the boundary), it is difficult to explain how reworked Maastrichtian nannofossils could carry a Tertiary isotope signature. This originally led to the suggestion that most Maastrichtian nannoplankton species survived the K/T boundary crisis (Perch-Nielsen, 1981; Perch-Nielsen *et al.*, 1982). More studies of this nature will be necessary to differentiate Cretaceous survivors from reworked populations in Tertiary sediments.

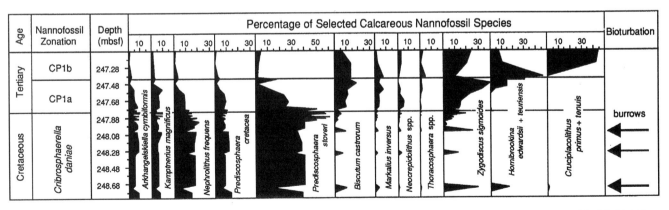

FIGURE 4.6 Abundance of dominant calcareous nannoplankton (percent) across the K/T boundary at ODP Site 690, Weddel Sea, Antarctic Ocean (data from Pospichal and Wise, 1990).

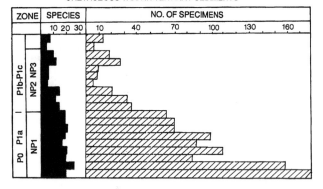

FIGURE 4.7 Relative abundance of Cretaceous calcareous nannofossils in basal Tertiary sediments at El Kef. Data based on counts of Cretaceous specimens encountered during a 10-min examination per sample with an optical microscope at 1000×.

POPULATION RESPONSE TO K/T DISTURBANCE

Relative abundance changes in dominant species groups of both planktic foraminifera and nannofossils yield a more reliable measure of environmental disturbance than systematic diversity. For instance, at the K/T boundary, virtually all planktic foraminiferal species that became extinct were numerically rare (<1%) in the latest Cretaceous ocean, and even relatively minor disturbances of the ecosystem may have caused their demise. In contrast, survivor species dominated the faunal assemblages. In this group, disturbance of the ecosystem is reflected in relative abundance changes and particularly the terminal abundance decline. In each of the sections examined (except Site 528), there are strong similarities in the timing of faunal change among planktic foraminifera, although faunal compositions vary from one region to another.

Among calcareous nannofossils the most commonly preserved taxa in the uppermost Maastrichtian are also still the most common in basal Tertiary deposits as also observed in planktic foraminifers (Figures 4.5 and 4.6). However, since these taxa did not give rise to the new Tertiary species and their presence in Tertiary deposits may be due to reworking, Jiang and Gartner (1986) labeled them "Cretaceous species" (Figure 4.5). Taxa they considered "survivors" or persistent species are very rare in the Maastrichtian and become dominant only in early Tertiary sediments of middle to high latitudes in the Atlantic (Denmark, Biarritz, Zumaya, DSDP Site 524) and Antarctic Oceans (Figure 4.6). These taxa include relatively large extant and new species of the genera *Biscutum*, *Cyclagelosphaera*, *Markalius*, *Neocrepidolithus*, *Zygodiscus*, and *Placozygus* (Figures 4.5 and 4.6). In contrast, such survivors are rare in lower latitudes of the Tethyan Sea (Caravaca and El Kef), where very small new species appear and soon dominate the assemblages. In addition to these floral changes there are increases in the genera *Thoracosphaera* and *Braarudosphaera* in both realms, although blooms of the latter are not observed in all sections. The faunal and floral turnovers at El Kef, Caravaca, Brazos, DSDP Site 528, and ODP Site 738 are discussed below.

El Kef, Tunisia

Figure 4.8 shows the relative abundances of survivor species of planktic foraminifera at El Kef along with changes in $\delta^{13}C$ values for planktic and benthic foraminifera (data from Keller, 1988; Keller and Lindinger, 1989). The K/T boundary, defined by the first appearance of Tertiary planktic foraminifera and nannofossils, coincides with an iridium anomaly and spinels (Kuslys and Krahenbuhl, 1983; Robin *et al.*, 1991), a lithological shift from tan colored marls to black clay with a thin basal red layer, a shift of –3.0‰ in $\delta^{13}C$ values, and elimination of the surface-to-deep $\delta^{13}C$ gradient, which suggests a drop in surface productivity (Zachos and Arthur, 1986; Keller and Lindinger, 1989). At El Kef the terminal decline of survivor species begins about 10 cm (or about 8000 to 10,000 yr) after the K/T boundary and leads the to near disappearance (<1%) of this group within a few thousand years.

The decline of the Cretaceous species and drop in $\delta^{13}C$ values are accompanied by the rise and dominance of *Guembelitria cretacea* (a Cretaceous survivor) in Zone P0 and the evolution of Tertiary species. Figure 4.9 illustrates a sequence of rapidly evolving and changing dominant faunal components in the earliest Tertiary Zones P0 and P1a that seem to be related to low surface productivity as indicated by the very low $\delta^{13}C$ values, also observed in numerous low-latitude sections (Zachos and Arthur, 1986; Barrera and Keller, 1990). Increased surface productivity (higher $\delta^{13}C$ values) at the top of Zone P1a (4 m above K/T) coincides with the disappearance of dominant earliest Tertiary species (*G. conusa, P. eugubina*), evolution of new species, and increased diversity (Keller, 1988, 1989b; Keller and Lindinger, 1989). This initial recovery of the ecosystem occurred about 250,000 to 350,000 yr after the K/T boundary (Keller, 1988, 1993).

Among calcareous nannofossils, semiquantitative data from Perch-Nielsen (1981) indicate that the first group to bloom in the basal Tertiary are species of *Thoracosphaera* followed by *Neobiscutum romeinii* and *N. parvulum*. The interval with common *N. parvulum* also contains common *Lanternithus minutus*, *Braarudosphaera bigelowii*, *Chiastozygus ultimus*, *Cruciplacolithus primus*, and *Placozygus sigmoides*. This increased floral diversification corresponds to the initial increase in surface productivity and first major increase in foraminiferal diversity

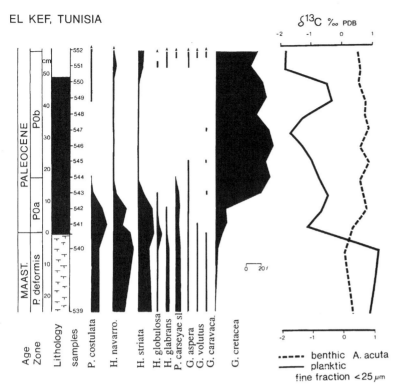

FIGURE 4.8 Abundance of Cretaceous planktic foraminiferal species (percent) across the K/T boundary at El Kef in relation to δ¹³C values of benthic foraminifera and sediment fine fraction. Note the terminal decline of Cretaceous taxa (except for *G. cretacea*) beginning 10 cm above the K/T boundary.

FIGURE 4.9 Abundance of dominant planktic foraminiferal taxa (percent) in the early Danian at El Kef in relation to benthic (*A. acuta*) and planktic (<25 mm sediment fraction) δ¹³C values. Note the increase in species diversity coincident with increased surface productivity at the top of Zone P1a.

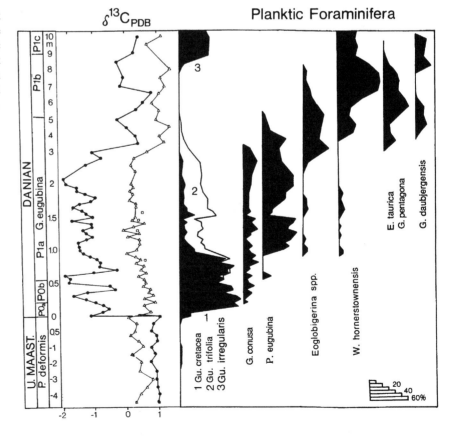

(top of P1a) that marks the initial recovery of the ecosystem at El Kef (Figure 4.9). The *N. parvulum* abundance peak is followed by abundance peaks in *Futyania petalosa* (*Toweius petalosus*), *Praeprinsius* of *P. dimorphesus*, and *Octolithus multiplus* (Perch-Nielsen, 1981).

Caravaca, Spain

The planktic foraminiferal assemblages that dominate during the latest Maastrichtian at Caravaca (as well as Agost) are similar to those at El Kef, although species abundance varies significantly (Figure 4.10; Canudo et al., 1991). The K/T boundary is characterized by the same species datums as well as lithological and geochemical markers (Robin et al., 1991), but the boundary clay is much thinner (7 cm) than at El Kef (50 cm). Species abundances are variable across the boundary. *Heterohelix navarroensis*, *H. glabrans*, and *H. globulosa* decline in abundance 5 to 10 cm below the boundary, whereas *H. globulosa*, *Pseudotextularia costulata*, *P. kempensis*, *Globigerinelloides aspera*, and *G. yaucoensis* remain the same or increase in the boundary clay. All Cretaceous species, except *G. cretacea*, decline to <2% immediately above the clay layer. As indicated earlier, this sharp faunal change marks a short hiatus (Figure 4.3), and part of the anomalous abundance increase in P0 may be due to reworking of Cretaceous taxa. Nevertheless, these relative abundance changes suggest ecological disturbances beginning prior to the K/T boundary and continuing into the earliest Tertiary (7-cm clay layer).

At Caravaca and El Kef, similar sequences of rapidly evolving and changing dominant planktic foraminiferal components are present in the earliest Tertiary, including the maximum abundances of *P. longiapertura*, *P. eugubina*, and *Woodringina* (Canudo et al., 1991). However, because the earliest Tertiary at Caravaca is more condensed

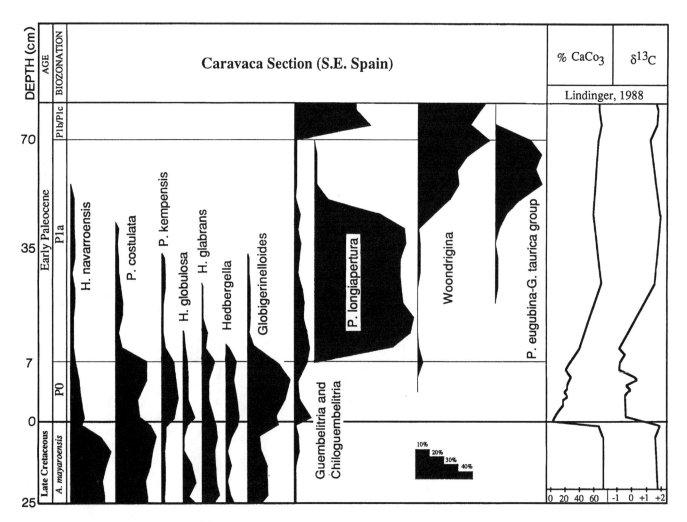

FIGURE 4.10 Percentage $CaCO_3$, $\delta^{13}C$ values, and abundance of dominant planktic foraminiferal species (percent) across the K/T boundary at Caravaca, SE Spain. Note the presence of abundant Cretaceous specimens in the boundary clay (Zone P0).

and contains two short hiatuses (Figure 4.3), the ranges are truncated. Similar to El Kef, $\delta^{13}C$ values at Caravaca also show low surface productivity through Zone P0 and the lower part of Zone P1a, with initial recovery in the upper part of this zone (Figure 4.10).

Calcareous nannofossil distribution at Caravaca is also similar to El Kef, with a succession of dominant *Thoracosphaera*, *B. bigelowii*, and *N. parvulum* parallel with the major decrease of Cretaceous nannofossils in the basal Tertiary (Romein, 1979; Romein and Smit, 1981). Differences in floral successions include the possible absence of a bloom of *N. romeinii*, which has not yet been studied, and the occurrence of common *Octolithus multiplus* preceding an abundance peak of *F. petalosa*.

Brazos, Texas

Brazos River sections do not have a well-defined boundary clay but contain a 2-mm thin red-brown layer that exhibits the iridium anomaly (Beeson et al., in preparation). The first appearance of Tertiary foraminifera and nannofossils places the K/T boundary at this thin red-brown layer and Ir anomaly (Jiang and Gartner, 1986; Keller, 1989a) and at the onset of the $\delta^{13}C$ shift in the Brazos Core section (Barrera and Keller, 1990) (Figure 4.11). The same group of Cretaceous survivors dominates foraminiferal assemblages, although because of the shallow neritic water depth, *H. globulosa* is most abundant. There is no major faunal change apparent below or at the K/T boundary. Above the boundary, Cretaceous species thrive well into Zone P1a (1 m above the K/T) similar to Nye Klov, Denmark (Keller et al., 1993), in contrast to Caravaca and El Kef where Cretaceous survivors become rare within or immediately above Zone P0. However at Brazos, *H. globulosa* populations also begin their terminal decline in Zone P0, parallel with a gradual decline in $\delta^{13}C$ values of *H. globulosa* and the benthic species *Lenticulina* (Figure 4.11; Barrera and Keller, 1990). This is the first time that the $\delta^{13}C$ shift has been observed as a gradual rather than a sudden change (Hsu et al., 1982; Zachos et al., 1986) and implies deteriorating environmental changes beginning within a few thousand years of the K/T boundary.

Above the K/T boundary the Cretaceous survivors *G. cretacea* and *G. trifolia* dominate (50 to 60%) through Zones P1a-P1b (Figure 4.11). Such prolonged dominance of this group has not been observed in other sections and seems to be the result of local ecological conditions. However, the evolutionary sequence of early Tertiary planktic foraminiferal species is similar to that of Nye Klov, ODP Site 738C, El Kef, and Caravaca, although *P. eugubina* is rare, presumably because of the shallow neritic environment. The $\delta^{13}C$ curve is similar to El Kef and Caravaca in that

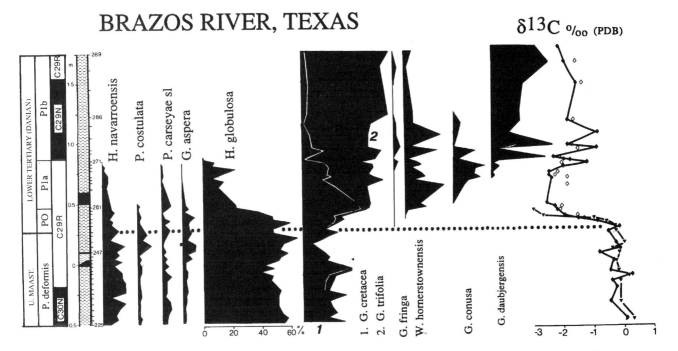

FIGURE 4.11 Abundnace of planktic foraminifera (percent) and $\delta^{13}C$ values of benthic (solid diamond = *Lenticulina sp.*; open diamond = analysis of single specimens) and planktic (triangles = *H. globulosa*) foraminifera across the K/T boundary in the Brazos Core section. Note the abundance of Cretaceous specimens in the basal Tertiary and their terminal decline parallel with the gradual decrease in $\delta^{13}C$ values of both planktic and benthic foraminifera. (Stable isotope data from Barrera and Keller, 1990.)

low surface productivity prevails during most of Zone P1a, and increasing but fluctuating values mark the top of Zone P1a (Figure 4.11; Keller, 1989a; Barrera and Keller, 1990).

Calcareous nannofossil abundance changes in the nearby Brazos-1 section show a similar pattern of rapid decline beginning near the K/T boundary and ending about 1 m above (Jiang and Gartner, 1986; Figure 4.5). The decline of the Cretaceous species is accompanied by increased abundance of *Thoracosphaera* spp. followed by *Braarudosphaera* and *N. romeinii*. As at El Kef and Caravaca there are some differences in the succession of *N. parvulum* and *F. petalosa* abundance distributions. At Brazos, abundance peaks of these taxa overlap with the maximum abundance of *C. primus*, whereas at Caravaca and El Kef the *C. primus* and *F. petalosa* peaks follow after the abundance peak of *N. parvulum*. It is possible that these differences are due primarily to differing taxonomic concepts among nannofossil workers especially for *N. parvulum*, a very small species that is difficult to identify with an optical microscope.

DSDP Site 528, South Atlantic

Relative abundances of planktic foraminiferal species are illustrated in Figure 4.12 for Site 528. Although the K/T boundary is located between two cores, a laminated boundary transition was recovered in the core catcher (D'Hondt and Keller, 1991). As noted earlier, Site 528 has a hiatus or interval of nondeposition at the K/T boundary and Zone P0, and the lower part of Zone P1a is missing (Figure 4.3). Moreover, nannofossil assemblages lack the characteristic basal Tertiary taxa *N. romeinii*, *N. parvulum*,

and *F. petalosa* (Manivit, 1984; Manivit and Feinberg, 1984). Instead, a typical high-latitude Zone NP1 assemblage with common *Thoracosphaera* and *Braarudosphaera* is present immediately above the K/T boundary. Thus, it is because of a hiatus, rather than catastrophic extinctions, that all planktic foraminiferal species truncate at the K/T boundary and Tertiary species suddenly appear. Because of this hiatus, which is present in virtually all deep-sea sections (Worsley, 1974; MacLeod and Keller, 1991a,b), the ecological response to the K/T boundary disturbance cannot be evaluated from such deep-sea sections. However, it is interesting to note that as in continental shelf sections, a similar species assemblage consisting of small cosmopolitan *heterohelicids*, *globigerinellids*, and *hedbergellids* dominates the late Cretaceous at Site 528, and there seems to be a trend toward decreased abundance in the *heterohelicids*, *globotruncanids*, and *globotruncanellids* in the terminal Cretaceous, indicating a changing environment.

ODP Site 738C, Indian Antarctic Ocean

Relative abundances of planktic foraminifera for Site 738C are illustrated in Figure 4.13. In this southern high-latitude section the K/T transition is preserved within a 15-cm-thick laminated layer (which has been analyzed at 1-cm intervals), with the K/T boundary and Ir anomaly 2 cm above the base of the laminated layer (Schmitz *et al.*, 1991; Thierstein *et al.*, 1991; Keller, 1993). Planktic foraminiferal changes are dramatic across the K/T transition, but do not coincide with the Ir anomaly and K/T boundary. No significant species extinctions coincide directly with

FIGURE 4.12 Abundance of dominant planktic foraminiferal species (percent) across the K/T boundary in DSDP Site 528 (data from D'Hondt and Keller, 1991). Note the sudden faunal change at the K/T boundary is due to a hiatus that encompasses Zone P0.

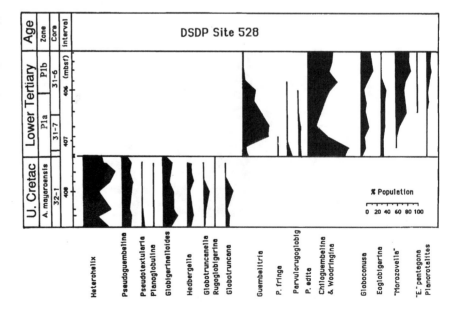

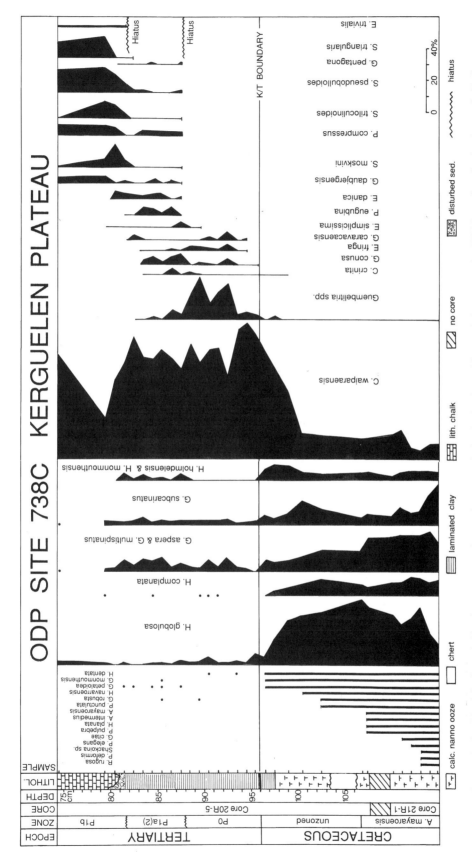

FIGURE 4.13 Planktic foraminiferal turnover at the centimeter-scale in the laminated interval that spans the K/T boundary in the Antarctic Indian Ocean ODP Site 738C. Note the absence of significant species extinctions at the K/T boundary, and the survivorship of all dominant Cretaceous species well into the Tertiary, but an onset of decline in their relative abundances beginning below the K/T boundary. Data from Keller (1993).

the K/T boundary; instead, there is a gradual elimination of more specialized and generally rare taxa below the K/T boundary. Moreover, all dominant species survive well into the Tertiary and die out as the diversity of new Tertiary species increases (Zones P1b and P1c)

The major faunal turnover, however, begins 5 cm below the K/T boundary with a sharp decline in abundances of *H. globulosa* from 40 to 4% by K/T boundary time, the decline and disappearance of *H. complanata*, and gradual declines in *G. aspera* and *G. subcarinatus* and hedbergellids. The most dramatic change is seen in *Chiloguembelina waiparaensis*, which increased from 15 to 90% in the 4 cm below the K/T boundary as triserial taxa (guembelitrids) increased in Zone P0 and new Tertiary species evolved. All Cretaceous taxa are dwarfed beginning 4 cm below the K/T boundary and through Zone P0, and reach maturity at less than half their usual size.

Indian Antarctic Ocean Site 738C thus illustrates a record of long-term environmental deterioration beginning well before and continuing well after the K/T boundary. It also shows that the demise of the cosmopolitan latest Maastrichtian assemblage cannot have been caused by a K/T boundary bolide impact, but was related to the environmental changes that resulted in a change of sediment deposition from calcareous ooze to laminated clay. Clearly, there was no sudden and catastrophic mass extinction at Site 738C (or in the northern-high latitude Nye Klov section; Keller *et al.*, 1993). In fact, it seems impossible to differentiate the effects of the proposed bolide impact from the continued environmental changes in progress. At most, the bolide impact may have hastened the demise of an already declining latest Cretaceous cosmopolitan fauna.

Stable isotope measurements of planktic and benthic foraminifera indicate that there is no negative $\delta^{13}C$ shift across the K/T boundary. In fact, $\delta^{13}C$ values remain stable or increase slightly (0.2‰, Barrera and Keller, 1994). This contrasts with the 3.0‰ negative shift observed in low latitudes and supports the interpretation that the effects of the K/T boundary event may have been limited to low latitudes.

MAGNITUDE OF K/T DISTURBANCE

Planktic Foraminifera

The magnitude of the K/T disturbance can be estimated from faunal and floral abundance changes. In Figure 4.14, planktic foraminiferal abundance data are summarized for all four sections. Species and their relative numerical abundances are grouped into (a) species extinct at or before the K/T boundary, (b) survivor species except *Guembelitria*, (c) *Guembelitria*, and (d) evolving Tertiary species. The most dramatic change among the five sections is seen in the relative abundances of all species extinct at or before the K/T boundary. For instance, in the Brazos Core (deposited in a middle to inner neritic environment), species extinctions account for less than 5% of the individuals in the total fauna, although one-third of the species disappear. At El Kef, Caravaca, and Site 528, which were deposited in outer neritic to upper bathyal and bathyal depths, respectively, only between 10 and 30% of the individuals are affected, whereas two-thirds of the species disappear. Moreover, Figure 4.14 shows that at El Kef and Site 528 there is a gradual decline in numerical abundance of the disappearing fauna beginning below the K/T boundary. In contrast to these low- and middle-latitude sites, almost all species survive the K/T boundary in the Indian Antarctic Ocean Site 738C (Keller, 1993). These data illustrate that a taxonomic census alone significantly overestimates the environmental effects of the K/T boundary crisis. However, these data also imply that the effect on faunal populations was variable between shallow nearshore and open marine environments, and between low or middle and high latitudes.

The numerical abundance of Cretaceous survivors (excluding *G. cretacea*) shows a similar pattern between El Kef and Caravaca, with a relatively rapid decline above the K/T boundary over about 10,000 yr and a more gradual decline over about 30,000 to 40,000 yr at Brazos. No data are available for Site 528 because of a hiatus. At Site 738C the decline of Cretaceous survivors is masked by the concurrent increase in *C. waiparaensis*. The Cretaceous survivor *Guembelitria* generally thrived after the K/T boundary as illustrated at El Kef, but is not present in great abundance at Caravaca in Zones P0 and basal P1a. This absence may be due, however, to a short hiatus at this interval at Caravaca (Figure 4.3). A peak in *Guembelitria* abundance within Zone P1a is present, however, in both Caravaca and Site 528 (Figure 4.13), as well as El Kef (Figure 4.9), but not at high-latitude Site 738C. At Brazos, *Guembelitria* is common throughout the Maastrichtian and increases in abundance in the early Tertiary.

Thus, Figure 4.14 illustrates unequivocally that only a small group of the planktic foraminiferal population became extinct at the K/T boundary, that the dominant group gradually declined and became extinct well after the boundary, and that the adverse effects of the K/T boundary event diminished toward high latitudes. The magnitude of the K/T boundary disturbance is thus significantly overestimated if based on taxic diversity alone or if low- to high-latitude differences are disregarded.

Calcareous Nannoplankton

Quantitative data to estimate the magnitude of the effect of the K/T boundary disturbances are available for El

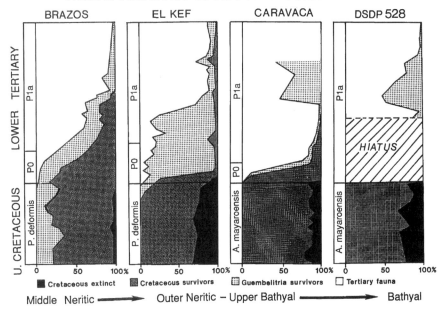

FIGURE 4.14 Faunal turnover across the K/T boundary in planktic foraminifera between midneritic and bathyal environments based on percentage of the population eliminated (number of individuals in all taxa that disappear at the K/T boundary) and percentage of the population surviving. Note the decreasing effect into shallower waters in percentage of fauna eliminated and the prolonged survivorship of the Cretaceous fauna.

Kef, Brazos River (Perch-Nielsen *et al.*, 1982; Jiang and Gartner, 1986), and ODP Sites 690, 738, and 761 (Pospichal and Wise, 1991; Wei and Pospichal, 1991; Pospichal and Bralower, 1992). However, as noted earlier, their interpretation is not unambiguous and depends on whether the so-called vanishing Cretaceous taxa are assumed to have survived at least partly into the basal Tertiary, or whether they are considered to be reworked above the K/T boundary.

If we assume that the vanishing Cretaceous species survived into the early Tertiary, then their abundance rapidly decreases to less than 10% by the *N. romeinii* subzone in the Brazos section. This decline in Cretaceous nannofossils parallels that of Cretaceous planktic foraminifers in the basal 1 m of the Tertiary at Brazos. A similar rapid decline in vanishing Cretaceous species is observed at El Kef. In higher-latitude sections (including DSDP Site 524, ODP Sites 690C, 738, and 761, Biarritz (France) and the Danish sections), vanishing Cretaceous species seem to remain relatively common (10 to 30%) into the upper part of NP1 (Perch-Nielsen, 1979a,b; Pospichal and Wise, 1991; Wei and Pospichal, 1991; Pospichal and Bralower, 1992). Their decrease is accompanied by a rapid increase in *Thoracosphaera* and, in high latitudes, also by an increase in persistent species.

If we assume that all vanishing Cretaceous species became extinct at the K/T boundary and their presence in Tertiary sediments is due to reworking, then the extinction rate exceeds 100 species per year, compared to an average of 1.5 species per million years (m.y.) and 5% per m.y. estimated by Perch-Nielsen (1986) and Roth (1987), respectively, for the Late Cretaceous. Given the uncertainties in this assumption, this estimate is not a satisfactory approximation of the magnitude of the K/T boundary event on the species level. Extinctions clearly occurred at a rate at least one order of magnitude higher than during the Late Cretaceous.

On the genus level about 18 of some 50 to 60 Cretaceous genera survived into the Tertiary (Perch-Nielsen, 1985, 1986). Among these, 4 genera are still extant; the remainder disappeared during the Paleocene (7 genera) and Eocene (7 genera) at an average rate of about 0.5 genus/m.y. This rate of extinction is similar to that of Paleocene to Middle Eocene floras in general. Because of the present ambiguity in determining whether Cretaceous species in basal Tertiary deposits are reworked or in situ, an accurate estimate of the effect of the K/T crisis based on nannoplankton cannot be obtained.

ARE SPECIFIC HABITATS SELECTIVELY DESTROYED?

Most planktic foraminifera live in the upper 200 to 400 m of the water column and within this interval can be grouped into surface, intermediate, or deep dwellers based on their oxygen and carbon isotope ranking and the assumption that they grew their shells in isotopic equilibrium with the sea water in which they lived (see Table 4.1; Douglas and Savin, 1978; Boersma and Shackleton, 1981; Stott and Kennett, 1989; Lu and Keller, 1993). Species living in warm surface waters have the lightest $\delta^{18}O$ and heaviest $\delta^{13}C$ values, whereas species living in deeper, cooler waters have successively heavier $\delta^{18}O$ and lighter $\delta^{13}C$ values. Variations in the distribution of these groups may indicate environmental changes related to climate, ocean circulation, and water chemistry.

Because stable isotope ranking of Cretaceous species is still preliminary and awaiting further analyses of individual species, Table 4.1 groups by genera. However, future isotope analyses may show that not all species within a genus inhabited the same depth environment, or that some species calcified their shells in disequilibrium with the seawater in which they grew. *Heterohelix globulosa* is one example of a species that calcified in isotopic disequilibrium or changed its habitat over time. In the Cretaceous, this species has been reported with relatively heavy $\delta^{18}O$ values in the open ocean (although other heterohelicid species have light values), indicating that it lived at thermocline depths (Boersma and Shackleton, 1981). However, in the shallow water (<100 m) Brazos sections (where this species is dominant throughout the Late Cretaceous and into the early Tertiary (Keller, 1989a), it apparently adapted to living in a shallower environment. The same pattern has been observed in the shallow water Stevns Klint and Nye Klov sections (Schmitz *et al.*, 1992; Keller *et al.*, 1993). In this study *H. globulosa* has therefore been grouped with surface dwellers.

In Figure 4.15, the Cretaceous fauna has been grouped into surface, intermediate, and deep dwellers based on two faunal parameters, the percentage of species in each group and the percentage of individuals in each group. The latter is a more sensitive parameter of environmental change because it weighs species according to their relative abundance in the total fauna, whereas the former gives each species the same significance regardless of whether it is represented by 1% or 80% in the faunal assemblage. These data are illustrated in Figure 4.15 along with the stable isotope record for each of the four sections, which are arranged from shallow (Brazos) to deep (Site 528).

TABLE 4.1 Oxygen Isotope Ranking of Cretaceous Species

Surface	Intermediate	Deep
Pseudoguembelina	*Globotruncana*	*Globotruncanella*
Rugoglobigerina	*Rugotruncana*	*Racemiguembelina*
Heterohelix		*Hedbergella*
Contusotruncana		*Shakoina*
Pseudotextularia		*Gublerina*
Globigerinelloides		*Planoglobulina*

Both faunal parameters in the four sections show a strong trend toward decreasing deep and intermediate dwellers and increasing surface dwellers. About 15 to 20% of the species are deep dwellers at Site 528 (midbathyal) and Caravaca (upper bathyal), and a decrease to 8 and 12%, respectively, occurs in the latest Maastrichtian (Figure 4.15). At El Kef (outer neritic) an average of 10% of the species are deep dwellers, and at Brazos 0% (midneritic). The relative abundance of individuals in this group shows the same trend, but they comprise a significantly smaller part of the population except at Site 528. Abundance of deep dwelling individuals decreases from 15 to 20% at Site 528, to 5 to 10% at Caravaca and El Kef, and 0% at Brazos. All deep dwellers disappear at the K/T boundary.

Intermediate dwellers show a similar trend. At Site 528, 5 to 10% of the species are in this group; at Caravaca 15 to 20%; at El Kef 10 to 15%; and at Brazos between 10 and 20% decreasing up-section and disappearing well below the K/T boundary (Figure 4.15). The relative abundance of individuals in this group is smaller. At Site 528, maximum abundance is 10%, decreasing to 2% well below the K/T boundary; at Caravaca maximum 5% decreasing up-section to <2%; at El Kef <2%; and at Brazos <1%. Like the deep water dwellers all intermediate dwellers disappear at the K/T boundary.

Surface dwellers are the largest component of the late Maastrichtian fauna. Their abundance increases from 75 to 85% in the open marine Site 528, to 85 to 90% at Caravaca and El Kef, and 99 to 100% at Brazos. In all sections the dominant surface dwellers are cosmopolitan taxa and survive the K/T boundary event; no data are available for Site 528 because the earliest Tertiary is missing. The relative abundance of dominant survivor species rapidly declines in the boundary clay at Caravaca and El Kef, but at Brazos, Cretaceous survivors thrive well into Zone P1a. Rapid expansion of the evolving Tertiary fauna accompanies the decline of Cretaceous survivors.

Figure 4.15 thus illustrates environmental conditions in four different geographic localities and at four different water depths. Some differences between these sections are due to changing water depths and some are due to other global environmental changes. For instance, the variations in the faunal assemblages among the four sites are due primarily to changing water depth and its effect on the thermocline and vertical stratification of species. Global sea-level changes and their effect on water mass stratification are indicated by the decrease and disappearance of deep and intermediate groups at each locality, especially in the middle to upper bathyal and outer neritic environments of Site 528, Caravaca, and El Kef. These gradual faunal changes indicate a changing environment beginning 100,000 to 300,000 yr before the end of the

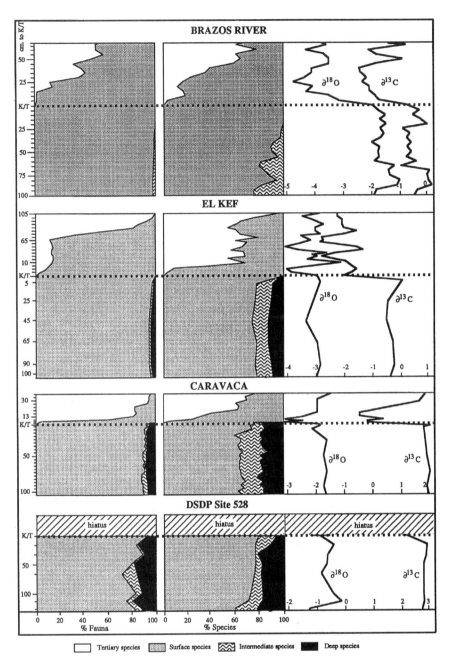

FIGURE 4.15 Cretaceous planktic foraminiferal species grouped into surface, intermediate, and deep water dwellers based on stable isotope ranking. Data are presented in terms of percentage of species and fauna (individuals) in each category and are arranged from shallow to deep. Stable isotope data from Lindinger (1988), Keller and Lindinger (1989), and Barrera and Keller (1990). Note: Species data overemphasize the magnitude of the K/T boundary effect when compared to the fauna affected; there is a decreasing effect from deep to shallow water environments; there is a gradual decline in the Cretaceous deep and intermediate dwellers and all become extinct at the K/T boundary, only surface dwellers survive.

Maastrichtian (Keller and Barrera, 1990). The terminal disappearance of all deep and intermediate dwellers near the K/T boundary, however, indicates accelerated environmental changes that favored the survival of cosmopolitan surface dwellers. The effect of these changes may have included global warming, disruption of the vertical water mass stratification and thermocline, and rapid reduction of marine primary productivity. Stable isotope data indicate that both warming and a drop in marine productivity beginning at the K/T boundary were restricted to low latitudes. However, through the latest Maastrichtian, stable isotope data indicate high marine productivity. The gradual faunal changes prior to the K/T disturbance imply Earth-derived causes and may be related to the latest Maastrichtian sea-level regression followed by a transgression across the K/T boundary clay (Schmitz et al., 1992; Keller et al., 1993), but large-scale volcanism and mantle plume activity may also have been contributing factors. However, the considerably smaller mass extinction in planktic foraminifera than generally assumed suggests that the effects of a bolide impact on marine plankton would have been significantly less catastrophic than proposed by current theory.

Moreover, recent studies of high-latitude sections show that the effects of the K/T boundary event were negligible in high latitudes.

DISCUSSION AND SUMMARY

We have examined the effect of global change on calcareous marine microplankton across the K/T boundary with respect to four critical topics: (1) the temporal completeness of boundary sequences; (2) the nature of species extinctions; (3) the effect on dominant species populations; and (4) the specific habitats affected. Our investigation has not uncovered any simple answers. The negative effects of global environmental changes across the K/T transition appear complex, prolonged, and unlikely to result from a single cause. We summarize the results below.

1. MacLeod and Keller (1991a,b) have demonstrated that nearly all K/T boundary sequences are temporally incomplete, but that the hiatus patterns may vary between deep sea and neritic environments. In deep-sea sections a short hiatus or interval of nondeposition that includes the basal Tertiary Zone P0 and part of Zone P1a is virtually always present. In continental shelf sequences, sediment deposition continued at this time, albeit at a reduced rate, and short hiatuses are usually present at the P0/P1a Zone boundaries. This hiatus pattern seems to be linked to the sea-level regression maximum just below the K/T boundary, followed by a rapid transgression across the K/T boundary and into Zone P0 (Schmitz et al., 1992; Keller et al., 1993), which trapped terrigenous sediment and organic carbon on continental shelves and temporarily deprived deep ocean basins of an inorganic sediment source and enhanced carbonate dissolution (Loutit and Kennett, 1981; Haq et al., 1987; Keller, 1988, 1989a, 1993; Keller et al., 1993). The two short hiatuses coincide with sea level lowstands. In view of this information, the sudden, near-complete mass extinction observed in deep-sea sections reflects artificial truncation of species ranges due to a hiatus as illustrated here for DSDP Site 528. Moreover, the abrupt changes in geochemical tracers (iridium, shocked quartz, soot, $\delta^{13}C$, $CaCO_3$, total organic carbon) observed in many deep-sea sections also need to be reevaluated, and the possible effects of elemental concentrations during a time of no carbonate or terrigenous sediment deposition must be considered (see Donovan et al., 1988).

2. The record of species extinctions and evolution in the low-latitude El Kef section, the temporally most complete K/T boundary sequence known to date, indicates that Cretaceous foraminiferal species extinctions occurred over an extended time period beginning well before and ending well after the K/T boundary. About one-third of the species disappeared before the boundary, nearly one-third disappeared at the boundary, and more than one-third survived and gradually disappeared during the early Tertiary. Moreover species extinctions appear systematic, affecting tropical complex, large, and highly ornamented morphologies first and favoring the survival of more cosmopolitan, smaller, and less ornamented morphologies. Furthermore, Cretaceous survivor species in Tertiary deposits are generally dwarfed. Evolution of Tertiary species begins immediately after the K/T boundary with the appearance of very small unornamented and primitive morphologies. Their increasing diversification and abundance seem to have contributed to the demise of the Cretaceous survivors. This species extinction pattern is representative of depositional sequences that are temporally complete across the K/T boundary in low latitudes (including Caravaca, Agost, Brazos) and illustrates the absence of a near-complete mass extinction as commonly reported from the deep-sea sections that contain a hiatus. Moreover, this gradual species extinction pattern cannot be reconciled with a single geologically instantaneous cause.

The species extinction pattern in high latitudes (Site 738C, Nye Klov) is quite different from that in low latitudes. In contrast to low latitudes, no significant species extinctions occur at the K/T boundary and nearly all species survive well into the Tertiary. Yet, as in low latitudes, extinctions occur over an extended time, beginning well below and extending well above the K/T boundary. From low to high latitudes, Cretaceous survivor taxa are primarily small cosmopolitan heterohelicids, hedbergellids, guembelitrids, and globigerinellids. These taxa were able to tolerate changing environmental conditions, whereas the more specialized tropical and subtropical taxa went extinct.

The emerging global pattern of species extinctions among planktic foraminifera suggests that the K/T boundary event may have been most severe in low latitudes but that its effect dimished into high latitudes. This interpretation is supported by the $-3.0‰$ shift of $\delta^{13}C$ in low latitudes and absence in high latitudes (Keller et al., 1993; Barrera and Keller, 1994).

3. Changes in the numerical abundances of species are an important factor in evaluating the magnitude of global change. For instance, the numerical abundance of all planktic foraminiferal species extinct at or before the K/T boundary is relatively small, less than 5% in midneritic environments (Brazos) and between 10 and 30% in outer neritic and bathyal depths (El Kef, Caravaca, and Site 528), and this group declined gradually during the late Maastrichtian. Cretaceous survivors dominate during the late Maastrichtian and decline rapidly in the Tertiary over about 10,000 to 40,000 yr in low latitudes when maximum negative conditions are reached accompanied by low primary marine productivity. Thus, only a small group of the planktic foraminiferal population became extinct at or be-

fore the K/T boundary. The major adverse effect of the global environmental change is seen in the rapid but gradual post-K/T faunal, floral, and $\delta^{13}C$ changes in low latitudes.

4. Changes in the marine environment are inferred from $\delta^{18}O$ and $\delta^{13}C$ depth stratification of Cretaceous planktic foraminiferal species. We demonstrate that faunal differences between neritic and bathyal depths are due primarily to changes in paleodepth and their effect on the thermocline and vertical stratification of species. Most surprising, however, is the total disappearance of deep and intermediate water dwellers (as well as some surface dwellers) at or before the K/T boundary in all sections. Only surface water dwellers survived the K/T boundary. This implies that the global environmental change preferentially eliminated deeper habitats first, favoring survival in surface waters, or that subsurface dwellers were less able to adapt to changing environmental conditions. Environmental changes, however, seem to have been gradual beginning 100,000 to 300,000 yr before the end of the Cretaceous; accelerating at the boundary; and reaching a negative maximum between about 10,000 and 40,000 yr after the boundary, coincident with maximum low primary marine surface productivity in low latitudes. Return to a more stable ecosystem and increased marine productivity does not occur until about 250,000 to 350,000 yr after the K/T boundary.

There is no single cause that can account for this prolonged environmental change. Hallam (1989) has consistently advocated sea-level changes as a major cause of faunal turnover. The latest Maastrichtian sea-level regression followed by a rapid transgression across the K/T boundary and into the early Tertiary (Brinkhuis and Zachariasse, 1988 Donovan et al., 1988, Schmitz et al., 1992; Keller et al., 1993), however, has largely been ignored as an important factor in the K/T faunal transition. The data presented here in terms of the differential nature of hiatus patterns in the deep-sea and neritic environment, the selective nature of pre-K/T species extinctions, and the elimination of all deep- and intermediate-dwelling planktic foraminifera at the K/T boundary all point toward sea-level change as a major contributor, if not causal factor. The late Maastrichtian sea level regression reached a maximum just before the K/T boundary, followed by a rapid transgression into the earliest Tertiary (Figure 4.3). Elimination of deeper-dwelling species could have occurred as a result of the breakdown in the water mass structure, a change in thermocline, expansion of the oxygen minimum zone, and decreased productivity. However, many sea-level changes of equal or greater magnitude have occurred during the past 100 m.y., and none has resulted in a complete change of marine plankton over a few 100,000 yr. Thus, sea-level change is probably only one, although perhaps the major, contributing factor to the K/T faunal transition. Other environmental changes (volcanism, bolide impact) may have accelerated the demise of a Cretaceous fauna already on the decline.

ACKNOWLEDGMENTS

We thank the reviewers H. Thierstein and W. Berger for their critical and helpful comments, and we gratefully acknowledge contributions from N. MacLeod, I. Canudo, S. D'Hondt, S. Gartner, and J. Pospichal. This research was supported by NSF Grants OCE 90-21338 and EAR 91-15044 to G.K.

REFERENCES

Alvarez, W., L. W. Alvarez, F. Asaro, and H. V. Michel (1980). Extraterrestrial cause for the Cretaceous-Tertiary extinction, *Science 208*, 1095-1108.

Barrera, E., and B. T. Huber (1990). Evolution of Antarctic Water during the Maastrichtian: Foraminifer oxygen and carbon isotope ratios, Leg 113, in *Proceedings of the Ocean Drilling Program, Scientific Results 113*, P. S. Barker, J. P. Kennett et al., eds., Ocean Drilling Program, College Station, Texas, pp. 813-827.

Barrera, E., and G. Keller (1990). Foraminiferal stable isotope evidence for gradual decrease of marine productivity and Cretaceous species survivorship in the earliest Danian, *Paleoceanography, 5*, 867-890.

Barrera, E., and G. Keller (1994). Productivity across the Cretaceous-Tertiary boundary in high latitudes, *Geological Society of America Bulletin*, in press.

Beeson, D., S. Gartner, G. Keller, N. MacLeod, J. Médus, and R. Rocchia (in preparation). Multidisciplinary stratigraphy and depositional environment across the Cretaceous-Tertiary boundary at the Brazos River, Falls County, Texas, *Palaios*.

Berger, W. H. (1970). Biogenous deep-sea sediments: Fractionation by deep-sea circulation, *Geological Society of America Bulletin 81*, 1385-1402.

Berger, W. H., and E. L. Winterer (1974). Plate stratigraphy and fluctuating carbonate line, *International Association of Sedimentology Special Publication 1*, 11-48.

Berggren, W. A. (1962). Some planktonic foraminifera from the Maastrichtian and type Danian stages of southern Scandinavia, *Stockholm Contributions in Geology 9*(1), 1-102.

Boersma, A., and N. J. Shackleton (1981). Oxygen and carbon isotope variations and planktonic foraminiferal depth habitats: Late Cretaceous to Paleocene, central Pacific, DSDP Sites 463 and 465, Leg 65, in *Initial Reports of the Deep Sea Drilling Project 65*, U.S. Government Printing Office, Washington, D.C., pp. 513-526.

Brinkhuis, W., and W. J. Zachariasse (1988). Dinoflagellate cysts, sea level changes and planktonic foraminifers across the Cretaceous-Tertiary boundary at El Haria, northwest Tunisia, *Marine Micropaleontology 13*, 153-191.

Canudo, I. J., G. Keller, and E. Molina (1991). K/T boundary extinction pattern and faunal turnover at Agost and Caravaca, SE Spain, *Marine Micropaleontology 17*, 319-341.

D'Hondt, S., and G. Keller (1991). Some patterns of planktic foraminiferal assemblage turnover at the Cretaceous-Tertiary boundary, *Marine Micropaleontology 17*, 77-118.

Donovan, A. D., G. R. Baum, G. L. Blechschmidt, T. S. Loutit, C. E. Pflum, and P. R. Vail (1988). Sequence stratigraphic setting of the Cretaceous-Tertiary boundary in central Alabama, *Society of Economic Paleontologists and Mineralogists Special Publication No. 42*, 300-307.

Douglas, R. G., and S. M. Savin (1978). Oxygen isotope evidence for depth stratification of Tertiary and Cretaceous planktic foraminifera, *Marine Micropaleontology 3*, 175-196.

Hallam, A. (1989). The case for sea-level change as a dominant causal factor in mass extinction of marine invertebrates, *Philosophical Transactions of the Royal Society of London, Series B 325*, 437-455.

Haq, B. U., J. Hardenbol, and P. R. Vail (1987). Chronology of fluctuating sea levels since the Triassic, *Science 235*, 1156-1166.

Herbert, T. D., and S. D'Hondt (1990). Environmental dynamics across the Cretaceous-Tertiary extinction horizon measured 21 thousand year climate cycles in sediments, *Earth and Planetary Science Letters 99*, 263-275.

Hofker, J. (1960). The foraminifera of the lower boundary of the Danish Danian, *Meddr. Dansk Geol. Forening 14*, 212-242.

Huber, B. T. (1991). Maastrichtian planktonic foraminifer biostratigraphy and the Cretaceous-Tertiary boundary at ODP Hole 738C (Kerquelen Plateau, Southern Indian Ocean), in *Proceedings of the Ocean Drilling Program, Scientific Results 119*, Ocean Drilling Program, College Station, Texas, pp. 451-466.

Jiang, M. J., and S. Gartner (1986). Calcareous nannofossil succession across the Cretaceous-Tertiary boundary in east-central Texas, *Micropaleontology 32*, 232-255.

Kauffman, E. G. (1984). The fabric of Cretaceous marine extinctions, in *Catastrophes and Earth History*, W. A. Berggren and J. A. Van Couvering, eds., Princeton University Press, Princeton, N.J., pp. 151-237.

Keller, G. (1988). Extinction, survivorship and evolution of planktic foraminifers across the Cretaceous-Tertiary boundary at El Kef, Tunisia, *Marine Micropaleontology 13*, 239-263.

Keller, G. (1989a). Extended Cretaceous/Tertiary boundary extinctions and delayed population change in planktonic foraminiferal faunas from Brazos River, Texas, *Paleoceanography 4*, 287-332.

Keller, G. (1989b). Extended period of extinctions across the Cretaceous/Tertiary boundary in planktonic foraminifera of continental shelf sections: Implications for impact and volcanism theories, *Geological Society of America Bulletin 101*, 1408-1419.

Keller, G., (1993). The Cretaceous Tertiary boundary transition in the Antarctic Ocean and its global implications, *Marine Micropaleontology*.

Keller, G., and E. Barrera, E. (1990). The Cretaceous-Tertiary boundary impact hypothesis and the Paleontological ecord, *Geological Society of America Special Paper 247*, 556-575.

Keller, G., and C. Benjamini (1991). Paleoenvironment of the eastern Tethys in the Early Paleocene, *Paleogeography, Paleoclimatology, Paleoecology 6*, 439-464.

Keller, G., and M. Lindinger (1989). Stable isotope, TOC and $CaCO_3$ record across the Cretaceous-Tertiary boundary at El Kef, Tunisia, *Palaeogeography, Palaeoclimatology, Palaeoecology 67*, 243-265.

Keller, G., C. Benjamini, M. Magaritz, and S. Moshkovitz (1990). Faunal, erosional and $CaCO_3$ events in the early Tertiary eastern Tethys, *Geological Society of America Special Paper 247*, 471-480.

Keller, G., E. Barrera, B. Schmitz, and E. Mattson (1993). Gradual mass extinction, species surviorship and long-term environmental changes across the Cretaceous-Tertiary boundary in high latitudes, *Geological Society of America Bulletin 105*, 979-997.

Kuslys, M., and U. Krahenbuhl (1983). Noble metals in Cretaceous/Tertiary sediments from El Kef, *Radiochimica Acta 34*, 139-141.

Lindinger, M. (1988). The Cretaceous/Tertiary boundaries of El Kef and Caravaca: Sedimentological, geochemical and clay mineralogical aspects, Ph.D. thesis, Swiss Federal Institute of Technology, Zurich, 254 pp.

Loper, D. E., and K. McCartney (1986). Mantle plumes and the periodicity of magnetic field reversals, *Geophysical Research Letters 13*, 1525-1528.

Loper, D. E., and K. McCartney (1988). Shocked quartz found at the K/T boundary: A possible endogenous mechanism, *EOS 69*, 971-972.

Loutit, T. S., and J. P. Kennett (1981). Australasian Cenozoic sedimentary cycles, global sea level changes and the deep sea sedimentary record, *Oceanologia Acta*, 46-63.

MacLeod, N. and G. Keller (1991a). Hiatus distribution and mass extinctions at the Cretaceous/Tertiary boundary, *Geology 19*, 497-501.

MacLeod, N., and G. Keller (1991b). How complete are K/T boundary sections? A chronostratigraphic estimate based on graphic correlation, *Geological Society of America Bulletin 103*, 1439-1457.

Manivit, H. (1984). Paleogene and upper Cretaceous calcareous nannofossils from Deep Sea Drilling Project Leg 73, in *Initial Reports of the Deep Sea Drilling Project 74*, U.S. Government Printing Office, Washington, D.C., pp. 475-500.

Manivit, H., and H. Feinberg (1984). Correlation of magnetostratigraphy and nannofossil biostratigraphy in upper Cretaceous and lower Paleocene sediments of the Walvis Ridge Area, in *Initial Reports of the Deep Sea Drilling Project 74*, U.S. Government Printing Office, Washington, D.C., pp. 469-474.

McLean, D. M. (1985). Mantle degassing induced dead ocean, in *The Cretaceous-Tertiary Transition*, Geophysical Monograph 32, American Geophysical Union, Washington, D.C., pp. 493-503.

Officer, C. B., and C. Drake (1985). Terminal Cretaceous environmental events, *Science 227*, 1161-1167.

Olsson, R. K. (1960). Foraminifera of latest Cretaceous and earliest Tertiary age in the New Jersey coastal plain, *Journal of Paleontology 34*(1), 1-58.

Olsson, R. K., and C. Liu (1993). Controversies on the placement of the Cretaceaus-Paleogene boundary and the K/T mass extinction of planktic foraminifera, *Palaios 8*, 127-139.

Perch-Nielsen, K. (1979a). Calcareous nanofossils at the K/T boundary in Tunisia, in *K/T Boundary Events*, W. A. Christensen and R. G. Bromley, eds., University of Copenhagen, Denmark, pp. 238-243.

Perch-Nielsen, K. (1979b). Calcareous nannofossils from the Cretaceous between the North Sea and the Mediterranean, *International Union of Geological Sciences, Ser. A 6*, 223-272.

Perch-Nielsen, K. (1981). Nouvelles observations sur les nannofossiles calcaires á la limite Crétacé/Tertiaire prés de El Kef, Tunisie, *Cah. Micropaléontol. 3*, 25-36.

Perch-Nielsen, K. (1982). Maastrichtian coccoliths in the Danian survivors or reworked "dead bodies," *Abstract IAS*.

Perch-Nielsen, K. (1985). Cenozoic calcareous nannofossils. Mesozoic calcareous nannofossils, in *Plankton Stratigraphy*, H. M. Bolli, J. B. Saunders, and K. Perch-Nielsen, eds., Cambridge University Press, Cambridge, pp. 329-554.

Perch-Nielsen, K. (1986). Geologic events and the distribution of calcareous nannofossils—Some speculations, *Bull. Centres Rech. Explor. Prod. Eif-Aquitaine 102*, 421-432.

Perch-Nielsen, K., J. McKenzie, and Q. He (1982). Biostratigraphy and isotope stratigraphy and the catastrophic extinction of calcareous nannoplankton at the Cretaceous/Tertiary boundary, *Geological Society of America Special Paper 190*, 353-371.

Percival, S. F., Jr. (1984). Late Cretaceous to Pleistocene calcareous nannofossils from the South Atlantic, Deep Sea Drilling Project Leg 73, in *Initial Reports of the Deep Sea Drilling Project 73*, U.S. Government Printing Office, Washington D.C., pp. 391-424.

Percival, S. F., and A. G. Fischer (1977). Changes in calcareous nannoplankton in the Cretaceous-Tertiary biotic crisis at Zumaya, Spain, *Evolutionary Theory 2*, 1-35.

Pospichal, J. J., and T. J. Bralower (1992). Calcareous nannofossils across the Creaceous/Tertiary boundary, Site 761, Northwest Australian margin, in *Proceedings of the Ocean Drilling Program, Scientific Results 122*, Ocean Drilling Program, College Station, Texas, pp. 735-751.

Pospichal, J. J., and W. Wise, Jr. (1990). Calcareous nannofossils across the K/T boundary, ODP Hole 690C, Maud Rise, Weddell Sea, in *Proceedings of the Ocean Drilling Program, Scientific Results 113*, Ocean Drilling Program, College Station, Texas, pp. 515-532.

Robin, E., D. Boclet, D. Benté, L. Froget, C. Jéhanno, and R. Rocchia (1991). The stratigraphic distribution of Ni-rich spinels in Cretaceous-Tertiary boundary rocks at El Kef (Tunisia), Caravaca (Spain) and Hole 761 (Leg 122), *Earth and Planetary Science Letters 107*, 715-721.

Romein, A. J. T. (1979). Lineages in early Paleogene calcareous nannoplankton, *Utrecht Micropaleontological Bulletins 22*, 1-231.

Romein, A. J. T., and J. Smit (1981). The Cretaceous/Tertiary boundary: Calcareous nannofossils and stable isotopes, *Koninkliijke Nederlandse Akademie van Wetenschapen Proceedings, Series B 84*(3), 295-314.

Roth, P. (1987). Mesozoic calcareous nannofossil evolution: Relation to paleoceanographic events, *Paleoceanography 2*, 601-611.

Schmitz, B., G. Keller, and O. Stenvall (1992). Stable isotope and foraminiferal changes across the Cretaceous/Tertiary boundary at Stevns Klint, Denmark: Arguments for longterm oceanic instability before and after bolide impact, *Palaeogeography, Palaeoclimatology, Palaeoecology 96*, 233-260.

Shaw, A. B. (1964). *Time in Stratigraphy*, McGraw-Hill, New York, 365 pp.

Sloan, R. E., J. K. Rigby, L. M. Van Valen, and D. Gabriel (1986). Gradual dinosaur extinction and simultaneous ungulate radiation in the Hell Creek Formation, *Science 232*, 629-633.

Smit, J. (1982). Extinction and evolution of planktonic foraminifera after a major impact at the Cretaceous/Tertiary boundary, *Geological Society of America Special Paper 190*, 329-352.

Smit, J. (1990). Meteorite impact, extinctions and the Cretaceous-Tertiary boundary, *Geologie en Mijnbouw 69*, 187-204.

Smit, J., and A. J. T. Romein (1985). A sequence of events across the Cretaceous-Tertiary boundary, *Earth and Planetary Science Letters 74*, 155-170.

Stott, L. D., and J. P. Kennett (1989). New constraints on early Tertiary paleoproductivity from carbon isotopes in foraminifera, *Nature 342*, 526-529.

Thierstein, H. R. (1982). Terminal Cretaceous plankton extinctions: A critical assessment, *Geological Society of America Special Paper 190*, pp. 385-399.

Thierstein, H. R., F. Asaro, W. U. Ehrman, B. Huber, H. V. Michel, H. Sakai, and B. Schmitz (1991). The Cretaceous/Tertiary boundary at Site 738, South Kerguelen Plateau, in *Proceedings of the Ocean Drilling Program, Scientific Results 119*, Ocean Drilling Program, College Station, Texas, pp. 849-868.

Wei, W., and J. J. Pospichal (1991). Danian calcareous nannofossil succession at Site 738 in the Southern Indian Ocean, in *Proceedings of the Ocean Drilling Program, Scientific Results 119*, Ocean Drilling Program, College Station, Texas, pp. 495-512.

Worsley, J. (1974). The Cretaceous-Tertiary boundary event in the ocean, in *Studies in Paleo-oceanography*, W. W. Hay, ed., Society of Economic Paleontologists and Mineralogists Special Publication 20, pp. 94-125.

Zachos, J. C., and M. A. Arthur (1986). Paleoceanography of the Cretaceous/Tertiary boundary event: Inferences from stable isotopic and other data, *Paleoceanography 1*, 5-26.

Terminal Paleocene Mass Extinction in the Deep Sea: Association with Global Warming

JAMES P. KENNETT
University of California, Santa Barbara

LOWELL D. STOTT
University of Southern California

ABSTRACT

The end of the Paleocene Epoch was marked by an abrupt, worldwide extinction of deep-sea benthic organisms. At about 55 Ma, between 30 and 50% of the benthic foraminifers suddenly became extinct, in association with comparable ostracode extinctions. Extinctions of planktonic taxa were insignificant. This extinction event is considered the largest of the past 90 million years (m.y.) in the deep sea. Although of major proportions, this biotic crisis was almost unknown before the past decade because it had little effect on shallow marine invertebrates and ocean plankton. This was a "bottom-up" extinction, compared with the "top-down" extinction that marked the Cretaceous/Tertiary boundary.

High-resolution stratigraphic studies in deep-sea sediments indicate that the extinction occurred in less than 3000 yr. Foraminiferal oxygen and carbon stable isotope changes, in combination with a distinct change in benthic fauna, indicate an abrupt but temporary warming and oxygen depletion of deep waters related to a fundamental change in oceanic circulation. The available data point to an ocean temporarily dominated by warm saline deep water whose source was probably in the middle latitudes.

Although the terminal Paleocene environmental changes were relatively brief, this transient global warming event affected both ocean and terrestrial spheres simultaneously and had a great influence on the course of global biotic evolution for the remainder of the Cenozoic.

INTRODUCTION

Much interest exists in the character and causes of major biotic extinctions in the geologic past. The stratigraphic record demonstrates that the biosphere, or parts of it, have experienced major disruptions in the geologic past, some involving mass extinctions. Mass extinctions involve a sudden, and short-lived, increase in extinction rates well above normal background levels, and can affect a great variety of biotic groups (Flessa, 1990). Numerous theories have been proposed to explain mass extinctions (for reviews, see Stanley, 1984, 1987; Hallam, 1989). All involve a response by the biosphere to radical changes in the environment on regional or global scales. However, a persistent problem requiring resolution is how to explain contemporaneous extinctions over broad areas of the Earth's surface and over a wide range of habitats. Why did certain groups of taxa, particularly those that were abundant over large areas of the Earth, suddenly cease to exist? What factors control the timing of mass extinctions and the rate of biotic turnover? Theories advanced to explain mass extinction are of two general categories. The first, and perhaps most popular, has invoked extraterrestrial causes, particularly massive global environmental change resulting from bolide impacts on Earth (Alvarez et al., 1980). The second involves intrinsic changes exclusively within the Earth's environment (Stanley, 1984). General popularity for an extraterrestrial cause of mass extinction stems in part from the fact that it seems to provide a mechanism for sufficiently large and rapid changes in the global environment. In contrast, it appears more difficult to explain how the Earth's environment might have changed intrinsically on a magnitude necessary to cause mass extinction. Have intrinsic changes in the global environment ever been large and rapid enough to cause biotic crises of this scale? Stanley (1984) argued that most mass extinctions have resulted from climatic change, particularly cooling. Debate continues, however, about the relative merits of extrinsic and intrinsic causes.

The purpose of this essay is to briefly summarize evidence for the possible cause of a mass extinction of deep-sea biota near the end of the Paleocene about 55 million years ago. Existing data suggest that the extinction event resulted from large, rapid changes within the Earth's environmental system without extraterrestrial forcing.

Evidence for mass extinction comes exclusively from the stratigraphic record, and the quality of the stratigraphic data, including their resolution, is usually the key to better understanding of causes. Critical information includes the rate of extinction; which sectors of the biosphere were involved; description of the taxa that did or did not become extinct; the sequence of extinctions in taxa; and relationships of the extinction event to a wide variety of paleoenvironmental proxies. Relatively few high-quality sediment sequences are available that are sufficiently fossiliferous and were deposited continuously at high enough rates of sedimentation to provide the required resolution. Numerous stratigraphic sections contain hiatuses of various duration contemporaneous with major extinctions events. Such disruption in the stratigraphic record probably resulted from sediment erosion related to changes in oceanic circulation and/or sea level change at times of major global environmental change.

During the past two decades, the Deep Sea Drilling Project (DSDP) and its successor, the Ocean Drilling Program (ODP), have provided fossil-rich ocean sediment sequences covering vast, otherwise inaccessible, areas of the Earth's surface, including the high latitudes. The problem of mass extinction has thus been assisted by the availability of a broader array of sequences that record such events and of critical new information about changes in the deep-sea environments and biota. The ocean ecosystem deeper than the continental shelf is vast, forming more than 90% by volume of the Earth's habitable environments (Childress, 1983). Questions about mass extinctions require information about any response or role played by the deep-sea habitat.

The Paleocene-Eocene transition has long been differentiated by stratigraphers based on significant biotic changes at the end of the Paleocene. However, knowledge of stratigraphic relationships between marine and terrestrial records has been hampered by poor chronostratigraphic control. The primary problem has been the discontinuous nature and poor biostratigraphic control of the classic Late Paleocene-Early Eocene European stratotype sections that form the foundation of global stratigraphic correlations. Terrestrial to marine correlations have been improved with the application of carbon isotope stratigraphy.

During the Paleocene and Early Eocene there were large, systematic patterns of $\delta^{13}C$ variability in the ocean that have been correlated globally (Shackleton and Hall, 1984; Stott et al., 1990; Pak and Miller, 1992; Stott, 1992; Zachos et al., 1993a,b). These carbon isotope ($\delta^{13}C$) variations reflect changes in the $^{12}C/^{13}C$ ΣCO_2 in the ocean. Because the ocean and atmospheric reservoirs of CO_2 tend to maintain approximate isotopic equilibrium, variations in the ocean's $\delta^{13}C$ composition will be transmitted to the terrestrial reservoirs of carbon via the atmosphere. Soil carbonates, freshwater fossils, and terrestrial biomass, therefore, also exhibit the large carbon isotopic variations of the marine fossil record. The absolute isotopic values differ among these various terrestrial and marine carbon reservoirs due to systematic differences in the fractionation of ^{12}C and ^{13}C. However, these fractionation patterns are known and can be used to predict isotopic stratigraphies in terrestrial sections. Hence, the large-scale patterns of $\delta^{13}C$ variability recorded in the marine sections across the Paleocene-Eocene boundary are now being discovered in terrestrial sections (Koch et al., 1992; Sinha and Stott, 1994).

With this new global stratigraphy it has become apparent that accelerated evolution in terrestrial mammals during the Late Paleocene coincided with the extinction and environmental changes recorded in the deep sea (Rea et al., 1990; Koch et al., 1992). Although this chapter is concerned only with the marine record of extinction, the terrestrial record cannot be viewed independently. The abrupt climatic warming at the end of the Paleocene may have stimulated the evolution of major new mammalian groups such as the artiodactyls, perissodactyls, and the primates (Koch et al., 1992). Within the next few years it should be possible to integrate the terrestrial and marine records of faunal change at sufficient stratigraphic resolution to provide perhaps the best example of biotic change associated with abrupt environmental change.

Originally, observations on biotic change in the Paleocene-Eocene transition were limited to shallow marine invertebrate and terrestrial fossils. It has been generally assumed that, as with other epoch boundaries, changes in the global environment produced the biospheric response. Until recently the character of these changes remained largely unknown. Abrupt global warming and associated environmental changes now known for the end of the Paleocene are clearly implicated as a cause of this biotic crisis.

TERMINAL PALEOCENE MASS EXTINCTION IN THE DEEP SEA

The oceanic deep sea sediment record of the Paleocene-Eocene transition is clearly marked by major deep-sea benthic foraminiferal extinctions, perhaps the largest of the past 90 m.y. (Thomas, 1990). This event profoundly affected oceanic benthic communities deeper than the continental shelf (>100 m; neritic zone) resulting in a 35 to 50% species reduction in benthic foraminifera (Thomas, 1990). Deep-sea benthic foraminiferal assemblages radically changed as a result of this event. Late Paleocene assemblages before the extinction are highly diverse and contain genera with long stratigraphic ranges through the Late Cretaceous and Paleocene. Indeed, as pointed out by Thomas (1990, 1992), deep-sea benthic foraminiferal assemblages were little affected during the massive extinctions at the Cretaceous-Tertiary boundary (K/T), about 8 m.y. earlier. Instead many cosmopolitan benthic foraminiferal taxa typical of late Mesozoic and Paleocene assemblages, including all species within certain genera, were eliminated at the end of the Paleocene.

Until recently, the Paleocene-Eocene boundary was not generally recognized as a time of major biotic crisis, because generic-level extinction rates were low (Raup and Sepkoski, 1984). These patterns of extinction, however, are from the shallow marine and terrestrial spheres, not the deep sea (Thomas, 1992).

The mass extinction that predates the Paleocene-Eocene boundary is located near the middle of a long reversed polarity interval, identified as Magnetochron 24 R (Miller et al., 1987; Stott and Kennett, 1990), within the latest Paleocene. The position of the Paleocene-Eocene boundary appears to be about 40,000 yr younger, but also within Chron 24 R (Kennett and Stott, 1991). Correlation of these respective levels between high- and low-latitude sequences is still inadequate to establish the precise age and prove synchroneity (for detailed discussion see Thomas, 1990; Kennett and Stott, 1991; Miller et al., 1992). Ages applied to the extinction differ among different sequences, ranging between ~57.33 and 58 m.y. ago (Ma) (Miller et al., 1992). However, for several reasons discussed below, we believe that the mass extinction was synchronous and that the different age assignments have resulted from the current inadequacy of biostratigraphic correlations using planktonic microfossils across different latitudes and oceans. This appears, in part, to be the result of diachronism in planktonic microfossil datums and/or insufficient resolution in the biomagnetostratigraphic schemes employed. The age that we previously adopted for the mass extinction was 57.33 Ma (Kennett and Stott, 1991). However, the time scale of the magnetostratigraphic sequence in the vicinity of the extinction is now considered to be 2 m.y. younger (Cande and Kent, 1992).

The mass extinction is marked by four types of change in benthic foraminiferal assemblages: (1) the extinction of many taxa, (2) diversity decrease, (3) changes in the relative abundance of taxa, and (4) a general decrease in test size of taxa within the assemblages. As a result, the benthic faunas were distinctly different before and after the mass extinction.

The extinction removed species within the genera *Stensioina, Neoflabellina, Bolivinoides, Pyramidinia, Pullenia, Aragonia, Tritaxia, Gyroidinoides, Neoeponides, Quadratobulimina, Stilostomella, Dorothia*, and many other forms making up what has been termed the *S. beccariiformis* assemblage. Many of the extinct forms were morphologically distinctive, and their loss from the deep-sea record adds to the conspicuousness of this datum level. Epifaunal forms were particularly devastated (Thomas, 1990).

The extinction was first recognized in upper bathyal marine sequences from Trinidad (Beckmann, 1960); the Richenhall and Salzburg Basins, Austria (von Hildebrandt, 1962); and northern Italy (Braga et al., 1975). With the recognition of the extinction in the deep sea by Tjalsma and Lohmann (1983), as well as Schnitker (1979), Kaiho (1988), Berggren and Miller (1989), Nomura (1991), and Thomas (1990, 1992), its global character was clearly established. It is also evident that a wide range of paleodepths were affected, ranging from bathyal to abyssal depths (Miller et al., 1987).

Despite the global distribution of this extinction horizon, stratigraphic resolution was too low to determine whether the event was geologically instantaneous throughout the oceans or diachronous. More recent biostratigraphic investigations in conjunction with stable isotope stratigraphy (Kennett and Stott, 1991; Thomas, 1992) have strengthened the concept that the extinction was synchronous throughout the oceans. Nevertheless, this still requires confirmation with high-resolution studies at numerous sequences that can be accurately correlated by using carbon isotopic, paleomagnetic, and biostratigraphic data (Sinha and Stott, 1994).

Immediately following the extinction, the benthic foraminiferal assemblages were dominated by small, thin-walled specimens (Thomas, 1990). Benthic taxa that survived the extinction included *Nuttallides truempyi*, which became a dominant component in the Eocene, as well as *Bulimina semicostata* and other taxa making up what has been termed the *Nuttallides truempyi* assemblage. This new, relatively low-diversity assemblage includes about six forms that dominated Early to Middle Eocene benthic foraminiferal assemblages (Tjalsma and Lohmann, 1983; Miller *et al.*, 1987). Faunal assemblages following the extinction are less cosmopolitan (Thomas, 1990).

Although the extinction event was abrupt, there is some evidence that the *S. beccariformis* assemblage became progressively restricted to shallower depths during the Paleocene (Tjalsma and Lohmann, 1983; Miller *et al.*, 1987). Also, the relative abundances of certain forms in this assemblage decreased during the Late Paleocene and were replaced by forms more typical of the *Nuttallides truempyi* assemblage of latest Paleocene to Early Eocene age (Miller *et al.*, 1987). These changes culminated at the mass extinction and suggest that some form of biological threshold was surpassed.

For several million years following the extinction there occurred a radiation of benthic foraminiferal taxa. These probably filled vacancies left by the latest Paleocene extinctions (Tjalsma and Lohmann, 1983; Miller *et al.*, 1987). The postextinction assemblages included long-ranging forms such as *Pullenia bulloides* and *Globocassidulina subglobosa* (Thomas, 1990). The radiation caused a diversity increase that peaked during the early Middle Eocene. Nevertheless, the high diversity values of the Late Cretaceous and Early Paleocene were never attained again (Thomas, 1990).

In contrast to the benthic assemblages, oceanic planktonic microfossil assemblages underwent no mass extinction at the end of the Paleocene, but did exhibit distinct change in the species composition in the Antarctic. A general increase in diversity marks the Late Paleocene high- to middle-latitude assemblages of planktonic foraminifera, calcareous nannofossils, and dinoflagellates (Premoli-Silva and Boersma, 1984; Oberhänsli and Hsü, 1986; Stott and Kennett, 1990; Pospichal and Wise, 1990). This increase in diversity stemmed, in part, from the incursion of lower-latitude groups into the Southern Ocean. The diversity increase at the end of the Early Paleocene was superimposed on a longer-term increase that began during the Paleocene, following the K/T boundary extinctions (Corfield, 1987). The plankton diversity increase may have been caused by the increased surface water temperatures at high- to middle-latitude regions. This increase in surface water temperatures was particularly pronounced in the Antarctic during the latest Paleocene, as reflected by the relatively brief appearance of the subtropical-tropical morozovellid group and a peak in discoaster abundance (Pospichal and Wise, 1990; Stott and Kennett, 1990). The emigration of these warm-loving planktonic microfossils to the Antarctic was particularly pronounced during the mass extinction. In one Antarctic site 32% of the planktonic foraminiferal species appeared for the first time in the latest Paleocene, 27% underwent major abundance changes, and only 13% were eliminated from the assemblages (Lu and Keller, 1993). Most new entries were surface dwellers. Of those that were eliminated, most were deeper dwellers such as the subbotinids (Lu and Keller, 1993). Coeval low-latitude planktonic assemblages underwent little change (Miller *et al.*, 1987; Miller, 1991) presumably because of the relatively stable sea surface temperatures (Stott, 1992).

ASSOCIATION BETWEEN MASS EXTINCTION AND OCEANIC WARMING

In earlier work (Kennett and Stott, 1990; Stott *et al.*, 1990) we discovered a dramatic negative oxygen and carbon isotopic excursion of brief duration (Figure 5.1) that coincided closely with the terminal Paleocene benthic foraminiferal extinction event in an Antarctic Paleogene sequence (ODP Site 690B) (Thomas 1989, 1990). This discovery stimulated a high-resolution study of the extinction event (Kennett and Stott, 1991). Results from that study demonstrated the intimate temporal relationship between the mass extinction and a large oxygen and carbon isotope excursion in both benthic and planktonic foraminifera. Planktonic values of $\delta^{18}O$ abruptly decreased by 1.0 to 1.5‰, and by about 2‰ in the benthics; values of $\delta^{13}C$ also decreased by 4‰ in surface-dwelling planktonic foraminifera, and ~2‰ in the deeper-dwelling planktonic and benthic forms. The planktonic foraminifer *Acarinina praepentacamerata* records the lowest oxygen and highest carbon isotope values within the excursion, consistent with an inferred near-surface habitat (Stott *et al.*, 1990; Kennett and Stott, 1991). The species of *Subbotina* record higher $\delta^{18}O$ and lower $\delta^{13}C$ values, indicating a deep water planktonic habitat and/or a preference for cooler months of the year (Stott *et al.*, 1990; Kennett and Stott, 1991). The highest $\delta^{18}O$ and lowest $\delta^{13}C$ values are exhibited by the benthic foraminifer *Nuttalides truempyi*, reflecting its habitat in relatively nutrient-rich high-latitude deep water.

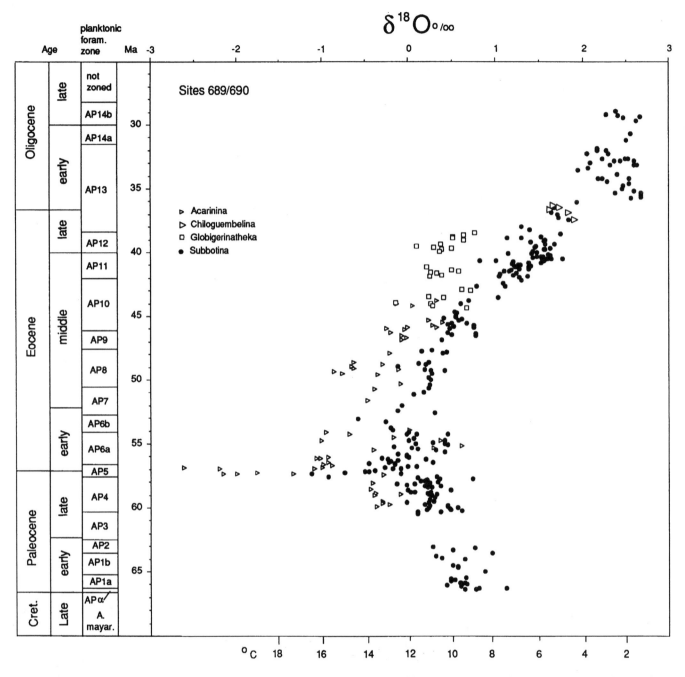

FIGURE 5.1 Composite oxygen isotopic record of planktonic foraminifers of ODP Sites 689 and 690 in the Antarctic Ocean (from Stott et al., 1990). Note the abrupt negative excursion near the Paleocene-Eocene boundary. Time scale follows that of Berrgren et al. (1985), although the Paleocene-Eocene boundary is now placed at 55 Ma (Cande and Kent, 1992). The temperature scale is based on the assumption of no significant ice sheet prior to the Early Oligocene and does not account for variation in surface water salinity.

During the latest Paleocene interval immediately preceding the extinction (before ~55.33 Ma), surface water temperatures estimated from oxygen isotopic values were ~13 to 14°C (Figure 5.2). Deep water temperatures were ~10°C at about 2100 m. Thus, before the extinction, little temperature difference existed between surface and deep water in the Antarctic.

The excursion began abruptly at 55.33 Ma, with conspicuous decreases in $\delta^{18}O$ and $\delta^{13}C$ values, followed by a return to values only slightly lower than those before the

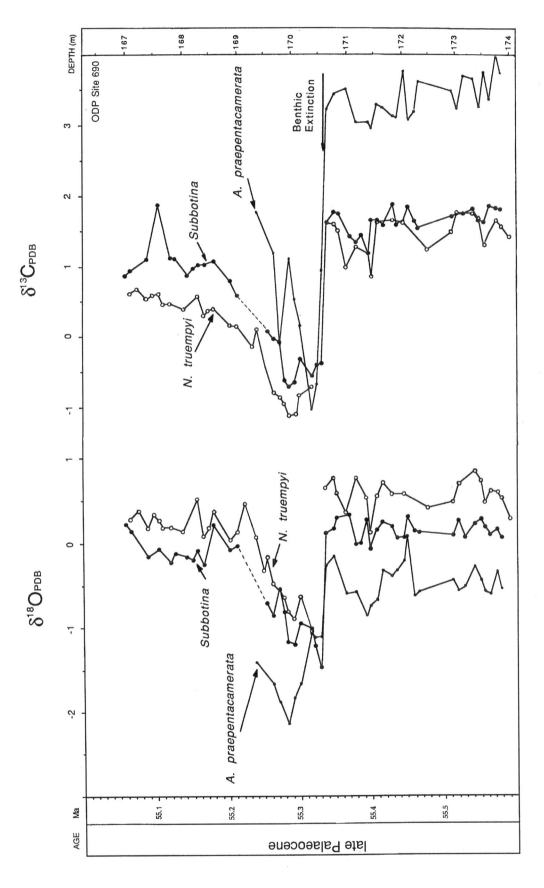

FIGURE 5.2 Changes in oxygen and carbon isotopic composition of planktonic foraminifera (*Acarinina praepentacamerata* and *Subbotina*) and a benthic foraminifer (*Nuttalides truempyi*) in the latest Paleocene (55.0 to 55.6 Ma) in relation to mass extinction in benthic foraminifera in ODP Site 690B, Maud Rise, Antarctica. These changes constitute the terminal Paleocene isotopic excursion. Note the abrupt negative isotopic shifts coinciding with the extinction. (Figure modified after Kennett and Stott, 1991.)

excursion (Figure 5.2). The $\delta^{18}O$ and $\delta^{13}C$ changes are reflected in all three foraminiferal taxa, although at different amplitudes (Figure 5.2). The significance of these differences is discussed by Kennett and Stott (1991). We limit our discussion here to the trends of critical paleoenvironmental importance. The largest $\delta^{18}O$ shift (2.0‰) at the beginning of the excursion is recorded by the benthic foraminifera, an intermediate shift (1.5‰) by deeper-dwelling planktonics, and the smallest shift (1.0‰) by shallow-dwelling planktonic forms (Figure 5.2). The initial oxygen isotopic shift exhibited by the surface-dwelling form, which coincided with the mass extinction, possibly reflects an increase in surface water temperatures from 14 to 18°C. This was followed by an additional $\delta^{18}O$ decrease of ~1.0‰, indicating a further possible temperature increase in surface waters to 22°C. The brief interval represented by the excursion was likely the warmest of the entire Cenozoic although, as discussed later, some fraction of the decrease in $\delta^{18}O$ values may have resulted from reduction in surface water salinity.

Of great significance, however, is the observation that the largest $\delta^{18}O$ change was recorded by the benthic, rather than the planktonic, foraminifera (Figure 5.2). Thus, deep waters warmed more than surface waters. For a brief interval, beginning at ~55.31 Ma, deep waters had warmed to such a degree that the temperature gradient between deep and surface waters was virtually eliminated at this location in the Antarctic region. The extinctions occurred at the beginning of the temperature excursion.

The initial, rapid temperature rise encompassed ~3000 yr and was followed by a more gradual decrease in ocean temperatures at all water depths. At the end of the excursion, the water column in this Antarctic region was only slightly warmer than it had been immediately before the excursion less than 100,000 yr earlier.

The magnitude of carbon isotopic change between the planktonic and benthic foraminfera was different from that of oxygen isotopes. Whereas the benthic (bottom dwellers) recorded the largest $\delta^{18}O$ change, it was the plankton that recorded the largest $\delta^{13}C$ change (4‰). The 4‰ shift in $\delta^{13}C$ is the largest so far known for the Cenozoic Period. The magnitude of the shift clearly underscores the significance of this event. During the brief interval at ~55.32 Ma when the vertical $\delta^{18}O$ gradient was eliminated, the previously large surface to deep water $\delta^{13}C$ gradient was also almost completely eliminated. The cause of the $\delta^{13}C$ change remains enigmatic. However, Stott (1992) presented evidence that the $\delta^{13}C$ of marine organic matter became more positive at the time of the excursion. If this was a global phenomenon, it would suggest that the negative $\delta^{13}C$ excursion recorded in foraminiferal calcite resulted from a redistribution of $\delta^{12}C$ between photosynthetic organic matter and the inorganic pool of carbon in the Late Paleocene oceans.

It is clear from this Antarctic record that the mass extinction coincided with the beginning of the sharp, negative shifts in $\delta^{18}O$ and $\delta^{13}C$. However, to determine whether all the extinctions occurred simultaneously and in conjunction with the initiation of the $\delta^{18}O$ and $\delta^{13}C$ change, Kennett and Stott (1991) increased the sample resolution to only 1-cm intervals (~800 yr) across the extinction interval in Site 690B (Figure 5.3). This was possible because the interval was not bioturbated (Kennett and Stott, 1991). Before the extinction, benthic foraminiferal assemblages (>150-μm fraction) were diverse, averaging about 60 species, or even more (Thomas, 1990). Assemblages included an abundance of forms interpreted to be of both infaunal and epifaunal habit (Corliss and Chen, 1988; Thomas, 1990). This included a high diversity of trochospiral and other coiled forms.

The extinction in Site 690B (Figure 5.3) involved a rapid drop in benthic foraminiferal diversity (>150 μm) from ~60 to 17 species, representing a diversity reduction of 72% within 3000 yr (4 cm). Many distinct taxa such as *Stensioina beccariiformis* and *Neoflabellina* disappeared early, during an interval of less than 1500 yr (Figure 5.3). Most trochospiral forms such as *Stensioina* had disappeared by the midpoint of the oxygen isotopic shift. The survival of a higher proportion of infaunal forms indicates some advantage over the epifaunal forms that lived at or close to the sediment-water interface. Nevertheless, the infaunal environment was not entirely unaffected since many of the taxa inferred to have been living there also disappeared. The abundance of benthic foraminifera (>150-μm fraction) was also severely reduced, although small individuals (<150 μm) remained abundant throughout. The ostracoda also exhibit a drastic decrease in diversity and abundance. The benthic foraminiferal assemblage was strongly depleted of coiled forms for several thousand years following the extinction. This left assemblages (>150 μm) temporarily dominated by relatively small, thin-walled, uniserial, triserial, and other forms more typical of an infaunal habitat (Corliss and Chen, 1988; Thomas, 1990). The relative increase in abundances of small benthic foraminiferal specimens, associated with a decrease in diversity, suggests conditions low in oxygen and higher in nutrients (Bernard, 1986; Thomas, 1990).

Following a brief interval of extremely low diversity and abundance in the fraction greater than 150 μm (Figure 5.3), diversity increased again to about 30 species on average—a diversity of about half that before the extinction. This increase seems to have resulted mainly from the reappearance of forms that had been temporarily excluded from the benthic foraminiferal assemblage. Nevertheless,

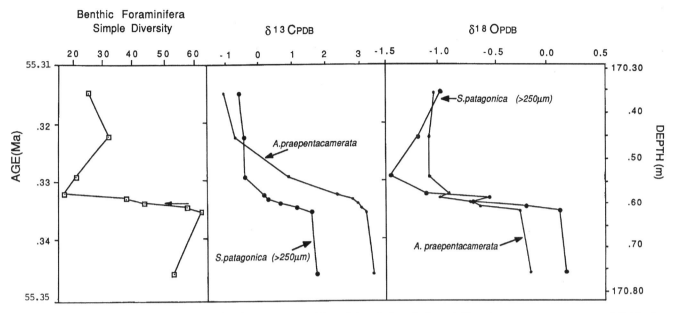

FIGURE 5.3 Changes in oxygen and carbon isotope composition of the planktonic foraminifers *A. praepentacamerata* and *Subbotina patagonica*, and in simple diversity of benthic foraminiferal assemblages at high stratigraphic resolution over the latest Paleocene mass extinction. The abrupt decrease in diversity reflects the mass extinction. Arrow at left indicates the last appearance of *Stensioina beccariiformis*, a distinctive foraminifer of the Paleocene. (Figure modified after Kennett and Stott, 1991.)

about 35% of the Late Paleocene species at Site 690B became completely extinct.

CAUSE OF MASS EXTINCTION IN DEEP SEA

It is clear that the mass extinction was restricted to the deep-sea biota deeper than the continental shelf or the thermocline. The lack of major extinctions in the oceanic planktonic and shallow-water benthic communities strongly suggests that the extinctions were not caused by an extraterrestrial impact with the Earth, as has been implicated for the terminal Cretaceous extinctions (Alvarez et al., 1980). An intrinsic oceanic cause is considered more likely (Kennett and Stott, 1991). The process that caused the mass extinction must have had the capacity to strongly affect the vast volume of the deep ocean in an interval of less than 3000 yr. Indeed, the extinctions may well have taken place at the rate of replacement time of the oceans, currently about 1000 yr, although this was possibly slightly slower during the early Paleogene.

The superposition of the abrupt, negative $\delta^{13}C$ and $\delta^{18}O$ shifts upon similar, more gradual trends during the Late Paleocene (~60 Ma) to Early Eocene (~55 Ma) (Figure 5.1) suggests the involvement of a climatic threshold event similar to the oxygen isotopic shift near the Eocene-Oligocene boundary, although in an opposite sense (Kennett and Shackleton, 1976). The speed and magnitude of the associated temperature increase imply global warming with strong positive feedback mechanisms, not just warming restricted to the oceans. Indeed, isotopic fluctuations in the marine carbonate record are closely tracked by the terrestrial records provided by paleosol carbonates and mammalian tooth enamel (Koch et al., 1992).

Three main hypotheses have been proposed to account for this mass extinction. These are (1) the rapid warming of deep waters (Miller et al., 1987); (2) an oxygen deficiency in deep waters resulting from the sudden warming and change in deep-sea circulation (Kennett and Barker, 1990; Kennett and Stott, 1990a; Thomas, 1990, 1992; Katz and Miller, 1991); and (3) a sharp drop in surface ocean biological productivity that reduced the supply of organic matter, the food source of deep sea benthic organisms, initiating a cascading trend of food chain collapse (Shackleton et al., 1985; Shackleton, 1986; Rea et al., 1990; Stott, 1992). If this had occurred, significant changes in oceanic plankton would be expected as well. There is no suggestion that this happened in the carbonate groups. It is also likely that reductions in abundance would have occurred in the infaunal deep-sea benthic foraminiferal assemblages. Indeed, Thomas (1990) showed that during the mass extinction, an increase occurred in the abundance of small infaunal species. Apparently these types of benthic foraminifera occur where there is availability of sedimentary organic carbon. Therefore, it is more likely that a greater abundance of these taxa resulted from an increase rather than a decrease in organic productivity, or a decrease in the oxygen content of deep waters, resulting in decreased oxidation of organic matter.

A general consensus exists that the mass extinction was caused by a rapid temperature increase of deep waters or the environmental effects associated with higher temperatures, including reduction in oxygen concentrations (Miller et al., 1987; Thomas, 1989, 1990, 1992; Kennett and Stott, 1990, 1991; Stott et al., 1990). Deep waters at the time of the excursion warmed to ~18°C. At such temperatures, deep waters would almost certainly have been depleted in oxygen even if oceanic primary production was lower and atmospheric oxygen levels were higher (Kennett and Stott, 1991). Widespread dysaerobism occurred in the deep ocean. However, there is no evidence that deep waters became completely anoxic, which would have caused an increase in accumulation of organic carbon during the excursion. Furthermore, infaunal benthic foraminifera remained abundant, which would not have been the case if there had been complete anoxia (Bernard, 1986). The dysaerobism was not as extreme as during the Cretaceous "oceanic anoxic events" (Schlanger and Jenkyns, 1976).

CAUSE OF OCEANOGRAPHIC AND CLIMATE CHANGE

The available data point to a mass extinction at the end of the Paleocene, resulting from physicochemical environmental changes in the deep sea, especially rapid warming and a decrease in oxygen concentrations. What ocean process could have created such widespread and rapid change? Consensus has developed (Miller et al., 1987; Thomas, 1990, 1992; Kennett and Stott, 1990a, 1991; Katz and Miller, 1991) that the rapid deep-sea warming resulted from a rapid change to near dominance in the deep ocean, of warm saline deep waters (WSDW) produced in the middle-latitude regions (see also Mead et al., 1993). At the same time it is believed that there was a severe reduction in the production of deep waters produced at high latitudes. Such ocean circulation, drastically different from that of the modern ocean, was termed Proteus by Kennett and Stott (1990b; Figure 5.4). In the present ocean, most bottom waters are formed at high latitudes where cold temperatures, in combination with moderately high salinities, cause waters to become dense and sink (for summary see Broecker and Peng, 1982). These waters are relatively oxygen rich. At the same time, warm saline dense waters are formed in the modern Mediterranean and Red Seas as a result of high net evaporation. Because of low buoyancy fluxes, these waters sink only to thermocline depths. These waters do not represent large volumes in the modern deep

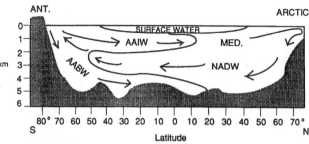

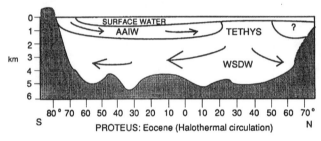

FIGURE 5.4 General model for deep and intermediate water circulation proposed for the time of the terminal Paleocene isotopic excursion and mass extinction. This model has been termed Proteus by Kennett and Stott (1990b). This is compared with the general circulation of the modern ocean (Oceanus). NOTE: WSDW = warm saline deep water; AAIW = Antarctic intermediate water; AABW = Antarctic bottom water; NADW = North Atlantic deep water; Med = Mediterranean. (Figure modified from Kennett and Stott, 1990b.)

ocean, although warm saline waters produced in the Mediterranean eventually become an important component of North Atlantic deep water (NADW) (Reid, 1979). Brass et al. (1982) suggested that this warm deep water (>10°C) of the Cretaceous and Early Paleogene reflected production of warm saline deep waters in middle-latitude areas.

The rapid ocean warming associated with the isotope excursion and mass extinction must have occurred from the deep ocean upward. We believe that this was caused by the elimination of deep water formation at high latitudes and the incursion of warm water from middle latitudes.

Synchroneity (Figure 5.3) of the mass extinction, the negative $\delta^{18}O$ shift, and the negative $\delta^{13}C$ shift in the deep-dwelling planktonic foraminifer *Subbotina patagonica* is apparent. Calculations of rates of sedimentation suggest that the mass extinction occurred in less than about 3000 yr, the oxygen isotopic shift in less than 4000 yr, and the carbon isotopic shift in less than 6000 yr. The mass extinction occurred at the beginning of the oceanographic and climatic changes that mark the excursion. Most of the large δ¹³C shift occurred slightly later in surface waters (Figure 5.3).

WHY THE EARLY PALEOGENE?

No such deep oceanic mass extinction and isotopic excursion has yet been discovered at any other time in the Cenozoic. This was an unusual, if not unique, event. Why did the event occur at about 55 Ma, early in the Cenozoic, rather than later? During the early Paleogene, different global geography and climate (Kennett, 1977; Haq, 1981; Hay, 1989) combined to make ocean circulation distinct from that of modern and, indeed, Neogene oceans (Kennett, 1977; Benson, 1979; Kennett and Stott, 1990a). Much evidence exists for relatively warm climates in the Antarctic region during the early Cenozoic (Kennett and Barker, 1990). Oxygen isotopic data suggest average Antarctic Ocean surface water temperatures of ~14°C during the Late Paleocene. Decreased meridional thermal gradients led to a decrease in global zonal wind intensity (Janecek and Rea, 1983; Hovan and Rea, 1992). Clay mineral assemblages in offshore sequences derived from the Antarctic continent were formed predominantly by chemical weathering under conditions of relative continental warmth and humidity (Robert and Kennett, 1992). Extensive coastal cool temperate rain forests dominated by *Nothofagus* indicate high continental rainfall and a lack of perglacial conditions at sea level (Case, 1988). Ice-rafted sediments are absent, as is other evidence for continental cryosphere of any extent (Kennett and Barker, 1990), although montane glaciation seems probable. The Antarctic Ocean was dominated by calcareous planktonic microfossil assemblages of high diversity rather than siliceous forms (Kennett, 1977). Faunas were cool to warm temperate in character. Deep waters in the global ocean were warm, averaging 10 to 12°C compared with ~2°C in the modern ocean (Shackleton and Kennett, 1975; Stott et al., 1990). The Earth was clearly in a "greenhouse" mode—a condition that appears to have been much exaggerated during the terminal Paleocene isotopic excursion. Relatively high precipitation in the Antarctic region at this time is inferred to have contributed to a large reduction in deep water production at high latitudes (Kennett and Stott, 1991).

At the same time, the extensive mid-latitude Tethys Seaway north of Africa was a likely location for the production of large volumes of warm saline deep waters (Kennett and Stott, 1990). Tectonic reconstructions (Dercourt et al., 1986) show that the Tethys Seaway in the early Cenozoic contained extensive shallow carbonate platforms with dolomites and evaporitic sediments. During the excursion, these various factors combined to form, through positive feedback responses, an extreme case of the Proteus Ocean, an ocean dominated by middle-latitude deep water production (Kennett and Stott, 1990, 1991). Climate model studies (Barron, 1987; Covey and Barron, 1988) suggest that large-scale meridional heat transport became more effective via the deep oceans relative to the atmosphere (Hovan and Rea, 1992).

The forcing mechanism of ocean warming and associated faunal turnover at the end of the Paleocene is not yet known. Rea et al. (1990) have suggested that the abruptness of environmental changes and the associated mass extinction were possibly triggered by rapid input of CO_2 into the atmosphere from volcanism and/or hydrothermal activity that was extensive over the Paleocene-Eocene transition. The warming of the oceans would have been the most obvious effect of enhanced greenhouse forcing resulting from this. Whether or not this would have been sufficiently rapid and large to explain the rapid rise in temperatures associated with the extinctions at 55 Ma remains to be tested. The triggering mechanism for the rapid climate change at the end of the Paleocene remains unknown.

In one attempt to test whether CO_2 might be implicated in the oceanic warming, Stott (1992) presented evidence that the Paleocene ocean-atmosphere system was indeed associated with higher levels of CO_2 compared to the present time. However, on the basis of the same data it appears that the extinction interval was actually associated with lower oceanic CO_2, not higher. How could an abrupt warming at the end of the Paleocene be associated with lower oceanic CO_2? The answer may lie in the way the ocean and atmosphere cycle CO_2. The problem is that the solubility of CO_2 in seawater decreases with increasing temperature. The exchange of CO_2 between the ocean and atmosphere was further complicated by changes in atmospheric circulation occurring at that time, which would have affected turbulence of the mixed layer ocean. This, together with changes in the biological pump (photosynthesis), constitute factors that are not yet well constrained for the Paleocene-Eocene. However, it is evident that with the high sea surface temperatures at the end of the Paleocene, particularly in regions of normal deep water advection (e.g., high latitudes), the oceans were probably less efficient in taking up CO_2 from the atmosphere.

IMPLICATIONS AND SUMMARY

A conceptual model of the possible chain of environmental events related to the terminal Paleocene mass extinction in the deep sea is shown in Figure 5.5. The extinction occurred in less than ~3000 yr and was associated with global deep-sea warming of similar rapidity (Figure 5.3). Both benthic foraminifera and ostracoda were se-

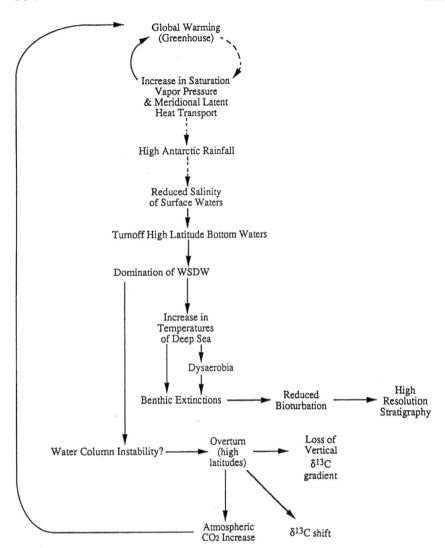

FIGURE 5.5 Conceptual model of possible chain of environmental events at the time of the mass deep-sea biotic extinction near the end of the Paleocene (55.33 Ma).

verely affected, although a synchronous decrease in bioturbation during the extinction at one location suggests that nonskeletal benthic organisms were also severely affected. Apparently the organisms that became extinct were unable to cope with the rapidity and magnitude of deep-sea warming and associated depletion in oxygen levels. Brief elimination of the vertical $\delta^{18}O$ and $\delta^{13}C$ gradients during the extinction event indicates vertical ocean mixing and homogenization of nutrient distributions over a large depth range at high latitudes (Figure 5.5). It seems that this was related to temporary instability of the water column and even ocean turnover at high latitudes. This did not seem to be the case in the tropics.

The character of the oxygen isotopic changes associated with the mass extinction indicates a temporary switch to almost total dominance of warm saline deep water and associated interruption in the production of deep waters at high latitudes (Figure 5.4). Surface waters warmed significantly at high latitudes but little in the tropics. The major $\delta^{13}C$ shift reflects large, rapid changes in the distribution of CO_2 and nutrients in the ocean during the extinction. The magnitude of the $\delta^{13}C$ shift in Antarctic surface waters suggests major changes in nutrients and partial pressure of CO_2 in surface waters. This may imply an associated increase in atmospheric CO_2 and resulting global greenhouse warming (Figure 5.5).

Rapid global warming would have led to increased transfer of heat from low to high latitudes, particularly via the oceans. Increased sea-surface temperatures would have caused an increase in atmospheric saturation vapor pressure. Latent heat transfer would have increased from the tropics to the poles, causing a significant increase in rainfall in the Antarctic region (Figure 5.5). Clay mineralogical evidence from the Antarctic suggests the occurrence of such an increase in rainfall. Increased rainfall in the Antarctic region would have reduced ocean surface water salinity. This in turn, would have contributed to a decrease in the production of Antarctic deep waters to the world ocean, thus reinforcing the dominance of warm saline deep waters (Figure 5.5).

A number of broad implications are suggested as a result of these discoveries. The extinction resulted from changes entirely within the Earth's environmental system, in the apparent absence of any extraterrestrial influences. Mass extinctions can be produced by large changes in the ocean. Such changes can occur very quickly under certain conditions. Mass extinction can be restricted to certain parts of the Earth's ecosystem and be effectively decoupled from other parts of the biosphere. In the case of the terminal Paleocene mass extinction, there was almost total decoupling between the deep and shallow marine ecosystems. Despite significant rapid warming in shallow and deep environments the mass extinction was limited to the deep sea. At particular times in the geologic past, broad sectors of the Earth's environmental system, such as the global ocean, were susceptible to major reorganizations on short time scales. Brief (10^2 to 10^3 yr), intense paleoenvironmental events can have large effects on the course of biotic evolution. At the end of the Paleocene, global climate change crossed a critical threshold, causing instability and mass extinction in the deep sea.

The potential for the early Paleogene ocean to shift between high- and middle-latitude dominated deep water production resulted from low vertical and meridional temperature gradients and multiple major sources of deep waters. Later in the Cenozoic, meridional and vertical temperature gradients strengthened, leading to decreased opportunities for such drastic switches in deep water sources. Thus, it is unlikely that the oceanographic changes and associated extinction event that occurred at the end of the Paleocene would have been repeated during the middle to late Cenozoic. The historic record reveals no such change.

ACKNOWLEDGMENTS

This contribution was supported by the National Science Foundation (Division of Polar Programs) DPP-9218720 to J.P.K. and (Ocean Sciences) OCE 9101662 to L.D.S.

REFERENCES

Alvarez, L. W., W. Alvarez, F. Asaro, and H. V. Michel (1980). Extraterrestrial cause for the Cretaceous-Tertiary extinction, *Science 208*, 1095-1108.

Barron, E. J. (1987). Eocene equator-to-pole surface ocean temperatures: A significant climate problem? *Paleoceanography 2*, 729-739.

Beckmann, J. P. (1960). Distribution of benthonic foraminifera at the Cretaceous-Tertiary boundary of Trinidad (West Indies), in *Report of the 21st Session, Norden, Part V, The Cretaceous-Tertiary Boundary*, International Geological Congress, pp. 57-69.

Benson, R. H. (1979). In search of lost oceans: A paradox in discovery, in *Historical Biogeography, Plate Tectonics, and the Changing Environment*, J. A. Gray and A. J. Boucot, eds., Oregon State University Press, Eugene, pp. 379-389.

Berggren, W. A., and K. G. Miller (1989). Cenozoic bathyal and abyssal calcareous benthic foraminiferal zonation, *Micropaleontology 35*, 308-320.

Berggren, W. A., D. V. Kent, J. J. Flynn, and J. A. Van Couvering (1985). Cenozoic geochronology, *Geological Society of America Bulletin 96*, 1407-1418.

Bernard, J. M. (1986). Characteristic assemblages and morphologies of benthic foraminifera from anoxic, organic-rich deposits; Jurassic through Holocene, *Journal of Foram. Research 16*, 207-215.

Braga, G., R. De Biase, A. Grunig, and F. Proto-Decima (1975). Foraminiferi bentonici del Paleocene e dell'Eocene della Sezione Possagno, *Schweizerische Palaeontologische Abhandlungen 97*, 85-111.

Brass, G. W., J. R. Southam, and W. H. Peterson (1982). Warm saline bottom water in the ancient ocean, *Nature 196*, 620-623.

Broecker, W. S., and T.-H. Peng (1982). *Tracers in the Sea*, Eldigio Press, Palisades, N.Y., 690 pp.

Cande, S. C., and D. V. Kent (1992). A new geomagnetic polarity time scale for the Late Cretaceous and Cenozoic, *Journal of Geophysical Research 97* (B10), 13,917-13,951.

Case, J. A. (1988). Paleogene floras from Seymour Island, Antarctic Peninsula, in *Geology and Paleontology of Seymour Island, Antarctic Peninsula*, Geological Society of America Memoir 169, pp. 523-530.

Childress, J. J. (1983). Oceanic biology: Lost in space? in *Oceanography: The Present and Future*, P. Brewer, ed., Springer, New York, pp. 127-135.

Corfield, R. M. (1987). Patterns of evolution in Palaeocene and Eocene planktonic foraminifera, in *Micropalaeontology of Carbonate Environments*, M. M. Hart, ed., Eflis Horwood, Chichester, 93-110.

Corliss, G. H., and C. Chen (1988). Morphotype patterns of Norwegian deep-sea benthic foraminifera and ecological implications. *Geology 16*, 716-719.

Covey, C., and E. J. Barron (1988). The role of oceanic heat transport in climatic change, *Earth Science Reviews 24*, 429-455.

Dercourt, J., *et al.* (1986). Geological evolution of the Tethys belt from the Atlantic to the Pamir since the Lias, *Tectonophysics 123*, 241-315.

Flessa, K. W. (1990). The "facts" of mass extinctions, *Geological Society of America Special Paper 247*, 1-7.

Hallam, A. (1989). Catastrophism in geology, in *Catastrophes and Evolution*, S. V. M. Clube, ed., Cambridge University Press, pp. 25-55.

Haq, B. U. (1981). Paleogene paleoceanography—Early Cenozoic oceans revisited, *Oceanologica Acta 4* (Suppl. to Vol. 4), Proceedings of the 26th International Geological Congress, pp. 71-82.

Hay, W. W. (1989). Paleoceanography: A review for GSA centennial, *Geological Society of America Bulletin 100*, 1934-1956.

Hovan, S. A., and D. K. Rea (1992). Paleocene/Eocene boundary changes in atmospheric and oceanic circulation: A Southern Hemisphere record, *Geology 20*, 15-18.

Janecek, T. R., and D. K. Rea (1983). Eolian deposition in the northeast Pacific: Cenozoic history of atmospheric circulation, *Geological Society of America Bulletin 94*, 730-738.

Kaiho, K. (1988). Uppermost Cretaceous to Paleogene bathyal benthic foraminiferal biostratigraphy of Japan and New Zealand: Latest Paleocene-Middle Eocene benthic foraminiferal species turnover, *Reviews Paléobiol., Vol. Spec. 2*, 553-559.

Katz, M. E., and K. G. Miller (1993). Early Paleogene benthic foraminiferal assemblage and stable isotope composition in the Southern Ocean, ODP Leg 114, in *Proceedings of the Ocean Drilling Program, Scientific Results 114*, P. F. Ciesielski and Y. Kristoffersen, eds., Ocean Drilling Program, College Station, Texas.

Kennett, J. P. (1977). Cenozoic evolution of Antarctic glaciation, the circumAntarctic Ocean, and their impact on global paleoceanography, *Journal of Geophysical Research 82*, 3843-3860.

Kennett, J. P., and P. F. Barker (1990). Latest Cretaceous to Cenozoic climate and oceanographic developments in the Weddell Sea, Antarctica: An ocean-drilling perspective, in *Proceedings of the Ocean Drilling Program, Scientific Results 113*, P. F. Barker, J. P. Kennett, et al., eds., Ocean Drilling Program, College Station, Texas, pp. 937-960.

Kennett, J. P., and N. J. Shackleton (1976). Oxygen isotopic evidence for the development of the psychrosphere 38 Myr. ago, *Nature 260*, 513-515.

Kennett, J. P., and L. D. Stott (1990a). Latest Cretaceous to Cenozoic climate and oceanographic developments in the Weddell Sea, Antarctica: An ocean-drilling perspective, in *Proceedings of the Ocean Drilling Program, Scientific Results 113*, P. F. Barker, J. P. Kennett, et al., eds., Ocean Drilling Program, College Station, Texas, pp. 865-880.

Kennett, J. P., and L. D. Stott (1990b). Proteus and Pro-oceanus: Ancestral Paleogene oceans as revealed from Antarctic stable isotopic results; ODP Leg 113, in *Proceedings of the Ocean Drilling Program, Scientific Results 113*, P. F. Barker, J. P. Kennett, et al., eds., Ocean Drilling Program, College Station, Texas.

Kennett, J. P., and L. D. Stott (1991). Abrupt deep-sea warming, palaeoceanographic changes and benthic extinctions at the end of the Palaeocene, *Nature 353*, 225-229.

Koch, P. L., J. C. Zachos, and P. D. Gingerich (1992). Correlation between isotope records in marine and continental carbon reservoirs near the Palaeocene/Eocene boundary, *Nature 358*, 319-322.

Lu, G., and G. Keller (1993) Climatic and oceanographic events across the Paleocene-Eocene transition in the Antarctic Indian Ocean: Inference from planktic foraminifera, *Marine Micropaleontology*.

Mead, G. A., D. A. Hodell, and P. F. Ciesielski (1993). Late Eocene to Oligocene vertical oxygen isotopic gradients in the South Atlantic: Implications for warm saline deep water, in *The Antarctic Paleoenvironment: A Perspective on Global Change 2*, J. P. Kennett and D. A. Warnke, eds., Antarctic Research Series, American Geophysical Union, Washington, D.C.

Miller, K. G. (1991). The Paleocene/Eocene boundary in the context of Paleogene global climate change. *Abstract, Geological Society of America Annual Meeting*, A141.

Miller, K. G., T. R Janecek, M. E. Katz, and D. J. Keil (1987). Abyssal circulation and benthic foraminiferal changes near the Paleocene/Eocene boundary, *Paleoceanography 2(6)*, 741-761.

Miller, K. G., M. E. Katz, and W. A. Berggren (1992). Cenozoic deep-sea benthic foraminifera: A tale of three turnovers, in *Benthos '90, Studies in Benthic Foraminifera*, Tokai University Press, pp. 67-75.

Nomura, R. (1991). Paleoceanography of upper Maastrichtian to Eocene benthic foraminiferal assemblages at ODP sites 752, 753 and 754, eastern Indian Ocean, in *Proceedings of the Ocean Drilling Program, Scientific Results 121*, J. W. Pierce et al., eds., Ocean Drilling Program, College Station, Texas, pp. 3-30.

Oberhänsli, H., and K. J. Hsü (1986). Paleocene-Eocene paleoceanography, in *Mesozoic and Cenozoic Oceans*, K. J. Hsü, ed., American Geophysical Union Geodynamics Series 15, Washington, D.C., pp. 85-200.

Pak, D. K., and K. G. Miller (1992). Paleocene to Eocene benthic foraminiferal isotopes and assemblages: Implications for deepwater circulation, *Paleoceanography 7(4)*, 405-422.

Pospichal, J. M., and S. W. Wise, Jr. (1990). Paleocene to Middle Eocene calcareous nannofossils of ODP sites 689 and 690, Maud Rise, Weddell Sea, in *Proceedings of the Ocean Drilling Program, Scientific Results 113*, P. F. Barker, J. P. Kennett, et al., eds., Ocean Drilling Program, College Station, Texas, pp. 613-638.

Premoli-Silva, I., and A. Boersma (1984). Atlantic Eocene planktonic foraminiferal historical biogeographic and paleohydrologic indices, *Palaeogeography, Palaeoclimatology, Palaeoecology 67*, 315-356.

Raup, D. M., and J. J. Sepkoski, Jr. (1984). Periodicity of extinctions in the geologic past, *Proceedings of the National Academy of Sciences USA 81*, 801-805.

Rea, D. K., J. C. Zachos, R. M. Owen, and P. D. Gingerich (1990). Global change at the Paleocene-Eocene boundary: Climatic and evolutionary consequences of tectonic events, *Palaeogeography, Palaeoclimatology, Palaeoecology 79*, 117-128.

Reid, J. L. (1979). On the contribution of Mediterranean Sea outflow to the Norwegian-Greenland Sea, *Deep-Sea Research 26*, 199-223.

Robert, C., and J. P. Kennett (1992). Paleocene and Eocene kaolinite distribution in the South Atlantic and Southern Ocean: Antarctic climatic and paleoceanographic implications, *Marine Geology 103*, 99-110.

Schlanger, S. O., and H. C. Jenkyns (1976). Cretaceous oceanic anoxic events: Causes and consequences, *Geologie en Mijnbouw 55*, 179-84.

Schnitker, D. (1979). Cenozoic deep-water benthic foraminifers, Bay of Biscay, in *Initial Reports of the Deep Sea Drilling*

Program 48, L. Montadert *et al.*, eds., U.S. Government Printing Office, Washington, D.C., pp. 377-414.

Shackleton, N. J. (1986). Paleogene stable isotope events, *Palaeogeography, Palaeoclimatology, Palaeoecology 57*, 91-102.

Shackleton, N. J., and M. A. Hall (1984). Carbon isotope data from Leg 74 sediments, in *Initial Reports of the Deep Sea Drilling Program 74*, U.S. Government Printing Office, Washington, D.C., pp. 613-619.

Shackleton, N. J., and J. P. Kennett (1975). Paleotemperature history of the Cenozoic and the initiation of Antarctic glaciation: Oxygen and carbon isotope analyses in DSDP sites 277, 279 and 281, in *Initial Reports of the Deep Sea Drilling Program 29*, J. P. Kennett, R. E. Houtz, *et al.*, eds., U.S. Government Printing Office, Washington, D.C., pp. 143-756.

Shackleton, N. J., R. M. Corfield, and M. A. Hall (1985). Stable isotope data and the ontogeny of Paleocene planktic foraminfera, *Journal of Foraminiferal Research 15*, 321-336.

Sinha, A. and L. D. Stott (1994). The transfer of ^{13}C change from the ocean to the atmosphere and terrestrial biosphere across the Paleocene/Eocene boundary: Criteria for terrestrial-marine correlations, in *Early Paleogene Correlation in NW Europe*, R. Knox, ed., Special Publication of the Geological Society, London (in press).

Stanley, S. M. (1984). Mass extinctions in the ocean, *Scientific American 250*, 64-72.

Stanley, S. M. (1987). *Extinction*, Scientific American Library, W. H. Freeman, New York.

Stott, L. D. (1992). Higher temperatures and lower oceanic PCO_2: A climate enigma at the end of the Paleocene epoch, *Paleoceanography 7*(4), 395-404.

Stott, L. D., and J. P. Kennett (1990). Antarctic Paleogene planktonic foraminifer biostratigraphy: ODP Leg 113, Sites 689 and 690, in *Proceedings of the Ocean Drilling Program, Scientific Results 113*, P. F. Barker, J. P. Kennett, *et al.*, eds., Ocean Drilling Program, College Station, Texas, pp. 548-569.

Stott, L. D., J. P. Kennett, N. J. Shackleton, and R. M. Corfield (1990). The evolution of Antarctic surface waters during the Paleogene: Inferences from the stable isotopic composition of planktonic foraminifera, ODP Leg 113, in *Proceedings of the Ocean Drilling Program, Scientific Results 113*, P. F. Barker, J. P. Kennett, *et al.*, eds., Ocean Drilling Program, College Station, Texas, pp. 849-864.

Thomas, E. (1989). Development of Cenozoic deep-sea benthic foraminiferal faunas in Antarctic waters, in *Origins and Evolution of the Antarctic Biota*, J. A. Crame, ed., Geological Society Special Publication, London 47, pp. 283-296.

Thomas, E. (1990). Late Cretaceous-Early Eocene mass extinctions in the deep sea, *Geological Society of America Special Paper 247*, 481-495.

Thomas, E. (1992). Cenozoic deep-sea circulation: Evidence from deep-sea benthic foraminifera, in *The Antarctic Paleoenvironment: A Perspective on Global Change*, Antarctic Research Series 56, American Geophysical Union, Washington, D.C., pp. 141-165.

Tjalsma, R. C., and G. P. Lohmann (1983). Paleocene-Eocene bathyal and abyssal benthic foraminifera from the Atlantic Ocean, *Micropaleontology Special Publication 4*, Micropaleontology Press, New York, 94 pp.

von Hildebrandt, A. (1962). *Akad. Wiss. (Wien), Math.-Naturw. Klasse, Abh. n. ser. 108*, 1-182.

Zachos, J., D. Rea, K. Seto, R., Nomura, and N. Niitsuma (1993a). Paleogene and early Neogene deepwater paleoceanography of the Indian Ocean as determined from benthic foraminifera stable isotope records, in *The Indian Ocean: A Synthesis of Results from the Ocean Drilling Program*, R.A. Duncan *et al.*, eds., Geophysical Monograph Series, American Geophysical Union, Washington, D.C.

Zachos, J. C., K. C. Lohmann, J. C. G. Walker, and S. W. Wise (1993b). Abrupt climate change and transient climates during the Paleogene: A marine perspective, *Journal of Geology 101*, 193-215.

6

Tropical Climate Stability and Implications for the Distribution of Life

ERIC J. BARRON
The Pennsylvania State University

ABSTRACT

The tropics are generally viewed as an environment in which the physicochemical factors are not undergoing major changes, and therefore the biotic composition and character are defined largely by biological competition. However, many equatorial species are also characterized by narrow environmental tolerances, which suggests that relatively small climate changes may result in a substantial biologic response. The climatic stability of the tropics is therefore a central issue in global change research. Evidence from the geologic record and from climate models suggests that a temperature variation of 3 to 5°C and a salinity variation of several parts per thousand from present values are plausible. The key question becomes the significance of changes of this magnitude for the distribution and character of tropical life. Data on the tolerances of tropical organisms and a case study for the mid-Cretaceous indicate that climate change may substantially influence tropical life. The changes in tropical organisms and their distribution through Earth history should be viewed as a rich, underutilized record that can provide new insights into the climate sensitivity of the tropics. This record provides the only major source of data on the biologic response to global change.

INTRODUCTION

The tropical biological environment is strongly associated with the notion of physical and chemical stability. However, there is also abundant evidence indicating that climate is a significant limiting factor in the distribution of life (e.g., Valentine, 1973; Stanley, 1984a,b). Tropical organisms may be sensitive to climate, in particular, because of their narrow environmental tolerances. Even small climate changes in the tropics can have a substantial impact on life. The question of tropical climate stability with respect to future global change therefore becomes a central issue of research (Crowley, 1991).

Oxygen isotopic paleotemperatures (Douglas and Savin, 1975; Savin, 1977; Shackleton, 1984) can be interpreted as evidence for large variations in tropical temperatures. However, these interpretations have been questioned by Matthews and Poore (1980) and Horrell (1990). Some simple physical arguments involving changes in evaporative cooling with warming (Newell *et al.*, 1978; Newell and Dopplick, 1979) and questions on the mechanisms of tropical temperature change (e.g., Horrell, 1990) have also been utilized to support the notion of tropical temperature stability. However, a number of more comprehensive model experiments suggest that variations within a limited, but significant, range cannot be ruled out (Washington and Meehl, 1984; Manabe and Bryan, 1985; Hansen *et al.*, 1988; Schlesinger, 1989). In combination, the isotopic data and the model studies support the hypothesis that past climate changes should have had a substantial impact on the character and distribution of life within the tropics. The climate stability of the tropics and its implications for the distribution and character of life are addressed here by (1) consideration of the oxygen isotopic records of low-latitude temperature variations; (2) discussion of the physical arguments for temperature stability within the tropics; (3) examination of climate model-derived tropical temperatures; (4) examination of model evidence for tropical salinity differences between different time periods in Earth history; (5) consideration of the climate tolerances of tropical organisms; and (6) consideration of a mid-Cretaceous case study in which simulated climate changes in the tropics can be compared with the biological record.

The primary conclusions are that (1) throughout Earth history there has been significant variation in tropical temperature (3 to 5°C differences from the present day) and salinity (several parts per thousand); (2) these variations are large enough to have substantial impact on life; and (3) greater study of the geologic record within the tropics will yield important insights into climate sensitivity and into the biologic response to global change.

OXYGEN ISOTOPIC RECORDS OF LOW LATITUDE TEMPERATURES

The oxygen isotope method of determining paleotemperatures has been widely utilized to study the Cenozoic and the Cretaceous (Savin, 1977). These paleotemperature determinations for the tropics suggest substantial variation.

Isotopic measurements on apparently unaltered planktonic foraminifera from the Shatsky Rise, near the equator during the mid-Cretaceous, yield temperature values of 25 to 27°C (Douglas and Savin, 1975), if an ice-free Earth is assumed. These values can be taken at face value and used to indicate little change in tropical temperatures or slightly lower temperatures than ar present (e.g., Horrell, 1990). However, several factors (regional variations, habitat, and selective preservation) must be considered in interpreting isotopic measurements on planktonic foraminifera. First, the isotopically lightest measurement (27°C) is likely to represent the shallowest dwelling foraminifera. Even present-day shallow-dwelling foraminifera give isotopic temperatures that are 3 to 5°C cooler than the surface. Further, selective dissolution of the more fragile, shallow-dwelling forms tends to bias estimates in the cold direction (Savin *et al.*, 1975). Consequently, a reasonable interpretation of the isotopic data within the Cretaceous tropics is surface temperatures of 27 to 32°C. The range of possible interpretation is from similar to the present day (28°C) to several degrees higher than at present (Figure 6.1).

Pre-Pleistocene Cenozoic isotopic temperatures are also substantially different from the present day. Shackleton (1984) presents data yielding isotopic paleotemperatures as low as 18°C for the low-latitude Pacific from the Maastrichtian to the Late Miocene. Values similar to the present day occurred only in the late Neogene. Early Eocene and Early Miocene values represent tropical ocean sea-surface temperature minima in the Shackleton (1984) analysis. The Early Eocene low-temperature values have received particular attention (Shackleton and Boersma, 1981). Recent synthesis and interpretation of these and other isotopic data (Sloan, 1990) suggest that at a maximum, Early Eocene tropical surface temperatures were about 24°C, about 3 to 5°C lower than present values.

Analysis of tropical sea-surface temperatures during the last glacial maximum also contributes to the notion of tropical temperature variation. Early estimates of tropical sea-surface temperatures from oxygen isotopes for ice age

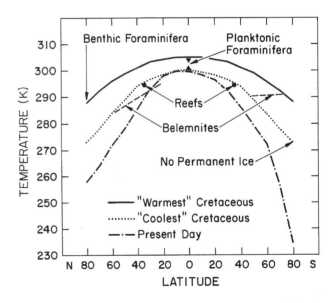

FIGURE 6.1 Cretaceous mean annual temperature limits in comparison with modern values (Barron, 1983). Some of the major constraints based on oxygen isotopes (benthic and planktonic foraminifera and bellemnites), reef distribution, and the absence of permanent ice. Solid line is "warmest" Cretaceous, dotted line is "coolest" Cretaceous, and dot-dashed line is present day.

climates were nearly 5°C lower than present-day values (Emiliani, 1970). However, these estimates were followed by a period of active debate over the relative role of temperature variation and ice volume changes in governing oxygen isotopic composition of foraminifera (e.g., Emiliani and Shackleton, 1974), with the general consensus that the signal was dominated by ice volume changes. CLIMAP (1981) reconstructions for 18,000 yr ago give tropical temperature values approximately 1 to 2°C lower than present-day values in most areas. The argument may still be unsettled, however; Rind and Peteet (1985) suggest that much cooler tropical temperatures are required to explain the snowline in low latitudes during the last glacial maximum. Rind and Peteet (1985) suggest that a 3 to 4°C lower tropical temperature in comparison with the present day would represent a better fit of the terrestrial observations.

Without critique, oxygen isotopic data suggest that tropical sea-surface temperatures have varied at least within ±5°C of present-day values. There is, however, substantial room for reinterpretation and debate.

ARGUMENTS FOR TROPICAL TEMPERATURE STABILITY

Three physically-based arguments, and additional evidence from biota, have been presented for essentially stable tropical temperatures and then utilized to question the validity of the oxygen isotopic analysis.

First, the conclusions of Newell *et al.* (1978) and Newell and Dopplick (1979) using a simple balance between radiative energy input and evaporation have been cited as an argument for tropical temperature stability. This argument is based on the fact that saturation vapor pressure, a measure of the amount of moisture contained within a parcel of air, is a strong nonlinear function of temperature. With increasing temperatures, evaporation should increase substantially, thus limiting any increase in surface temperature. Newell *et al.* (1978) and Newell and Dopplick (1979) argue that present-day sea-surface temperatures are at a maximum for present-day energy input from the Sun.

The problem with the argument presented by these authors is that the role of the atmosphere is basically ignored in their simple surface energy budget model. Higher temperatures and increased moisture should also lead to a greater greenhouse effect, thus promoting surface warming. In every major climate model study in which the atmosphere is included in detail and for which a forcing factor is incorporated that promotes warming (e.g., increased carbon dioxide), the tropical surface temperatures increase (e.g., Washington and Meehl, 1984; Manabe and Bryan, 1985; Hansen *et al.*, 1988; Schlesinger, 1989). These results reflect the increased radiative forcing from carbon dioxide and the importance of water vapor feedback at warmer temperatures. The arguments by Newell and others are not supported by these more comprehensive experiments and should not be a basis for assuming tropical temperature stability.

A second major argument centers on whether cool tropical temperatures (e.g., for the Eocene) could be explained plausibly by increased poleward heat transport, as suggested by Shackleton and Boersma (1981) and Barron (1987). Rind and Chandler (1991) and Barron *et al.* (1993) demonstrate that reasonable changes in ocean poleward heat transport can explain the equator-to-pole surface temperature distribution for time periods of past warm climates such as the Cretaceous or the Eocene. However, using a simple energy balance climate model, Horrell (1990) suggests that the heat transport required to maintain very low tropical temperatures (e.g., 15°C) would be a factor of two or three times the present-day total poleward heat flux. This magnitude of increase would be excessive and unlikely, especially since the Eocene equator-to-pole temperature gradient was small.

The question of plausibility in this case depends entirely on the magnitude of the tropical cooling that is proposed. A 3 to 5°C decrease in tropical temperatures in comparison with the present day is quite plausible. Sloan (1990) calculated the total poleward heat transport in an atmospheric general circulation model (GCM), with sea-surface temperatures specified at 3 to 5°C lower than present in the tropics but substantially warmer polar regions in accordance with observations. The atmospheric heat transport in the model decreased from present-day values, as expected because of the decreased temperature gradient from equator to pole and because of the cooler tropical temperatures. The total poleward heat transport increased, which implies a greater role for the ocean in order to achieve warmer poles and cooler tropics. However, the change was not substantial. Covey and Thompson (1989) examined explicitly the role of increased ocean heat flux on the total poleward heat transport and on the latitudinal distribution of surface temperatures. In a case for doubled oceanic poleward heat transport, the total poleward heat transport increased slightly (about 12% at the maximum in midlatitudes), while the role of the atmosphere declined substantially. The tropical sea-surface temperatures decreased by 5°C. Barron *et al.* (1993a) found similar results (2-3°C decrease) for the Cretaceous for increases of 15 to 30% of observed ocean heat transport. Therefore, relatively small changes in total poleward heat transport can substantially influence tropical sea-surface temperatures. Rind and Chandler (1991) provide a different perspective by calculating the ocean heat transport required to achieve a specific sea-surface temperature distribution. They conclude that perturbations in ocean heat transport, if strong enough to alter sea-ice distributions, may also be self-sustaining in terms of radiative balance. Global warming of near 6°C could be achieved by about 50% increases in poleward ocean heat transport.

A greater role for the oceanic thermohaline circulation may be a plausible mechanism for an increased role of the oceans, because of their heat capacity. Today, deep water forms in geographically restricted regions, for which a density contrast from the main ocean occurs through interaction with the atmosphere. The source of deep water depends on the buoyancy flux (Brass et al., 1982), which is a function of the density contrast and the volume flux. Brass et al. (1982) demonstrate the plausibility of a thermohaline circulation very different from today, including the potential for subtropical deep water formation. Interestingly, ocean GCM simulations completed for several periods during the Cenozoic predict warm saline deep water formed within the subtropics during the Eocene (Barron and Peterson, 1990). Modeling work by Ogelsby and Saltzman (1990) gives added support to the concept of warm salty bottom-water formation. Given the large volume fluxes of the deep circulation in the ocean and the evidence for the possibility of warm saline deep water formation, the thermohaline circulation is a plausible candidate for an increased role by the oceans in poleward heat transport. Much additional research is required to examine the potential role of changes in the thermohaline circulation.

In summary, the calculation of Horrell (1990) may render unlikely scenarios in which tropical ocean sea-surface temperatures are as low as 15°C. This conclusion is also supported by the recent analysis of Crowley (1991). However, decreases in tropical sea-surface temperatures of 3 to 5°C are plausible.

Third, Matthews and Poore (1980) argue, in part based on the conclusions of Newell and others described above, that the low-latitude surface ocean had a stable temperature very close to present-day values. The difference between the isotopic temperature and the actual temperature is explained by changes in the oxygen isotopic composition of the oceans due to the storage of isotopically light waters as snow and ice in ice caps. The argument of Matthews and Poore (1980) essentially proposes extensive ice on Antarctica during much of the Cenozoic. The occurrence of ice on Antarctica is a matter of substantial debate, however, and Shackleton (1984) takes exception to the interpretation that the difference in tropical values is due to ice volume. Shackleton points out that such an interpretation would require more ice on Antarctica during the warm middle Miocene interval than today and would also require very large fluctuations in ice volume in the Early Miocene. Such a result appears unlikely and is unsubstantiated.

The arguments for Cenozoic tropical sea-surface temperatures near 15°C have also been challenged by Adams et al. (1990), based on biotic evidence. Adams and others note that the existence of Eocene mangroves, corals with zooxanthellae, and larger foraminifera preclude such low temperatures. These forms have minimum temperature limits closer to 18 to 20°C. If the temperature tolerances estimated for these organisms are accurate, then very cool tropical temperatures can be rejected. However, temperature variations within 5°C of modern values cannot be eliminated by these data.

In conclusion, none of the discussions presented above provides convincing arguments for tropical temperature stability, only limits to temperature variation. Tropical temperatures near 15°C can probably be rejected based on heat transport arguments and the biotic composition within the tropics during the Cenozoic. However, tropical temperature variations within 3 to 5°C of present-day values are not eliminated by any of the physical or biological arguments proposed to date.

MODEL-DERIVED TROPICAL TEMPERATURES

Interestingly, much of the emphasis on the interpretation of tropical temperatures and the evaluation of estimates using geologic data described above has focused on cases in which tropical temperatures may have been lower than at present. Unfortunately, there is a notable failure of atmospheric GCMs to simulate tropical climates with lower temperatures than the present day (Barron, 1987). If the conclusions from the experiments of Covey and Thompson (1989) are correct, and an increased role by the ocean in poleward heat transport is required to achieve cooler tropical temperatures, then this problem is explained by the lack of an explicit ocean formulation in current atmospheric climate models. At present, the debate over tropical cooler temperatures cannot be addressed explicitly by current climate models. The results from ocean GCM experiments (Barron and Peterson, 1991) for an ocean driven by an Eocene atmospheric simulation in an uncoupled mode, which produced deep water within the subtropics, are suggestive of a different role for the oceans. However, to date, climate models have not simulated reduced tropical sea-surface temperatures based solely on physical processes incorporated within the model. The conditions for reduced tropical temperatures in models remain problematic.

The prospect of higher tropical ocean surface temperatures during warm climates, perhaps the most interesting case for future global change projections, has received much less attention than the "cool" tropics cases. Barron and Washington (1985) noted that higher carbon dioxide climates proposed for the mid-Cretaceous might result in tropical temperatures too high for many organisms. Crowley (1991) calls attention to the issue of tropical temperatures during warm climates in general.

As stated earlier, a wide variety of atmospheric GCMs (Washington and Meehl, 1984; Manabe and Bryan, 1985; Hansen et al., 1988; Schlesinger, 1989) predict increased tropical sea-surface temperatures for a doubling of carbon dioxide. Schlesinger and Mitchell (1987) illustrate results from models of the National Center for Atmospheric Research (NCAR), the NOAA Geophysical Fluid Dynamics Laboratory, and the Goddard Institute for Space Studies. In each, the doubling of carbon dioxide resulted in a 2 to 4°C increase in tropical sea-surface temperatures.

In the geologic record, tropical warming may be the product of changes in carbon dioxide (e.g., as proposed by Berner et al., 1983; Berner, 1990), and by changes in geography. Barron and Washington (1985) specifically examine the warmth of the mid-Cretaceous utilizing a version of the NCAR Community Climate Model. The specification of Cretaceous geography without polar ice resulted in a 2 to 3°C increase in tropical sea-surface temperatures. However, the global warming was insufficient to explain most of the geologic observations at higher latitudes. The addition of four times the present-day atmospheric carbon dioxide concentration produced a climate with temperatures high enough to satisfy most geologic observations. In this case, tropical sea surface temperatures are more than 5°C higher than present-day values (Figure 6.2). Similar experiments with a full seasonal cycle using the GENESIS GCM also produced tropical temperature increases of 3 to 4°C for 4× present day CO_2 (Barron et al., 1993b). These model predictions for the mid-Cretaceous are within the interpretations proposed based on the oxygen isotopic data.

Although still limited in scope, the results from comprehensive climate models supported by the oxygen isotopic data provide the best case for a working hypothesis on tropical temperature variation during Earth history.

EVIDENCE FOR TROPICAL SALINITY DIFFERENCES

Much of the discussion of tropical climates has centered on temperature analyses. However, salinity is also a major control on the distribution of organisms. Unfortunately, little or no information on salinity has been derived from either geochemical or biological paleoclimatic indices. Only recently (Barron and Peterson, 1989; 1990) have ocean GCM been utilized to derive ocean salinity maps for different periods in Earth history that provide a basis for examining the potential importance of salinity variations. Figure 6.3 illustrates salinity predictions for the mid-Cretaceous, Paleocene, Eocene, Miocene, and Present day continental geometries utilizing the ocean GCM. Substantial ranges in salinity are projected, largely as a result of changes in the area of the oceans within the subtropical arid zone, the restriction of the tropical and subtropical basins and the degree of warmth. In the Eocene and the mid-Cretaceous, salinity predictions for substantial areas of the subtropics exceed 38 parts per thousand (‰) and a range of several parts per thousand is evident within the tropics throughout the Cenozoic.

The results from the ocean GCM studies are highly preliminary, but suggest that large salinity variations are also plausible in response to climate and geographic changes. The salinity variations projected are sufficient to influence the distribution of organisms. Interestingly, the high salinities for some time periods (e.g., the Eocene) would also serve to increase the isotopic temperature for the tropics by approximately 2°C (J. Zachos and L. Sloan, personal communication).

SUMMARY OF TROPICAL CLIMATE EXTREMES

In summary, a combination of model sensitivity studies and isotopic temperature analyses supports the conclusion that the tropics have been subjected to substantial climatic variation during Earth history. A temperature range of 3 to 5°C and a salinity range of several parts per thousand are reasonable hypotheses for variation within the tropics during the Mesozoic and Cenozoic. The case for warmer, and potentially more saline, tropical and subtropical oceans presents interesting prospects for biogeography and the response of tropical organisms to global warming. This case, perhaps exemplified by the mid-Cretaceous, is par-

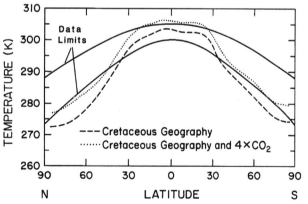

FIGURE 6.2 Cretaceous zonally averaged surface temperature (K) limits in comparison with Cretaceous model-derived surface temperatures for the geography and geography plus CO_2 quadrupuling experiments.

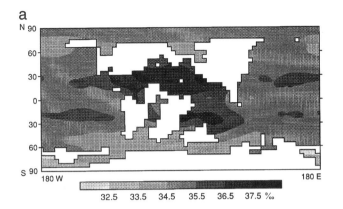
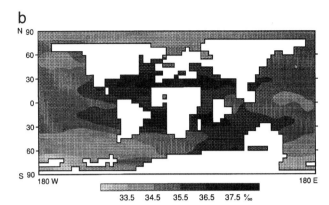
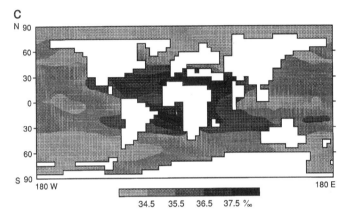
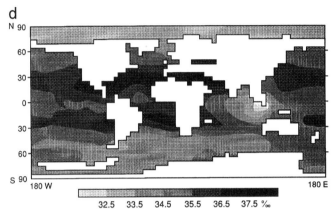
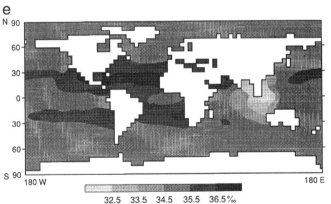

FIGURE 6.3 Simulated surface salinity (‰) with contour interval of 1‰: (a) mid-Cretaceous, minimum 34.5‰; (b) Paleocene, minimum 32.5‰; (c) Eocene, minimum 33.5‰; (d) Miocene, minimum <32.5‰; (e) present day, minimum <32.5‰.

ticularly interesting for global change studies and for considering the tropical response to increased levels of carbon dioxide in the near future. The key question becomes the significance of increases of 3 to 5°C in surface temperature and of several parts per thousand in salinity on the distribution and character of tropical organisms.

CLIMATE TOLERANCES OF TROPICAL ORGANISMS

Abundant evidence exists demonstrating that climate is a significant limiting factor in the distributions of organisms (e.g., Valentine, 1973; Stanley, 1984a,b). The tropics are generally viewed as an environment in which the physicochemical constraints are not undergoing major changes, resulting in an environment that is limited largely by biological competition. However, many equatorial species are also characterized by narrow environmental tolerances, suggesting that relatively small temperature and salinity changes could result in a substantial biologic response. Stanley (1988) suggests that tropical cooling could result in extinction if the temperature decrease removed the warm climate conditions required by many tropical organisms. In the case of hermatypic coral diversity, Stehli and Wells (1971) describe the importance of tropical temperature by noting the decrease in diversity with temperature from 50 genera at a mean temperature of 27°C to half that number at only 24°C. Tropical warming, on the other hand, might expand the latitudinal range of some organisms or displace organisms outside the tropics if some upper limit in temperature or salinity were exceeded.

Quantitative data on tolerances of organisms from the geologic record are obviously very limited. Experiments with modern organisms provide some guides. Vaughn and Wells (1943) performed a number of tank experiments on hermatypic corals for different temperatures and salinities. Tolerance limits were defined by evidence for notable damage or death due to exposure of only 12 to 24 hr. The optimum temperature range for corals was found to be 25 to 30°C, with minimum and maximum tolerances of 18 to 20°C and 35 to 37°C, respectively. The optimum salinity for growth was found to be 36‰, with minimum and maximum tolerances of 17 to 28‰ and 40 to 48‰, respectively. Since, the minimum and maximum tolerances were defined by damage over a very short time (24 hr), logically longer exposures could well result in narrower temperature limits or competitive disadvantage. Knowledge of organism tolerances provides crucial data for examining the question of tropical temperature variation.

A MID-CRETACEOUS CASE STUDY

Four lines of evidence—isotopic paleotemperatures, climate model results, the distribution of climate-sensitive organisms, and quantitative estimates of tropical tolerances—provide the basis for case studies of tropical climate sensitivity and the response of organisms during Earth history. From the viewpoint of both the availability of all lines of evidence and the interest in warm geologic climates for future global change considerations, the mid-Cretaceous is an interesting and valuable case study. Mid-Cretaceous atmospheric GCM experiments have been completed for Cretaceous geography with the present level of atmospheric CO_2 and Cretaceous geography with four times the present-day levels of atmospheric CO_2 (Barron and Washington, 1984). These atmospheric simulations were utilized to drive an ocean GCM (Barron and Peterson, 1990) that simulates surface ocean temperatures and salinities. Figure 6.4 illustrates the tropical areas that exceed the temperature optimum (30°C) and the salinity optimum (37‰) for each of the two simulations. A substantial portion of the Tethyan Ocean exceeds the optimum conditions for corals as described by Vaughn and Wells (1943). There are substantial differences between the two simulations. In the high CO_2 simulation, the optimum is exceeded over much of the tropical latitudinal

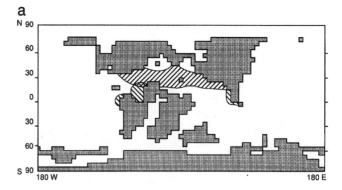

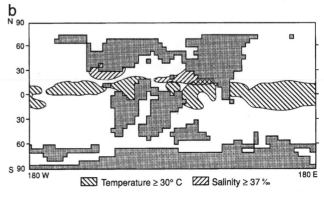

FIGURE 6.4 Ocean GCM predictions above the optimum conditions for coral growth for temperature (≥30°C) and salinity (≥37‰): (a) mid-Cretaceous simulation; (b) mid-Cretaceous simulation with atmospheric CO_2 concentrations at four times the present value.

band. In the simulation with present-day CO_2 levels, the optimum is exceeded largely within Tethys.

The region of warm temperatures and higher salinities corresponds closely to the Supertethys zone of Kauffman and Johnson (1988), which was dominated by rudistid bivalves. The history of reef-building rudists has been documented by numerous authors (Kauffman and Sohl, 1974; Scott, 1988; Kauffman and Johnson, 1988; Scott et al., 1990). During the Early Cretaceous, rudists became important in shallow water communities, tending to be more important in restricted environments such as lagoons and intrashelf basins. From the Hauterivian to the Albian, rudists became increasingly successful as reef dwellers and became the dominant tropical reef organism by the late Albian. Regionally, corals and rudists coexisted (e.g., Texas, Arizona margin of Tethys in the mid-Cretaceous), but rudists displaced corals in large measure. Corals decreased substantially within the tropics, and stromatoporoids almost disappeared. Many hypotheses have been offered, based on both competition and environment, to explain the displacement of corals by rudists. The model results presented here fit well with the speculation by Barron (1983) and the more comprehensive analysis of Kauffman and Johnson (1988) that the rise of rudist-dominated reefs corresponded to increased tropical warmth and higher salinities.

The rise of rudist-dominated reefs may well be one example of the response of tropical organisms to global warmth and increased tropical sea-surface temperatures and changes in surface salinity.

DISCUSSION AND CONCLUSIONS

Three major conclusions can be derived from the lines of evidence presented in this study. First, substantial variation in tropical temperatures and salinities during Earth history is plausible. Oxygen isotopic data, climate model studies, ocean heat transport experiments, and biologic data support a range of variation within 3 to 5°C of the present-day surface temperatures. Limited ocean model studies further suggest that salinities differing by several parts per thousand from present-day values are also reasonable. Although all the sources of data are characterized by uncertainty, and the individual sources of data are probably insufficient to describe these variations quantitatively through time, the data are sufficient to conclude that the tropics are sensitive to global change.

Second, the variations in temperature and salinity are large enough to have a substantial impact on tropical organisms. This conclusion is based on relatively limited experiments on the temperature tolerances of living organisms. The case study of the changes in reef communities from coral-dominated to rudist bivalve-dominated during the Cretaceous is perhaps one major example of the tropical response to global warming. Importantly, the emersion experiments on corals are likely to overestimate the optimum environmental range of tropical marine organisms, and such experiments fail to take into account changes in competitive advantage with changes in environmental conditions. More research within this area is required to be able to assess tropical response to global change.

Third, greater study of the tropics and tropical biota in Earth history may well yield substantial additional insights into global change research. The sensitivity of the tropics to external forcing factors is a subject of considerable debate, with as yet few data to verify or validate the model simulations. The rise of rudist-dominated reefs is likely to be just one example of the response of tropical organisms to change. The geologic record contains a wealth of other case studies of tropical changes.

The mid-Cretaceous case study also suggests that the fabric of the biologic changes contains much more information than evidence of warmth. For example, the fact that corals and rudists were able to compete or coexist at some localities (Scott *et al.*, 1990) has a number of additional implications. First, this may suggest that the temperatures and salinities did not exceed the limit of corals, but rather were likely to be outside the optimum conditions. Second, the model simulations and knowledge of the nature of environments suggest substantial spatial variations in temperature and salinity. Global warming may well cause restricted tropical regions to exceed temperature and salinity tolerances more readily than open ocean regions. Given the large temperature changes with depth, the changes in coral versus rudistid dominance may well exhibit a depth control. Further, the climate model simulations do not suggest a simple belt of above-optimum salinities and temperatures. Rather there is considerable spatial structure, with some areas exceeding only the temperature optimum, others exceeding only the salinity optimum, and still others exceeding both temperature and salinity optima. There may well be substantial structure in the rudist and coral communities that can be tied to the spatial characteristics of the model results. Finally, there are large differences between the Cretaceous and the Cretaceous simulation with high carbon dioxide. The degree of warming, the mechanism of warming, and the history of global warmth throughout the Cretaceous may well be described within the changes in tropical communities during this period.

The generally held view that the tropics are an environment in which the physicochemical constraints are not undergoing major changes should not be translated to a view that the tropics are stable to external forcing factors (e.g., increases in atmospheric carbon dioxide). Because of the narrow environmental tolerances of many tropical organisms, tropical biota may be very sensitive to global change. For this reason, changes in tropical organisms and their distribution should be viewed as a rich, underutilized record that can provide many new insights into the climate sensitivity of the tropics. This record provides the only major source of data on the biologic response to global change.

REFERENCES

Adams, C. G., D. E. Lee, and B. R. Rosen (1990). Conflicting isotopic and biotic evidence for tropical sea-surface temperatures during the Tertiary, *Palaeogeography, Palaeoclimatology, Palaeoecology* 77, 289-313.

Barron, E. J. (1983). A warm, equable Cretaceous: The nature of the problem, *Earth-Science Reviews* 19, 305-338.

Barron, E. J. (1987). Eocene equator-to-pole surface ocean temperatures: A significant climate problem? *Paleoceanography* 2, 729-739.

Barron, E. J., and W. H. Peterson (1989). Model simulation of the Cretaceous ocean circulation, *Science* 244, 684-686.

Barron, E. J., and W. H. Peterson (1990). Mid-Cretaceous ocean circulation: Results from model sensitivity studies, *Paleoceanography* 5(3), 319-337.

Barron, E. J., and W. H. Peterson (1991). The Cenozoic ocean circulation based on Ocean General Circulation Model results, *Palaeogeography, Palaeoclimatology, Palaeoecology 83*, 1-28.

Barron, E. J., and W. M. Washington (1984). The role of geographic variables in explaining paleoclimates: Results from Cretaceous climate model sensitivity studies, *Journal of Geophysical Research 89*, 1267-1279.

Barron, E. J., and W. M. Washington (1985). Warm Cretaceous climates: High atmospheric CO_2 as a plausible mechanism, in *The Carbon Cycle and Atmospheric CO_2: Natural Variations Archean to Present*, American Geophysical Union, Washington, D.C., pp. 546-553.

Barron, E. J., W. H. Peterson, D. Pollard, and S. Thompson (1993a). Past climate and the role of ocean heat transport: Model simulations for the Cretaceous, *Paleoceanography 8*, 785-798.

Barron, E. J., P. J. Fawcett, D. Pollard, and S. Thompson (1993b). Model simulations of Cretaceous climates: The role of geography and carbon dioxide, *Palaeoclimate and Their Modelling with Special Reference to the Mesozoic Era, Philosophical Transactions of the Royal Society of Biological Sciences 341*, 307-316.

Berner, R. A. (1990). Atmospheric carbon dioxide over Phanerozoic time, *Science 249*, 1382-1386.

Berner, R. A., A. C. Lasaga, and R. M. Garrels (1983). The carbonate-silicate geochemical cycle and its effect on atmospheric carbon dioxide over the last 100 million years, *American Journal of Science 283*, 641-683.

Brass, G. W., J. R. Southam, and W. H. Peterson (1982). Warm saline bottom waters in the ancient ocean, *Nature 296*, 620-623.

CLIMAP Project Members (1981). Seasonal reconstruction of the Earth's surface at the last glacial maximum, *Geological Society of America Map and Chart Series MC-36*.

Covey, C., and S. L. Thompson (1989). Testing the effects of ocean heat transport on climate, *Global and Planetary Change 75*, 331-341.

Crowley, T. J. (1991). Past CO_2 changes and tropical climate, *Paleoceanography 6*, 387-394.

Douglas, R. G., and S. M. Savin (1975). Oxygen and carbon isotope analyses of Tertiary and Cretaceous microfossils from Shatsky Rise and other sites in the North Pacific, *Initial Reports of the Deep-Sea Drilling Project 32*, 509-520.

Emiliani, C. (1970). Pleistocene paleotemperatures, *Science 168*, 822-825.

Emiliani, C., and N. J. Shackleton (1974). The Brunhes Epoch: Isotopic paleotemperatures and geochronology, *Science 183*, 511-514.

Hansen, J. E., et al. (1988). Global climate changes as forecast by Goddard Institute for Space Studies three-dimensional model, *Journal of Geophysical Research 93*, 9341-9364.

Horrell, M. A. (1990). Energy balance constraints on ^{18}O based paleo-sea surface temperature estimates, *Paleoceanography 5*, 339-348.

Kauffman, E. G., and C. C. Johnson (1988). The morphological and ecological evolution of Middle and Upper Cretaceous reef-building rudistids, *Palaios 3*(2), 194-216.

Kauffman, E. G., and N. F. Sohl (1974). Structure and evolution of Antillean Cretaceous rudist frameworks, *Verhandlungen Naturforschende Gesellschaft 84*(1), Basel, 399-467.

Manabe, S., and K. Bryan (1985). CO_2-induced change in a coupled ocean-atmosphere model and its paleoclimatic implications, *Journal of Geophysical Research 90*, 11,689-11,708.

Matthews, R. K., and R. Z. Poore (1980). Tertiary ^{18}O record and glacioeustatic sea-level fluctuation, *Geology 8*, 501-504.

Newell, R. E., and T. G. Dopplick (1979). Questions concerning the possible influence of anthropogenic CO_2 on atmospheric temperature, *Journal of Applied Meteorology 18*, 822-825.

Newell, R. E., A. R. Navato, and J. Hsiung (1978). Long term global sea surface temperature fluctuations and their possible influence on atmospheric CO_2 concentrations, *Journal of Pure and Applied Geophysics 116*, 351-371.

Ogelsby, R. J., and B. Saltzman (1990). Extending the EBM: The effect of the deep ocean temperatures on climate with applications to the Cretaceous, *Global and Planetary Change 2*, 237-259.

Rind, D., and D. Peteet (1985). Terrestrial conditions at the last glacial maximum and CLIMAP sea-surface temperature estimates: Are they consistent?, *Quaternary Research 24*, 1-22.

Savin, S. (1977). The history of the Earth's surface temperature during the past 100 million years, *Annual Review of Earth and Planetary Science 5*, 319-355.

Savin, S., R. Douglas, and F. Stehli (1975). Tertiary marine paleotemperatures, *Geological Society of America Bulletin 86*, 1499-1510.

Schlesinger, M. E. (1989). Model projections of the climatic changes induced by increased atmospheric CO_2, in *Climate and Geo-Sciences*, A. Berger, S. H. Schneider, and J.-C. Duplessy, eds., Kluwer, Dordrecht, Netherlands, pp. 375-415.

Schlesinger, M. E., and J. F. B. Mitchell (1987). Climate model simulations of the equilibrium climatic response to increased carbon dioxide, *Reviews of Geophysics 25*(4), 760-798.

Scott, R. W. (1988). Evolution of Late Jurassic and Early Cretaceous reef biotas, *Palaios 3*(2), 184-193.

Scott, R. W., P. A. Fernández-Mendiola, E. Gili, and A. Simó (1990). Persistence of coral-rudist reefs into the Late Cretaceous, *Palaios 5*(2), 98-110.

Shackleton, N. J. (1984). Oxygen isotope evidence for Cenozoic climatic change, in *Fossils and Climate*, P. J. Brenchley, ed., Wiley, Chichester, pp. 27-34.

Shackleton, N. J., and A. Boersma (1981). The climate of the Eocene ocean, *Journal of the Geological Society of London 138*, 153-157.

Sloan, L. C. (1990). Determination of Critical Factors in the Simulation of Eocene Global Climate, with Special Reference to North America, Ph.D. Dissertation, The Pennsylvania State University, University Park, Pa.

Stanley, S. M. (1984a). Marine mass extinction: A dominant role for temperature, in *Extinctions*, M. H. Nitecki, ed., University of Chicago Press, Chicago, pp. 69-117.

Stanley, S. M. (1984b). Temperature and biotic crises in the marine realm, *Geology 12*, 205-208.

Stanley, S. M. (1988). Climatic cooling and mass extinction of Paleozoic reef communities, *Palaios 3*(2), 228-232.

Stehli, F. G., and J. W. Wells (1971). Diversity and age patterns in hermatypic corals, *Systematic Zoology* 20, 115-126.

Valentine, J. W. (1973). *Evolutionary Paleoecology of the Marine Biosphere*, Prentice-Hall, Englewood Cliffs, N.J., 511 pp.

Vaughn, T., and J. Wells (1943). Revision of the suborders, families and genera of the Scleractinia, *Geological Society of America Special Paper 44*, 363 pp.

Washington, W. M., and G. A. Meehl (1984). Seasonal cycle experiment on the climate sensitivity due to a doubling of CO_2 with an atmospheric general circulation model coupled to a simple mixed-layer ocean model, *Journal of Geophysical Research* 89, 9475-9503.

7

Neogene Ice Age in the North Atlantic Region: Climatic Changes, Biotic Effects, and Forcing Factors

STEVEN M. STANLEY
Johns Hopkins University

WILLIAM F. RUDDIMAN
University of Virginia

ABSTRACT

Long-term climatic trends culminated in the recent ice age of the Northern Hemisphere. As late as mid-Pliocene time, however, many sectors of the North Atlantic region remained substantially warmer than today. Oxygen isotope ratios for marine microfossils indicate that a pulse of cooling occurred relatively suddenly at high and middle latitudes at ~3.2 to 3.1 million years ago (Ma) and that large ice sheets formed ~2.5 Ma, when more severe cooling and regional drying of climates occurred. Cycles of glacial expansion and contraction reflected orbital forcing at periodicities of ~41,000 yr until about 0.9 Ma and ~100,000 yr thereafter. Aridification in Africa at ~2.5 Ma resulted in the extinction of many forest-dwelling species of mammals and, soon thereafter, in the origins of numerous species adapted to savannas. Mammalian extinction intensified closer to 2 Ma in North America and was weaker in Europe, where forests changed in floral composition but remained widespread. Beginning at ~2.5 Ma and continuing into mid-Pleistocene time, life occupying shallow seafloors in the North Atlantic region suffered heavy extinction from climatic cooling, leaving an impoverished, eurythermal Recent fauna. Long-term climatic trends in the North Atlantic region during Neogene time probably resulted primarily from tectonic events, notably closure of the Straits of Panama and uplift of the Tibetan Plateau and other regions. A decrease in atmospheric CO_2 and consequent weakening of a greenhouse effect also appears to be required, perhaps due to increased weathering that accompanied the uplifting of plateaus.

INTRODUCTION

In this chapter, we discuss climatic events that marked the onset of the recent ice age, their impact on biotas, and their likely causes. Our focus is on the North Atlantic region because this part of the world, which became bordered in the north by major ice caps, was the scene of more severe environmental and biotic changes than occurred in other areas of the globe.

The detailed chronology for major glacial events of the past 3 million years (m.y.) or so has come primarily from deep-sea deposits. Many deep-sea cores provide relatively continuous records, for which magnetic reversals and biostratigraphic data yield key dates. Also especially useful are changes in the isotopic composition of microfossils; the proportion of ^{18}O increased in these forms at times of glacial maxima, both because isotopic partitioning during skeletal secretion varies with temperature and because ^{16}O is preferentially evaporated from the oceans, transported in water vapor, and sequestered in glacial ice.

Fossils in shallow marine and terrestrial sequences also record climatic changes of the recent ice age. Some biotic changes represent clear evidence of climatic transitions, but others, which will be discussed separately, can only be interpreted a posteriori as reflecting these environmental changes.

Historically, a great variety of hypotheses have been invoked to explain the onset of Plio-Pleistocene glaciation in the Northern Hemisphere. Although the issue is complex, forcing factors resulting from tectonic events were probably responsible. We also review ways in which ice sheets have been influenced by periodicity in the Earth's orbital motion and have themselves behaved as climatic forcing factors in a complex feedback system.

CLIMATIC EVENTS

The climatic changes that have occurred in the North Atlantic region since the ice age began about 2.5 Ma can be understood only in the context of events that were under way millions of years earlier. The climatic changes of the past 2.5 m.y. have been cyclical, relatively rapid, and associated with orbital-scale variations in ice volume in the Northern Hemisphere. In contrast, the major changes prior to 2.5 Ma represented net trends that were relatively gradual. For the most part, they resulted from tectonic events that are discussed in the final section of this chapter.

In the present section, we first evaluate climatic events prior to the Late Pliocene, which began at 3.4 Ma; next, events at 3.2–3.1 Ma that preceded the start of the ice age; and finally, events that marked the development of vast ice sheets close to 2.5 Ma.

Responses Prior to the Late Pliocene

Climatic responses to tectonic events of Miocene and Early Pliocene time were regionally complex. Although in general these climatic trends developed gradually over many millions of years, they were at times interrupted by more dramatic "steps," or brief intervals when critical thresholds in the system were exceeded and large-amplitude responses were triggered. Both the slow climatic drift and the steps altered the distributions of plants and animals, and may also have affected their evolution. Some marine records appear to be relatively continuous, but the discontinuous nature of most continental sedimentation precludes fully adequate resolution of many long-term climatic trends. The basic patterns that can be detected involve widespread cooling, especially at high and middle latitudes, and a mosaic of more regional trends toward wetter and drier climates.

The cooling trend is particularly evident at higher latitudes and elevations. Thick ice appeared on Antarctica early in the Cenozoic (Barron *et al.*, 1989). Deposits from small mountain glaciers are first recorded between 10 and 5 Ma in the Coast Range of Alaska (Denton and Armstrong, 1969) and the Andes of South America (Mercer, 1983). Pollen data from high-altitude sites in Iceland indicate significant cooling by 10 Ma (Mudie and Helgason, 1983). Traces of ice-rafted sand in Norwegian Sea sediments by 4 Ma (Henrich *et al.*, 1989) suggest at least the sporadic presence of mountain glaciers along the east coast of Greenland or the west coast of Scandinavia at this time.

Another major late Cenozoic trend was a progression toward more highly differentiated regional extremes of wet and dry climate. During the past 15 m.y., deserts have formed or expanded into new terrain in Asia (Wolfe, 1979), North Africa (Tiedemann *et al.*, 1989), and North America (Axelrod, 1950), while monsoonal climates have persisted or intensified in the Indo-Asian subtropics and South American tropics.

Several regions of west-central North America became markedly drier during the late Cenozoic. On the Northern Plains, prairie savanna gave way to grasses and herbs after 15 Ma (Thomasson, 1979). During the same interval, vegetation adapted to summer drought gradually came to dominate the California coast (Axelrod, 1966).

The Pliocene Prior to 2.5 Ma

Most of the profound climatic changes in the Northern Hemisphere during the Pliocene Epoch prior to about 3.1 Ma were regional in scale. Forest yielded to scrub vegetation in the rainshadow of the Cascades around 4 Ma, for example (Leopold and Denton, 1987), and desert vegetation expanded in the Great Basin near 4 to 3 Ma (Axelrod,

1950). Saharan dust fluxes also increased abruptly near 4.2 Ma (Tiedemann *et al.*, 1989). In general, climates of the Northern Hemisphere prior to about 3.2 Ma were warmer than those of the past 2.5 m.y., including climates during glacial minima. To set the scene for discussion of the ice age, we review climatic indicators of the widespread warmth that preceded it.

The terrestrial floras that have thus far been used most effectively to document continent-wide climatic changes during the Pliocene Epoch are those of Europe (Figure 7.1). Early in Pliocene time, dense coastal forests fringed the northwestern coast of the Mediterranean Sea. Dominance here of species belonging to the cypress family indicates that moist and relatively warm conditions persisted throughout the year (Suc and Zagwin, 1983).

The Mediterranean Sea itself was characterized by a marginally tropical, or at least warm subtropical, thermal regime during Early Pliocene time. This is indicated, for example, by fossil occurrences of numerous species of the generally tropical gastropod genus *Conus* (Marasti and Raffi, 1979). Furthermore, 5% of the polysyringian bivalve mollusks of the Mediterranean that survive from the Early Pliocene are restricted to tropical seas along the west coast of Africa today, apparently being unable to tolerate subtropical conditions (Raffi *et al.*, 1985). Quantitative assessment of the history of extant lineages of planktonic foraminifera reveals that surface waters of the Early Pliocene Mediterranean were characterized by equable thermal conditions, with winter temperatures usually higher than those of the present (Thunell, 1979).

The North Sea was also warmer than today: 20% of the extant polysyringian bivalve species from the lower Pliocene of the North Sea region today live only south of the North Sea, and 10% occur only in waters at least as warm as subtropical. The reproductive requirements of the warm-adapted survivors suggest that for 3 or 4 months, mean temperatures reached 20°C or more (Raffi *et al.*, 1985).

Marine biotas reveal that in the Western Atlantic region, as well, climates prior to about 3 Ma were warmer than today. Fossiliferous strata representing a highstand of sea level between about 3.5 and 3.0 Ma are exposed along the Atlantic Coastal Plain from Virginia to Florida. They were deposited seaward of the Orangeburg scarp, a conspicuous topographic feature from North Carolina to Florida cut by wave erosion during a stillstand coinciding with the maximum advance of the shoreline, when sea level stood 35 ±18 m above its present level (Dowsett and Cronin, 1990). At the border between North and South Carolina, the scarp lies about 150 km inland from the present shoreline. Fossil ostracods and microplankton from the Duplin Formation, which extends eastward from the scarp in this region, indicate an age of about 3.5 to 3.0 Ma, and the thermal tolerances of surviving ostracods suggest that nearshore bottom-water temperatures ranged from 26°C in August to 18°C in February (Dowsett and Cronin, 1990). By way of comparison, bottom temperatures along the continental shelf of this region today at a depth of 30 m drop to about 12°C in February. Rich fossil molluscan faunas represent the mid-Pliocene highstand from Virginia to southern Florida. In southern Florida they are associated with coral reefs at latitudes about 150 km north of the limit of reef growth today (Meeder, 1979).

Although climates in the southeastern United States were generally warmer than today throughout the 3.5 to 3.0 Ma interval, they became especially warm toward the end of the interval. Ostracod occurrences also suggest a rise of temperatures during the mid-Pliocene transgression. In Virginia, changing ostracod and molluscan faunas indicate a shift from warm temperate to subtropical conditions during deposition of the uppermost Yorktown For-

FIGURE 7.1 Examples of warm-adapted genera of plants that disappeared from northwestern Europe at about 2.3 Ma. Left: *Pseudilarix* (golden larch); center: *Liquidambar* (sweet gum); right: *Zelkova* (Caucasian elm).

mation (Hazel, 1971; Stanley and Campbell, 1981). This pulse of warming has been taken to reflect a strengthening of the Gulf Stream with the closure of the Isthmus of Panama. The temporary disappearance of the planktonic foraminiferan *Pulleniatina* from the Caribbean at 3.1 Ma suggests that this was the time of complete closure (Keigwin, 1978), as does evidence of strongly increased winnowing along the Yucatan Channel (Brunner, 1984). The interval of warming in the southeastern United States was short-lived. As discussed below, ice-age cooling affected the region profoundly within a few hundred thousand years.

Thermal conditions in the Caribbean have not been assessed for the first half of the Pliocene, but in the Bahamas, reefs flourished to a greater extent than today, and the Bahama Banks comprised atolls (Beach and Ginsburg, 1980). The molluscan fauna of southern Florida differed markedly from that of the Caribbean, in sharp contrast to the situation today, when Florida shares nearly all of its strictly tropical shallow-water species with the Caribbean. The distinctive Caribbean Pliocene fauna even characterized the Bahamas, which are separated from the Florida peninsula only by the narrow Straits of Florida (McNeill *et al.*, 1988). Possibly during the highstand of sea level of mid-Pliocene time, upwelling around Florida formed a biogeographic barrier of relatively cool water (Stanley, 1986). This may explain why, whereas hermatypic corals and many molluscan species of tropical affinities thrived in a lagoonal setting as far north as Sarasota, Florida, there was very little carbonate mud production by calcareous algae here even in the near absence of terrigineous clays. It appears that this region resembled the relatively cool but thermally stable tropical region of the modern Eastern Pacific, where upwelling also prevails (Stanley, 1986).

While the Panamanian straits connecting the Atlantic and Pacific became obstructed during mid-Pliocene time, the Bering Strait opened, to connect the two large oceans across the Arctic. This new connection apparently resulted largely from the global elevation of sea level that ultimately inundated the eastern United States along the Orangeburg scarp. Sediments of mid-Pliocene age have been recognized at elevations in Alaska as high as 35 m above present sea level (Carter *et al.*, 1986). The breaching of the Bering land bridge left its mark in the stratigraphic record of northeastern Iceland, where a host of Pacific mollusk species appear abruptly in mid-Pliocene sediments (Einarsson *et al.*, 1967).

Fossil occurrences and biogeographic distributions of molluscan species in the modern world indicate that the exchange of species between the Atlantic and Pacific was asymmetric. At least 125 species migrated from the Pacific to the Atlantic, whereas only 16 species are known to have moved in the opposite direction (Hopkins, 1967). The reason for this disparity is unclear (Vermeij, 1989), but it is significant that many groups that participated in the exchange are unable to live in most areas of the Arctic Ocean today (Carter *et al.*, 1986). This fact, and the occurrence of a fossil sea otter on the north slope of Alaska, indicate that temperatures in the Arctic were warmer in mid-Pliocene time than they are now. At least some fringes of the Arctic were apparently ice free during at least part of the year.

The First Pulse, 3.2 to 3.1 Ma

The first events presaging the onset of the Plio-Pleistocene ice age occurred about 3.2 to 3.1 Ma, but they did not include the buildup of major ice sheets. At this time, in many low-latitude areas fossil planktonic foraminifera exhibit a shift to heavier $\delta^{18}O$ values that was sustained for only about 100,000 yr. In contrast, a similar increase for deep-sea benthic foraminifera persisted, with fluctuations, to the present day (Prell, 1984). This divergence of values is taken to indicate a sustained cooling of the deep sea where, in fact, an interval of scouring by descending cold water is also recorded (Ledbetter *et al.*, 1978). This implies climatic cooling at high latitudes.

The temporary nature of the isotopic perturbation for planktonic fossils indicates that there was no permanent buildup of large ice sheets. Nonetheless, the oldest Pliocene tillites in northeastern Iceland occur just above a basaltic unit dated at 3.1 ± 0.1 Ma by the potassium-argon method (McDougall and Wensink, 1966) and just below the base of the Mammoth interval of reversed magnetism, dated at 3.15 Ma (Mankinen and Dalrymple, 1979). Climatic changes extended at least as far equatorward as the northwestern Mediterranean region. Here the coastal forest, dominated by cypresses, gave way to oaks and other forms of vegetation adapted to relatively drier, more seasonal climates (Suc, 1984). Statistical analysis of changes in the composition of the planktonic foraminiferal fauna of the Mediterranean reveal that a pulse of cooling between about 3.2 and 3.0 Ma reduced mean annual temperature by 2 to 4°C (Keigwin and Thunell, 1979). In northwestern Europe, several subtropical species of land plants, including palms, disappeared at the end of the Branussumian interval, about 3.2 Ma (van der Hammen *et al.*, 1971).

Onset of the Ice Age at 2.5 to 2.4 Ma

Warmer temperatures returned temporarily to at least some regions between 3 and 2.5 Ma. Fluctuations in the composition of the planktonic foraminiferal fauna of the Mediterranean indicate oscillating temperatures for this interval (Thunell, 1979). Fossils in transgressive shallow marine strata on the north slope of Alaska reveal that sea otters were again present along the margin of the Arctic

Ocean slightly before 2.4 Ma. Fossil pollen in the same Alaskan deposits point to the presence of tundra close to sea level, however, which means that by this time climates were deteriorating once again (Repenning et al., 1987).

The cooling of climates between about 2.5 and 2.4 Ma marked the true transition to the modern ice age. Most climatic changes in the Northern Hemisphere subsequent to ~2.4 Ma reflect the pervasive influence of ice sheets. Oxygen isotopic records from deep-sea cores provide the best proxy of global ice volume (Shackleton and Opdyke, 1976), with ~50 to 70% of the late Pleistocene fluctuations linked directly to ice volume, primarily in North American and Eurasia. The remaining part of these signals was determined largely by local temperature changes, some of which tracked ice volume closely, whereas others varied independently, depending on locale.

An oxygen isotopic record covering the entire span of significant glaciation in the Northern Hemisphere is shown in Figure 7.2. Several features of this record are worth noting. First, regular cycles of rather small amplitude began at least as early as 2.7 Ma, and scattered ice-rafted debris in the North Atlantic and Norwegian Sea confirms that small ice sheets existed between 2.7 and 2.4 Ma (Jansen et al., 1988; Raymo et al., 1989). The 2.4 to 2.3 Ma interval included the inception of much larger cycles, in both the isotopic (Shackleton et al., 1984) and the ice-rafting (Zimmerman, 1984) signals, marking the first appearance of really substantial ice sheets. For the next 1.5 m.y. (2.3 to 0.9 Ma), isotopic cycles varied mainly at the 41,000-yr period (Raymo et al., 1989; Ruddiman et al., 1989), and the maximum isotopic values during glacial climatic extremes generally did not exceed those near 2.4 to 2.3 Ma, except toward the end of the interval.

During the past 0.9 m.y., several isotopic cycles attained amplitudes at glacial maxima that were considerably larger than for any previous cycle, and the dominant tempo of glaciation shifted to 100,000 yr (Shackleton and Opdyke, 1976). In addition, very rapid deglaciations called "terminations" (Broecker and van Donk, 1970) began to occur at the end of the major glacial cycles.

Several other records that are continuous and span large portions of the past 3 m.y. confirm the basic climatic trends indicated by oxygen isotopes. These include sea-surface temperature records and ice-rafted fluxes in the subpolar North Atlantic Ocean (Ruddiman et al., 1989); loess deposits in eastern Europe (Kukla, 1977) and south-

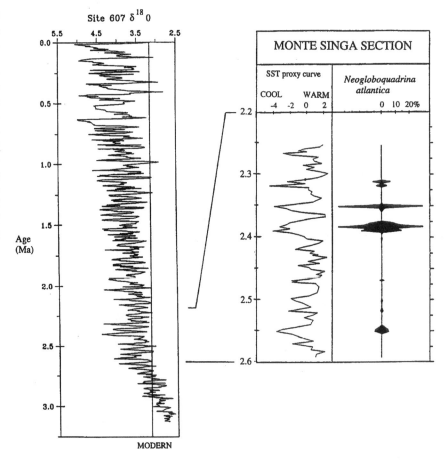

FIGURE 7.2 Benthic foraminiferal $\delta^{18}O$ from North Atlantic Site 607 (after Raymo et al., 1990) and events between 2.6 and 2.2 Ma in the Mediterranean (Zachariasse et al., 1990). This record spans the entire Northern Hemisphere ice age, at an average sampling interval of 3500 yr. The record shows the Late Pliocene initiation and mid-Pleistocene intensification of Northern Hemisphere glaciation. The sea-surface temperature (SST) proxy curve is the product of a principal component analysis of occurrences of planktonic foraminifera, based on evidence of their thermal tolerances. Relative abundance of *Neogloboquadrina atlantica*, a species adapted to cool conditions, increased markedly just after 2.4 Ma and again at 2.35 Ma.

ern China (Heller and Liu, 1982); and pollen sequences in the Netherlands, northern Italy, and Macedonia (van der Hammen *et al.*, 1971). Prominent features in these records include a large initial step toward glacial-like climates at or near 2.4 Ma, cycles of modest size and relatively high frequency until near the Jaramillo magnetic reversal (~0.9 Ma), and larger cycles displaying a 100,000-yr periodicity during the late Pleistocene.

Regions located well away from the Northern Hemisphere ice sheets might be expected to show climatic variations during the Late Pliocene and Pleistocene in response to other factors. Within middle and high latitudes of the Northern Hemisphere, however, the relatively continuous records available all appear to fluctuate sympathetically with the ice sheets. Some discrepancies may occur in the 2.3 to 0.9 Ma interval, where the isotopic record appears to detect more cycles than the loess or pollen sequences, but this may reflect different thresholds of sensitivity in the detection of small-amplitude 41,000 yr cycles.

Less continuous or well-dated stratigraphic sequences representing shallow marine and terrestrial environments yield fossils that document important climatic events. For example, the record of vegetational change along the northwestern margin of the Mediterranean at about 2.3 Ma points to a shift to drier conditions. Forests shrank and steppe vegetation, including sagebrush and weedy plants, expanded. Generally contemporaneous changes in the composition of some European floras reflect cooling (Suc, 1984). At about this time (the Praetiglian interval), what remained of the subtropical Malayan element disappeared from the flora of northwestern Europe (van der Hammen *et al.*, 1971; also see Figure 7.1). In Africa, palynofloras reveal sweeping climatic change at about 2.5 Ma., although the precise timing remains to be determined. In the Ethiopian uplands, climates became cooler and drier than today, whereas they had previously been warmer and moister (Bonnefille, 1985). Apparently this pulse of climatic change was soon partly reversed. The fact that even today, during a glacial minimum, Africa is cooler and drier than it was prior to 2.5 Ma may, however, indicate that this region has experienced a fundamental climatic change unrelated to glacial maxima.

In the Western Atlantic region and the Americas, climatic changes are less well documented, but in the high plain of Bogota, Columbia, fossil pollen records indicate cooling at about 2.5 Ma (van der Hammen, 1985). Cores obtained by drilling in the Bahamas document dramatic change a bit earlier. Here, magnetostratigraphy has provided a detailed chronology for a relatively complete stratigraphic sequence (McNeill *et al.*, 1988). Fossils in the core reveal that sea level dropped briefly at about 2.65 Ma and San Salvador ceased to grow as a coral atoll. The fact that the biotic crisis in the Bahamas occurred as early as 2.65 Ma reflects the fact that a transition to the ice age was a complex event that spanned more than 200,000 yr (Figure 7.2). The precise chronology of climatic change remains to be established for many geographic areas, but the magnetic reversal separating the Gauss and Matuyama Chrons at 2.5 Ma should prove useful here.

BIOTIC CONSEQUENCES

The Pliocene climatic changes that affected terrestrial biotas of the North Atlantic region included a general cooling of climates, with increased seasonality in many areas and widespread aridification on the land. Among the results were the migration of many species to favorable habitats; the extinction of species unable to escape intolerable new conditions through migration; and the origin of new species adapted to the new conditions.

Terrestrial Biotas

Some of the biotic consequences of the climatic changes, in fact, represent key evidence documenting the changes. Foremost of these were previously described transformations of terrestrial floras, which we review in discussing the general topic of biotic change.

Fossil pollen have provided most of our information about Pliocene terrestrial floras of the North Atlantic region. Although the evidence assembled to date is patchy, it indicates that drying of climates was at least as significant as thermal change. Often it was an accentuation of the dry season that had the greatest impact on floras. In general, grasslands expanded at the expense of forests, which require moist conditions. Climatic changes had their primary effects on terrestrial mammals indirectly, through their influence on vegetation.

Africa

In some areas, orographic effects of regional tectonic activity had a more pronounced impact on biotas than did global climatic change. Africa, however, experienced aridification so pervasive that it could not have resulted from small-scale tectonics. Fossil pollen reveals that savannas spread at the expense of forests, not only throughout northern Africa, where the trade winds prevail, but also in Kenya (Bonnefille, 1976). The dramatic change that occurred at about 2.5 Ma entailed both cooling and drying of climates. While a slight cooling probably had some effect on African floras, aridification was the most influential climatic change. Low rainfall is the primary factor that separates grasslands from forested areas today.

The onset of the ice age had a greater impact on mammals in Africa than in Europe or North America. Presum-

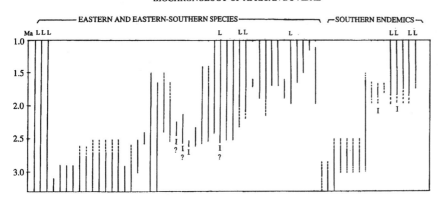

FIGURE 7.3 Stratigraphic ranges of antelope species in sub-Saharan Africa (I: immigrants; L:living; for identities of species, see Vrba, 1988). Most of the extinctions close to 2.5 Ma were of forest-adapted species. Most species that appeared at this time or shortly thereafter were savanna dwellers.

ably, this resulted from two conditions: (1) many species of mammals were endemic to Africa, and (2) the vegetational changes were continental in scale. Numerous forest-adapted species of antelopes died out close to 2.5 Ma, but there was a positive evolutionary response to the climatic change as well (Vrba, 1985). Several new species of antelopes adapted to savannas appeared through speciation events at about 2.5 Ma, or within 200,000 to 300,000 years thereafter (Figure 7.3). At the same time, several species adapted to grasslands invaded from Eurasia, apparently because forest barriers disappeared, at least temporarily. These species and others that evolved slightly later constitute the modern antelope fauna of the African savannas. Micromammals appear to have changed in a parallel fashion (Wesselman, 1985); in Ethiopia, numerous forest-dwelling species died out and were replaced by species adapted to drier conditions. African monkeys experienced a high rate of evolutionary turnover at about the same time, although the relationship of this change to the vegetational transition remains obscure (Delson, 1985). The dramatic effects on the history of the human family are discussed in Chapter 14 of this volume.

Europe

We have already noted the floral change along the northwestern border of the Mediterranean, where the decline of cypress-dominated forests beginning slightly before 3 Ma apparently resulted primarily from aridification but may have entailed climatic cooling as well. This region is marginal to the trade wind belt and is far more arid today than northwestern Europe. In the latter, more northerly region, cooling of climates and also divergence of seasonal extremes of temperature apparently had a stronger effect on biotas than did aridification.

We have previously noted the disappearance of the subtropical Malayan component from the flora of northwestern Europe in two pulses centered at about 3.1 and 2.5 Ma. This was not simply a matter of regional extermination: numerous endemic species suffered extinction (Leopold, 1967). As Reid and Reid (1915) noted long ago, at the time of the climatic change, southern Eurasia was spanned from east to west by a barrier of seas, deserts, and mountains unbreached by river valleys. At the same time, the increasingly dry climate of central Asia blocked eastward migration. The result was a lethal trap for many warm-adapted species, including palms. Other species, which were more widely distributed before climatic changes began, disappeared from Europe but survived in North America or China.

The floral change began earlier in northwestern Europe than in Poland (Leopold, 1967). Extinction continued into the Pleistocene Epoch, especially in Poland, but ceased almost entirely during the latter part of the Pleistocene. Floral compositions have continued to fluctuate markedly with shifts of climate, however, as a result of oscillatory migration. In Poland, for example, temperate elements have been more conspicuous during glacial minima (including the Holocene) and arctic-alpine elements during glacial maxima. In other words, the heaviest extinction occurred during the Pliocene, when numerous vulnerable species were present, and declined into the Pleistocene. By the latter part of the Pleistocene, vulnerable species had disappeared and floras consisted of species that could tolerate climatic fluctuations. Today, as a result of extinction and regional extermination, the European flora is markedly impoverished.

Mammals suffered much less heavy extinction in Europe than in Africa during the Pliocene. The strongest pulse of extinction in Europe appears to have occurred shortly after 2.0 Ma, before the existence of the faunas labeled "Tegelen"; about 15% of all known mammal species died out (Kurtén, 1968). The relatively high survivorship for the entire Pliocene Epoch probably reflected the nature of the climatic and floral changes in Europe. Whereas throughout the African continent the forest biome was constricted to a degree that was lethal for many kinds of animals, in Europe, forests changed in composition and distribution because of temperature changes but nonetheless persisted over broad areas because aridification was less pervasive. Evidently, many species of European mammals were insensitive to changes in the composition of forests. Perhaps exemplifying the greater specificity of extinction in Europe was the disappearance of *Paraliurus anglicus*, a close relative of the extant lesser panda, which feeds exclusively on bamboo (Kurtén, 1968).

Eastern North America

Unfortunately, Pliocene terrestrial floras of eastern North America have not been well dated or extensively studied. Floras of the Great Plains, however, reveal that climates were becoming drier during the Miocene Epoch, long before the global climatic changes of mid-Pliocene time. By the Late Miocene (10 to 5 Ma), grasslands were widespread in central North America, while forests were greatly restricted (Axelrod, 1985; Leopold and Denton, 1987). As discussed later, the early spread of grasslands can be related to tectonic events in the American West. Global climatic changes of the Pliocene presumably compounded these effects, but had a less severe impact on floras in central North America than in Europe or, especially, Africa because aridification had already progressed quite far during the Miocene.

Similarly, late Cenozoic events of mammalian extinction in North America spanned a considerable interval of time: there were six such events during the past 10 m.y. (Webb, 1984). One of these was a minor episode that occurred early in Pliocene time. A more severe event took place near the end of the Pliocene, eliminating about 55 genera. This one was approximately coincident with the heaviest pulse of Pliocene extinction in Europe, but it is not clear that they shared the same cause. A striking pattern of the North American event was that many species of large mammals, among them peccaries and capybaras, survived by retreating to low latitudes. For some species the geographic shift occurred in stages. Extinction and emigration resulted in a severe impoverishment of modern mammalian faunas in the temperate zone of North America. The northward decline in diversity is especially pronounced today at latitudes above 38° (Webb, 1984).

Marine Biotas

The onset of the ice age had profound consequences for shallow-water marine life of the North Atlantic, especially adjacent to North American and in the Caribbean Sea. Losses were relatively minor for planktonic species, most of which retained access to suitable biogeographic provinces, but heavy extinction left an impoverished bottom-dwelling fauna that has never recovered its diversity. Changes in sea-surface temperatures were the primary agent of extinction.

We have noted that relatively warm marine climates with low seasonality characterized mid-Pliocene waters adjacent to the eastern United States, with temperatures reaching the subtropical range as far north as Virginia near the end of Yorktown deposition, when the Gulf Stream was strengthened by closure of the Isthmus of Panama about 3.1 Ma. Thus, benthic faunas of the Western Atlantic were warm adapted and stenothermal, which means that they were highly vulnerable to climatic deterioration. The Pliocene fate of the bivalve mollusks, which has been studied in some detail (Stanley, 1986), presumably typifies the history of the benthic fauna in general.

Whereas in California and Japan, mid-Pliocene bivalve faunas contain about 70% extant species, in the Atlantic Coastal Plain only about 20% of mid-Pliocene species survive to the present. Although it might otherwise be tempting to attribute the heavy Western Atlantic extinction to the lowering of sea level and reduction of shallow seafloor that accompanied glacial expansion during Late Pliocene time, the fact that Pacific faunas experienced no major pulse of extinction rules out eustatic change as a primary agent of extinction. In fact, a broad depositional ramp borders the west coast of Florida, whereas only narrow shelves fringe the Pacific coast of North America. Even during glacial maxima, a large area of shallow seafloor was available for colonization west of Florida; yet extinction here was as heavy as along the Atlantic coast and much heavier than in California. If we take the rate of extinction shared by California and Japan to represent a normal or "background" rate, then the Western Atlantic crisis removed about 65% of all mid-Pliocene species, whereas a much smaller fraction died out through normal attrition (Stanley, 1986).

There is strong evidence that the cooling and accompanying increase in seasonality of shallow waters in the Western Atlantic described earlier constituted the dominant cause of heavy extinction. The mid-Pliocene bivalve fauna of west-central Florida, like any marginally tropical biota, included some species restricted to very warm climates and others that ranged into the temperate zone. As it turns out, however, every one of the nearly 60 species surviving from this fauna today range into the temperate zone: around the Gulf coast to Texas or northward along the Atlantic coast at least to the Carolinas. Thus, a thermal filter removed all stenothermal species, leaving a modern fauna dominated by forms with broad thermal tolerances (see Figure 7.4).

The Plio-Pleistocene strata exposed along the Atlantic Coastal Plain represent only high stands of sea level. Although discontinuous, the record here is consistent with the hypothesis that steps of extinction occurred during

FIGURE 7.4 Elongate species (scale bars are 1 cm) of the family Anadaridae that lived in shallow tropical seas of southern Florida at about 3 Ma. Only two of these species (**E** and **K**) survive today, and they range into temperate waters. **A**: *Arca williamsi*; **B**: *Arca wagneriana*; **C**: *Barbatia floridana*; **D**: *Barbatia irregularis*; **E**: *Barbatia dominigensis*; **F**: *Barbatia leonensis*; **G**: *Barbatia taeniata*; **H**: *Anadara notoflorida*; **I**: *Anadara campsa*; **J**: *Anadara improcera*; **K**: *Anadara lienosa*; **L**: *Anadara propatula*.

lowstands representing glacial maxima. Extinction had ended by late Pleistocene time, however. By this time, forms unable to tolerate conditions during the pronounced glacial maxima that began about 0.9 Ma had died out (Stanley, 1986). This pattern of change paralleled that for plants in Europe. What remains is an impoverished bivalve fauna of largely eurythermal species. Many range from the temperate zone to the tropics. In addition, many bivalve species that occupied lagoonal settings before the onset of the ice age are restricted to offshore shelf areas today. Apparently these species cannot tolerate the increased seasonality (in particular, the colder winter temperatures) that now characterize nearshore waters. Only a modest number of new bivalve species evolved during Late Pliocene and Pleistocene time. The extent to which climatic changes may have initiated some speciation in the Bivalvia remains to be investigated, but the onset of the ice age has been credited with triggering the origin of several new species of Western Atlantic ostracods (Cronin, 1988).

The molluscan fauna of the Caribbean, which was largely distinct from that of eastern North America, experienced a decline that was more or less as severe, although the details remain to be brought to light. In contrast, the molluscan fauna on the Pacific side of the Isthmus of Panama has maintained its very high diversity to the present day. It may be that some of the Caribbean extinctions resulted from reduced upwelling and productivity following the uplift of the isthmus (Vermeij and Petuch, 1986), but the Caribbean fauna suffered losses as far north as the Bahamas at 2.65 Ma (McNeill *et al.*, 1988). This occurrence and the extinction during early Pleistocene time of three species of planktonic foraminifera that were narrowly adapted to the tropical Caribbean (Stanley *et al.*, 1988) suggest that cooling was a major cause of extinction here, as it was in the Western Atlantic. While we do not know the temporal pattern of cooling in the Caribbean, the CLIMAP study showed that during the most recent glacial maximum the sea-surface temperature in the central Caribbean dropped to a level about 4°C below that of today (Prell and Hays, 1976).

On the eastern side of the Atlantic, molluscan faunas and (by inference) marine life in general experienced significant but less severe extinction. As noted earlier, patterns of extinction of bivalves—and geographic distributions of surviving species—point to climatic cooling as the dominant agent of extinction in the Mediterranean and North Sea basins, where only 54% of the total number of Early Pliocene species survive (Raffi et al., 1985). The incidence of extinction in both regions was reduced by the ability of species to survive in the southern parts of their ranges Most species restricted to either the North Sea or the Mediterranean during Early Pliocene time died out, whereas 60 of 64 species present in both basins survive today. Even so, the total fauna declined markedly in diversity. Today it includes only 198 polysyringian bivalve species, yet 323 Early Pliocene species are known.

FORCING OF LATE CENOZOIC CLIMATIC CHANGES

In the first section of this chapter, we summarized two major regimes of late Cenozoic climatic change in and around the North Atlantic area: (1) the long-term cooling (and regional drying) that preceded Northern Hemisphere glaciation, and (2) the ice-age cycles of the past 2.5 to 3 m.y. Here we provide a brief overview of some of the possible causes for these changes, with particular attention to processes affecting the North Atlantic Ocean and surrounding continents.

Tectonic Forcing of Climate (pre-2.5 Ma)

The most likely causes of climatic trends persisting for millions of years are tectonic changes in the configuration of the solid Earth that underpins the climate system, particularly changes in geography related to plate-tectonic processes. These include changes in plate position, sea level, mountain elevations, and narrow "gateways" (sills and isthmus connections) that constrict ocean circulation. Most such tectonic changes are so gradual that it is difficult to demonstrate that they provide strong climatic forcing in the late Cenozoic. Two of these changes that may be especially relevant to climatic changes in and around the North Atlantic are the narrowing and final closing of the Isthmus of Panama and the relatively rapid uplift of plateaus and mountains in Asia and North America.

Closure of the Straits of Panamanian Isthmus

Final formation of the Isthmus of Panama occurred near 3 Ma, but was probably preceded by a long interval of gradually shallowing sill depth (Keigwin, 1982). Experiments with ocean general circulation models (OGCMs) indicate that closure should have led to a dramatic increase in the salinity of North Atlantic waters because the prior subsurface flow of low-salinity waters into the Atlantic would slow and then cease (Maier-Reimer et al., 1990). Modeling also simulates two other related changes: (1) increased formation of North Atlantic deep water (NADW), and (2) decreased formation of sea ice, resulting in a warming of circum-Atlantic waters at middle and high latitudes.

Geologic evidence confirms that a long-term increase in rates of NADW formation occurred over the past 10 or 15 m.y. (Woodruff and Savin, 1989), in agreement with the isthmus experiment. It is unclear, however, what ramifications increased NADW would have for global climate (via effects on the large oceanic carbon reservoir and thus potentially on CO_2). On glacial-interglacial time scales, increased NADW formation correlates with increased, rather than decreased, levels of atmospheric CO_2.

The simulated circum-Atlantic warming, resulting from closure of the Straits of Panama, matches neither the generally observed Northern Hemisphere trend toward cooler climates, nor the conclusion that some cooling of the North Atlantic sea surface occurred prior to glaciation (Dowsett and Poore, 1990). It does, however, match the evidence for very warm Early Pliocene ocean temperatures along the southeastern seaboard of the United States (Hazel, 1971; Stanley and Campbell, 1981; Cronin, 1988). It is also pertinent that an already-formed isthmus cannot account for the additional mid-Pleistocene cooling that led to larger glaciations over the past 1 m.y.

Plateau Uplift

Geologic data summarized by Ruddiman et al. (1989) suggest major late Cenozoic uplift of the Tibetan Plateau in southern Asia and uplift across a broad region of high terrain in the American West centered on the Colorado Plateau (although the latter is contested by Molnar and England, 1990). Experiments with global circulation models (GCMs) show that uplift of rock masses on the scale of several million square kilometers can alter the basic planetary circulation of the atmosphere, by repositioning and intensifying meanders in the midlatitude surface westerlies and jet stream flow, and creating the strong monsoonal circulations of the subtropics (Kutzbach et al., 1989). Of the many large-scale changes due to uplift that are simulated by the model (Ruddiman and Kutzbach, 1989), two are particularly pertinent to the North Atlantic region: (1) strong winter cooling over east-central North America, because prevailing winds turn from westerly toward northwesterly; and (2) increased summer (and annual) evaporation over the Mediterranean and Eastern Atlantic, due to increased subsidence of dry air and outflow of dry air from the Asian interior (Figure 7.5).

Winter cooling over east-central North America agrees with scattered fossil evidence of the nature of Neogene continental vegetation from the Northern Plains (Thomasson, 1979) and the eastern seaboard (Rachelle, 1976; Fredericksen, 1985; Omar et al., 1987). It appears to disagree with the peak shallow marine warmth attained during mid-Pliocene time in the southeastern United States (Hazel, 1971; Stanley and Campbell, 1981; Cronin, 1988).

The simulated increase in summer evaporation over the Mediterranean and Eastern Atlantic agrees with the early to middle Pliocene shift to drier summer climates recorded by North African dust (Tiedemann et al., 1989) and Mediterranean vegetation (Suc, 1984). Although the ocean was not an interactive part of the uplift experiments, a drier Mediterranean and Eastern Atlantic should increase North Atlantic salinity, and, in combination with other simulated changes, could lead to increased NADW formation (Ruddiman and Kutzbach, 1989).

Most geologic data from the Northern Hemisphere indicate a large, progressive late Cenozoic cooling trend prior to Northern Hemisphere glaciations. This cooling trend appears, however, to be somewhat muted in Europe (van der Hammen et al., 1971) and, as we have seen, may even be contradicted along some margins of the Atlantic.

One possible explanation for this complexity is the more regional effect of plateau uplift and the Panamanian closure on the Atlantic Ocean. Both factors seem likely to cause a stronger northward flux of salty water through the late Cenozoic. Because salty water aids deep water formation but suppresses sea-ice formation, it should increase the release of heat from the ocean to the atmosphere in winter, thereby moderating climate over and around the North Atlantic. It thus seems possible that a localized tendency toward warming around the Atlantic might attenuate the effects of an otherwise "global" cooling trend. In addition, as noted earlier, a strengthened Gulf Stream seems to account for a warming along the southeastern margin of the United States. This effect was only tempo-

FIGURE 7.5 Uplift-induced changes in January atmospheric circulation over North America, based on GCM experiments summarized in Ruddiman and Kutzbach (1989). *Top left*: winds in no-mountain (NM) experiment. *Bottom left*: winds in full-mountain (M) experiment. Wind strength keyed to vectors at bottom left. *Top right*: changes in surface temperature due to uplift (M-NM difference), with cooler regions shaded. *Bottom right*: changes in precipitation due to uplift (M-NM difference), with wetter areas shaded. Regions in which temperature and precipitation changes are significant at the 99% confidence level are indicated by diagonal dashes.

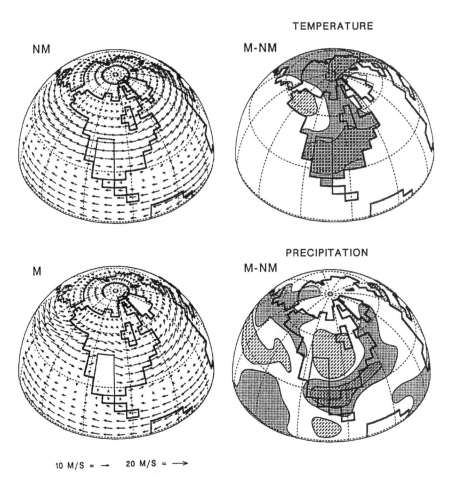

rary, however; since polar regions cooled dramatically in Late Pliocene time, polar outbreaks *have* produced very cold winter temperatures in the southeastern United States.

In summary, simulation experiments using GCMs to test the climatic effects of factors such as plateau uplift and isthmus connections have yielded numerous regional trends that agree with observed climatic changes, as well as some that do not. Basically, however, these experiments support the major conclusion from earlier simulations (Barron, 1985) that changes in geography alone are inadequate to explain the large amplitude of late-Cenozoic cooling. An additional factor is needed, and the most likely explanation is a long-term decrease in atmospheric CO_2 levels. Two factors are important in determining long-term CO_2 levels: (1) rates of input from volcanoes, and (2) rates of removal by chemical weathering of silicate rocks on land.

Sea-Floor Spreading and CO_2

One hypothesis links long-term CO_2 changes mainly to rates of seafloor spreading at midocean ridge crests (Berner *et al.*, 1983). Globally averaged spreading rates are proposed to control rates of CO_2 emission from island-arc volcanos, where ocean crust and sediments are consumed and destroyed in ocean trenches. However, over the past 30 m.y., global mean rates of seafloor spreading have been roughly constant and so cannot explain the pronounced climatic cooling.

Chemical Weathering and CO_2

A second hypothesis invokes increased chemical weathering due to uplift of plateaus and mountains as the mechanism for drawing down atmospheric CO_2 levels (Raymo *et al.*, 1988; Raymo and Ruddiman, 1992; see also Chamberlin, 1906). As noted above, plateau uplift produces a monsoonal increase in rainfall. Uplift also exposes fresh silicate rock on faulted slopes to attack by weathering, which consumes more atmospheric CO_2 than is subsequently released to the ocean during secretion of carbonate in plankton. Whereas the Berner *et al.* (1983) model assumed that chemical weathering is a function of both temperature and global sea level, the uplift hypothesis takes into account the fact that orography has a powerful control on weathering rates (Edmond, 1987). Steep slopes also flush away the products of chemical erosion, keeping weathering rates high.

An additional factor is the partitioning of carbon and alkalinity between the deep and the intermediate portions of the ocean. Because the ocean is the largest reservoir of carbon on Earth, this partitioning is potentially a critical factor in regulating atmospheric CO_2. At present, however, the partitioning even in the late Pleistocene (last glacial) ocean is poorly understood.

Whatever the ultimate cause, falling CO_2 levels during the late Cenozoic must eventually have passed through a series of critical climatic thresholds. At some point, sea ice began to expand across larger areas of the Arctic Ocean; GCM experiments show that its thickness and extent must have oscillated considerably in response to orbitally driven changes in insolation (Kutzbach and Gallimore, 1988). Modeling studies also indicate that one critical effect of more extensive sea ice would have been to increase the intensity of winter outbreaks of polar air masses across east-central North America and the rate of extraction of heat from the western North Atlantic (Raymo *et al.*, 1990).

Ice-Sheet Forcing of Climatic Change

Variations in the Earth's orbit have affected climate throughout the history of the Earth. Changes in the Earth's inclination (or tilt) relative to the plane of the ecliptic alter seasonal insolation at high latitudes at a period of 41,000 yr. Precession of the equinoxes around the elliptical orbit influence seasonal insolation at all latitudes, mainly at periods of 23,000 and 19,000 yr. Changes in the eccentricity of the orbit at periods near 100,000 and 410,000 yr do not directly result in variations in seasonal insolation at these rhythms, but instead modulate the amplitude of the precessional signal.

Most of the clear evidence for orbital control of climate has come from continuously deposited, well-dated marine sediments of Pleistocene age (Hays *et al.*, 1976). Although orbital variations have no doubt influenced climate through all of geologic time, once ice sheets appeared in the Northern Hemisphere, they provided a particularly effective means of amplifying insolation forcing from within the climate system. For the first 1.5 m.y. or more of the Northern Hemisphere ice age, these ice sheets varied mainly at periods of 41,000 and 23,000 yr, in response to direct orbital variations in summer insolation (Raymo *et al.*, 1989). Ice-volume changes lag several thousand years behind the direct insolation forcing at 41,000- and 23,000-yr periods, because of the inherently slow time constants of ice response.

For the larger ice sheets of the past 0.9 Ma, the dominant rhythm of ice-sheet change has been centered near 100,000 yr, despite the lack of direct insolation forcing at this period (Hays *et al.*, 1976; Shackleton and Opdyke, 1976, 1977). This rhythm of change appears to reflect either a highly nonlinear response to insolation forcing (linked to the modulation of the precession signal by eccentricity) or else some kind of natural resonance that has recently developed within the climate system and is paced by insolation.

The ice sheets in turn act as a source of forcing for other parts of the climate system, through several mechanisms. Like rock plateaus, the rise of these large domes of ice can rearrange the basic circulation of the atmosphere, with major effects that are sent far downstream to the south and east. For example, both marine geologic data and GCM experiments show that cold winds from the ice sheets can chill and freeze the surface of the North Atlantic Ocean at high latitudes (Ruddiman and McIntrye, 1984; Manabe and Broccoli, 1985). GCM experiments also show that chilling and freezing of the North Atlantic can in turn cool climates substantially over maritime regions of Eurasia and southward into the Mediterranean and even northern Africa (Rind et al., 1986). Loess records from China indicate that even in far southeastern Asia, aridity cycles follow the basic tempo of change in the size of ice sheets farther north (Heller and Liu, 1982; Kukla, 1987).

In addition, formation of deep water in the North Atlantic was suppressed when ice sheets were large (Boyle and Keigwin, 1985), probably because of changes in salinity created by the altered wind field (and possibly because of meltwater fluxes). Changes in rate of formation of NADW may also influence subsurface and surface circulation in the Southern Ocean (Weyl, 1968). Redistribution of nutrients and alkalinity in this region may, in turn, affect atmospheric CO_2 levels and thus global climate (Broecker and Peng, 1989). To some extent, however, CO_2 has also varied somewhat independently of Northern Hemisphere ice sheets over the past 150,000 yr (Barnola et al., 1987).

SUMMARY AND CONCLUSIONS

The abrupt climatic changes that marked the onset of the Recent ice age during the Pliocene Epoch followed a long interval of widespread cooling and accentuation of contrasts between wet and dry regional climates. Nonetheless, prior to about 3.1 Ma, climates in many areas of the North Atlantic region remained warmer and less seasonal than today. The Atlantic and Pacific Oceans exchanged temperate species of mollusks by way of the Arctic, the Mediterranean Sea was marginally tropical, and shallow seas off Virginia became briefly subtropical.

Oxygen isotope ratios for marine microfossils offer the best chronological record of the present ice age. They reveal that the deep sea became permanently cooler at about 3.2 to 3.1 Ma, which implies that there was increased cooling at high latitudes. At this time, several subtropical species of plants disappeared from northwestern Europe, and a brief pulse of cooling extended at least as far south as the Mediterranean, where changes in the planktonic foraminiferal fauna record an estimated reduction in mean annual temperature of 2 to 4°C.

Climates warmed again in at least some areas during the 3.0 to 2.5 Ma interval, but at 2.5 to 2.4 Ma the modern ice age began. Mountain glaciers had been present earlier, but the expansion of large ice sheets at this time accounted for the strong isotopic signal recorded in deep-sea cores. Subsequent isotopic cycles in these cores and in loess reveal an orbitally forced periodicity of ~41,000 yr for about 1.5 m.y. and then of ~100,000 yr between 0.9 Ma and the present.

Terrestrial biotas underwent major changes in the vicinity of the North Atlantic during Late Pliocene time, with aridification playing at least as large a role as cooling. In most areas, the greatest changes took place at about 2.5 Ma, or a bit later. Throughout Africa, forests contracted and many species of mammals that had adapted to them died out; speciation soon produced many new species adapted to the expanding savannas. In Europe forests changed their character, in part through the disappearance of subtropical taxa, but they remained widespread so that mammals were little affected. Details of floral change during the Pliocene remain poorly known for many areas of North America; grasslands were widespread before the end of Miocene time, but about 55 genera of mammals disappeared near the end of the Pliocene.

Beginning slightly before 2.5 Ma, heavy extinction of shallow marine life occurred in the Western Atlantic and Caribbean. Some species survived by migrating offshore to relatively stable thermal regimes, but nearly all strictly tropical species of Florida died out. Heavy extinction ended in late Pleistocene time, after the onset of severe climatic oscillations at ~0.9 Ma. Moderately heavy extinction also occurred in the Mediterranean and the North Sea. In general, shallow-water benthic faunas of the North Atlantic region are impoverished today as a result of Plio-Pleistocene extinction.

Tectonic forcing has probably been responsible for persistent, longer-term climatic trends in the North Atlantic region during late Neogene time. Formation of the Isthmus of Panama slightly before 3 Ma should have increased the formation of deep water in the North Atlantic and decreased the formation of sea ice by reducing the influx of low-salinity water to the Atlantic. In OGCMs, closure of the Straits of Panama elevates temperatures in the southeastern United States, as actually happened, but fails to produce the observed general cooling at high latitudes. General circulation models suggest that late Neogene uplift of the Tibetan Plateau and other regions should have intensified meanders in westerlies and jet stream flow, producing aridification of the Mediterranean region and reduction of winter temperatures in the southeastern United States. Both predictions match actual events, although the second change is reflected in the fossil record only after mid-Pliocene cooling at high latitudes. It appears, however, that changes in geography are insufficient to account for general climatic cooling during the late Neogene time.

A trend of decreasing atmospheric CO_2 is the most likely cause of long-term climatic cooling. Spreading rates at midocean ridges have not changed enough during the past 30 m.y. to have been the primary factor. A likely cause is elevation of mountains and plateaus, which increased rates of CO_2 uptake by weathering. Since extensive sea ice and glaciers first formed at high latitudes, changes in the Earth's orbital behavior have caused their volumes to oscillate periodically and further influence climate.

REFERENCES

Axelrod, D. I. (1950). Evolution of desert vegetation in western North America, in *Studies in Late Tertiary Paleobotany*, Carnegie Institution of Washington, Publication 590, pp. 217-306.

Axelrod, D. I. (1966). The Pleistocene Soboba flora of southern California, *University of California at Berkeley Publications in Geological Sciences 60*, 1-79.

Axelrod, D. I. (1985) Rise of the grassland biome, central North America, *Botan. Reviews 51*, 163-201.

Barnola, J. M., D. Raynaud, Y. S. Korotkevich, and C. Lorius (1987). Vostok ice core provides 160,000-year record of atmospheric CO_2, *Nature 329*, 408-414.

Barron, E. J. (1985). Explanations of the Tertiary global cooling trends, *Palaeogeography, Palaeoclimatology, Palaeoecology 50*, 45-61.

Barron, E. J., *et al.*, eds. (1989). *Proceedings of the Ocean Drilling Program, Scientific Results 119*, Ocean Drilling Program, College Station, Texas, 942 pp.

Beach, D. K., and R. N. Ginsburg (1980). Facies succession of Pliocene-Pleistocene carbonates, northwestern Great Bahama Banks, *American Association of Petroleum Geologists Bulletin 64*, 1634-1642.

Berner, R. A., A. C. Lasaga, and R. M. Garrels (1983). The carbonate-silicate geochemical cycle and its effect on atmospheric carbon dioxide over the last 100 million years, *American Journal of Science 283*, 641-683.

Bonnefille, R. (1976). Palynological evidence for an important change in the vegetation of the Omo basin between 2.5 and 2 million years ago, in *Earliest Man and Environments in Lake Rudolf Basin*, Y. Coppens, F. C. Howell, G. L. Isaac, and R. E. F. Leakey, eds., University of Chicago Press, Chicago, pp. 421-431.

Bonnefille, R. (1985). Evolution of the continental vegetation: The palaeobotanical record from East Africa, *South African Journal of Science 81*, 267-270.

Boyle, E. A., and L. D. Keigwin (1985). Comparison of Atlantic and Pacific paleochemical records for the last 215,000 years: Changes in deep ocean circulation and chemical inventories, *Earth and Planetary Science Letters 76*, 135-150.

Broecker, W. S., and T.-H. Peng (1989). The cause of glacial to interglacial atmospheric CO_2 change: A polar alkalinity hypothesis, *Global Biogeochemical Cycles 3*, 215-239.

Broecker, W. S., and J. Van Donk (1970). Insolation changes, ice volumes, and the [18]O record in deep-sea cores, *Reviews of Geophysics and Space Physics 8*, 169-198.

Brunner, C. A. (1984). Evidence for increased volume transport of the Florida current in the Pliocene and Pleistocene, *Marine Geology 54*, 223-235.

Carter, L. D., J. Brigham-Grette, L. Marincovich, V. L. Pease, and J. W. Hillhouse (1986). Late Cenozoic Arctic Ocean sea ice and terrestrial paleoclimate, *Geology 14*, 675-678.

Chamberlin, T. C. (1906). On a possible reversal of deep-ocean circulation and its influences on geologic climates, *Journal of Geology 14*, 363-373.

Cronin, T. M. (1988). Evolution of marine climates during the past four million years, *Philosophical Transactions of the Royal Society of London B318*, 661-678.

Delson, E. (1985). Neogene African catarrhine primates: Climatic influence on evolutionary patterns, *South African Journal of Science 81*, 273-274.

Denton, G. H., and R. L. Armstrong (1969). Miocene-Pliocene glaciations in southern Alaska, *American Journal of Science 267*, 1121-1142.

Dowsett, H. J., and T. M. Cronin (1990). High eustatic sea level during the middle Pliocene: Evidence from the southeastern U.S. Atlantic coastal plain, *Geology 18*, 435-438.

Dowsett, J. M, and R. Z. Poore (1990). A new planktic foraminifer transfer function for estimating Pliocene-Recent paleoceanographic conditions in the North Atlantic, *Marine Micropaleontolgy 16*, 1-23.

Edmond, J. M. (1987). Hydrothermal fluxes in the oceanic geochemical budgets, *EOS 68*, 1209.

Einarsson, T., D. M. Hopkins, and R. R. Deoll (1967). The stratigraphy of Tjörnes, northern Iceland, and the history of the Bering Land Bridge, in *The Bering Land Bridge*, D. M. Hopkins, ed., Stanford University Press, Stanford, Calif., pp. 312-325.

Fredericksen, N. O. (1985). Stratigraphic, paleoclimatic, and palaeogeographic significance of Tertiary sporomorphs from Massachusetts, *U.S. Geological Survey Professional Paper 1308*, 1-25.

Hays, J. D., J. Imbrie, and N. J. Shackleton (1976). Variations in the Earth's orbit: Pacemaker of the ice ages, *Science 194*, 1121-1132.

Hazel, J. E. (1971) Paleoclimatology of the Yorktown Formation (upper Miocene and lower Pliocene) of Virginia and North Carolina, *Centre de Recherches Pan-SNPA Bulletin 5*(Supplement), 361-375.

Heller, F., and T. S. Liu (1982). Magnetostratigraphical dating of loess deposits in China, *Nature 300*, 431-433.

Henrich, R., T. C. Wolf, G. Bohrmann, and J. Thiede. (1989). Cenozoic paleoclimatic and paleoceanographic changes in the Northern Hemisphere revealed by variability of coarse fraction composition in sediments from Voring Plateau—ODP Leg 104 drill sites, *Ocean Drilling Program 105*, Ocean Drilling Program, College Station, Texas, pp. 75-188.

Hopkins, D. M. (1967). Quaternary Marine transgressions in Alaska, in *The Bering Land Bridge*, D. M. Hopkins, ed., Stanford University Press, Stanford, Calif., pp. 47-90.

Jansen, E. U., Bleil, R. Henrich, L. Kringstad, and B. Slettemark (1988). Paleoenvironmental changes in the Norwegian Sea and the northeast Atlantic during the last 2.8 m.y.: Deep Sea Drilling Project/Ocean Drilling Program sites 610, 642, 643, and 644, *Paleoceanography 3*, 563-581.

Keigwin, L. D. (1978). Pliocene closing of the Isthmus of Panama, based on biostratigraphic evidence from nearby Pacific Ocean and Caribbean Sea cores, *Geology 6*, 630-634.

Keigwin, L. D. (1982). Isotopic paleoceanography of the Caribbean and East Pacific: Role of Panama uplift in late Neogene time, *Science 217*, 350-353.

Keigwin, L. D., and R. C. Thunell (1979). Middle Pliocene climatic change in western Mediterranean from faunal and oxygen isotopic trends, *Nature 282*, 294-296.

Kukla, G. (1977). Pleistocene land-sea correlations, *Earth-Science Reviews 13*, 307-374.

Kukla, G. (1987). Loess stratigraphy in central China and Correlation with an extended oxygen isotope stage scale, *Quaternary Science Reviews 6*, 191-219.

Kurtén, B. (1968). *Pleistocene Mammals of Europe*, Aldine Publishing Co., Chicago.

Kutzbach, J. E., and R. G. Gallimore (1988). Sensitivity of a coupled atmosphere/mixed layer ocean model to changes in orbital forcing at 9000 years B.P., *Journal of Geophysical Research 93*, 803-821.

Kutzbach, J. E., P. J. Guetter, W. F. Ruddiman, and W. L. Press (1989). Sensitivity of climate to late Cenozoic uplift in southern Asia and the American West: Numerical experiments, *Journal of Geophysical Research 94*, 18,393-18,407.

Ledbetter, M. T., D. F. Williams, and B. B. Ellwood (1978). Late Pliocene climate and south-west Atlantic abyssal circulation, *Nature 272*, 237-239.

Leopold, E. B. (1967). Late-Cenozoic patterns of plant extinction, in *Pleistocene Extinctions: The Search for a Cause*, P. S. Martin and H. W. Wright, eds., Yale University Press, New Haven, Conn.

Leopold, E. B., and M. F. Denton (1987). Comparative age of grasslands and steppe east and west of the northern Rocky Mountains, *Annals of the Missouri Botanical Gardens 74*, 841-867.

Maier-Reimer, E., U. Mikolajewica, and T. Crowley (1990). Ocean general circulation model sensitivity experiment with an open Central American Isthmus, *Paleoceanography 5*, 349-366.

Manabe, S., and A. J. Broccoli (1985). The influence of continental ice sheets on the climate of an ice age, *Journal of Geophysical Research 90*, 2167-2190.

Mankinen, E. A., and G. B. Dalrymple (1979). Revised geomagnetic polarity time scale for the interval 0-5 m.y. B.P., *Journal of Geophysical Research 84*, 615-626.

Marasti, R., and S. Raffi (1979). Observations on the paleoclimatic and biogeographic meaning of the Mediterranean Pliocene molluscs, state of the problem, VII International Congress on the Mediterranean Neogene, Athens, *Ann. Géol. Pays Hellén Tome Hors Série 1979 2*, 727-734.

McDougall, I., and H. Wensink (1966). Plaeomagnetism and geochronology of the Pliocene-Pleistocene lavas in Iceland, *Earth and Planetary Science Letters 1*, 232-236.

McNeill, D. F., R. N. Ginsburg, S. B. R. Chang, and J. L. Kirschvink (1988). Magnetostratigraphic dating of shallow-water carbonates from San Salvador, Bahamas, *Geology 16*, 8-12.

Meeder, J. F. (1979). *A Pliocene Fossil Reef of Southwest Florida*, Miami Geological Society Field Trip Guide, January 20-21, 1979.

Mercer, J. H. (1983). Cenozoic glaciation in the Southern Hemisphere, *Annual Review of Earth and Planetary Science 11*, 99-132.

Molnar, P., and P. England (1990). Late Cenozoic uplift of mountain ranges and global climate change: Chicken or egg? *Nature 346*, 29-34.

Mudie, P. J., and J. Helgason (1983). Palynological evidence for Miocene climatic cooling in eastern Iceland about 9.8 Myr ago, *Nature 303*, 689-692.

Omar, G., K. R. Johnson, L. J. Hickey, P. B. Robertson, M. R. Dawson, and C. W. Barnosky (1987). Fission-track dating of Haughton Astrobleme and included biota, Devon Island, Canada, *Science 231*, 1603-1605.

Prell, W. L. (1984). Covariance patterns of foraminiferal $\delta^{18}O$: An evaluation of Pliocene ice volume changes near 3.2 million years ago, *Science 226*, 692-694.

Prell, W. L., and J. D. Hays (1976). Late Pleistocene faunal and temperature patterns of the Columbia Basin, Caribbean Sea, *Geological Society of America Memoir 145*, 201-220.

Rachelle, L. O. (1976). Palynology of the Lexler Lignite: A deposit in the Tertiary Cohansey Formation of New Jersey, U.S.A., *Review of Paleobotany and Palynology 22*, 225-252.

Raffi, S., S. M. Stanley, and R. Marasti (1985). Biogeographic patterns and Plio-Pleistocene extinction of Bivalvia in the Mediterranean and southern North Sea, *Paleobiology 11*, 368-388.

Raymo, M. E. (1994). The initiation of Northern Hemisphere glaciation, *Annual Review of Earth and Planetary Sciences 22*, 353-383.

Raymo, M. E., and W. F. Ruddiman (1992). Tectonic forcing of late cenozoic climate, *Nature 359*, 117-122.

Raymo, M. E., W. F. Ruddiman, and P. N. Froelich (1988). The influence of late Cenozoic mountain building on oceanic geochemical cycles, *Geology 16*, 649-653.

Raymo, M. E., W. F. Ruddiman, J. Backman, B. M. Clement, and D. G. Martinson (1989). Late Pliocene variation in Northern Hemisphere ice sheets and North Atlantic deep water circulation, *Paleoceanography 4*, 413-446.

Raymo, M. E., D. Rind, and W. F. Ruddiman (1990). Climatic effects of reduced Arctic sea ice limits in the GISS II general circulation model, *Paleoceanography 5*, 367-382.

Reid, C., and E. M. Reid (1915). *The Pliocene Floras of the Dutch-Prussian Border*, Mededeel. van de Rijisopsp. van Delfstoffen, no. 6, The Hague.

Repenning, C. A., E. M. Brouwers, L. D. Carter, L. Marincovich, and T. A. Ager (1987). The Beringian ancestry of *Phenacomys* (Rodentia: Cricetidae) and the beginning of the modern Arctic Ocean borderland fauna, *U.S. Geological Survey Bulletin 1687*.

Rind, D. Y., D. Peteet, W. S. Broecker, A. McIntyre, and W. F. Ruddiman (1986). The impact of cold North Atlantic sea-surface temperatures on climate: Implications for the Younger Dryas cooling (11-10 Ka), *Climate Dynamics 1*, 3-34.

Ruddiman, W. F., and J. E. Kutzbach (1989). Forcing of late Cenozoic Northern Hemisphere climates by plateau uplift in southern Asia and the American West, *Journal of Geophysical Research 94*, 18,409-18,427.

Ruddiman, W. F., and A. McIntyre (1984). Ice-age thermal response and climatic role of the surface North Atlantic Ocean, 40° to 63°N, *Geological Society of America Bulletin 95*, 381-396.

Ruddiman, W. F., W. L. Prell, and M. E. Raymo (1989). Late Cenozoic uplift in southern Asia and the American West: Rationale for general circulation modeling experiments, *Journal of Geophysical Research 94*, 18-379-18,391.

Shackleton, N. J., and N. D. Opdyke (1976). Oxygen isotope and paleomagnetic stratigraphy of Pacific core V28-239: Late Pliocene to latest Pleistocene, in *Investigation of Late Quaternary Paleoceanography and Paleoclimatology*, R. M. Cline and J. D. Hays, eds., Geological Society of America Memoir 145, pp. 449-464.

Shackleton, N. J., and N. D. Opdyke (1977). Oxygen isotope and paleomagnetic evidence for early Northern Hemisphere glaciation, *Nature 270*, 216-239.

Shackleton, N. J, J. Backman, H. Zimmerman, D. V. Kent, M. A. Hall, D. G. Roberts, D. Schnitker, J. G. Baldauf, A. Desprairies, R. Homrighausen, P. Huddlestun, J. B. Keene, A. J. Kaltenback, K. A. O. Krumsiek, A. C. Morton, J. W. Murray, and J. Westberg-Smith (1984). Oxygen isotope calibration of the onset of ice-rafting and history of glaciation in the North Atlantic region, *Nature 307*, 620-623.

Stanley, S. M. (1986). Anatomy of a regional mass extinction: Plio-Pleistocene decimation of the Western Atlantic bivalve fauna, *Palaios 1*, 17-36.

Stanley, S. M., and L. D. Campbell (1981). Neogene mass extinction of western Atlantic molluscs, *Nature 293*, 457-459.

Stanley, S. M., K. L. Wetmore, and J. P. Kennett (1988). Macroevolutionary differences between the two major clades of Neogene planktonic foraminifera, *Paleobiology 14*, 235-249.

Suc, J.-P. (1984). Origin and evolution of the Mediterranean vegetation and climate in Europe, *Nature 307*, 429-432.

Suc, J.-P., and W. H. Zagwin (1983). Plio-Pleistocene correlations between the N-W Mediterranean and N-W Europe according to recent biostratigraphic and paleoclimatic data, *Boreas 12*, 153-166.

Thomasson, J. R. (1979). Late Cenozoic grasses and other angiosperms from Kansas, Nebraska, and Colorado: Biostratigraphy and relationships to living taxa, *Kansas Geological Survey Bulletin 218*, 1-67.

Thunell, R. C. (1979). Climatic evolution of the Mediterranean Sea during the last 5.0 million years, *Sedimentary Geology 23*, 67-79.

Tiedemann, R., M. Sarnthein, and Stein R. (1989). Climatic changes in the western Sahara: Aeolo-marine sediment record of the last 8 million years (Sites 657-661), *Initial Reports Deep Sea Drilling Project 108*, 241-278.

van der Hammen, T. (1985). The Plio-Pleistocene climatic record of the tropical Andes, *Journal of the Geological Society of London 142*, 483-489.

van der Hammen, T., T. A. Wijmstra, and W. H. Zagwijn (1971). The floral record of the late Cenozoic of Europe, in *The Late Cenozoic Glacial Ages*, K. K. Turekian, ed., Yale University Press, New Haven, Conn., pp. 392-424.

Vermeij, G. J., and E. J. Petuch (1986). Differential extinction in tropical American molluscs: Endemism, architecture, and the Panama land bridge, *Malacologia 27*, 29-41.

Vermeij, G. J. (1989). Invasion and extinction: The last three million years of North Sea pelecypod history, *Conservation Biology 3*, 274-281.

Vrba, E. S. (1985). African Bovidae: Evolutionary events since the Miocene, *South African Journal of Science 81*, 263-266.

Vrba, E. S. (1988). Late Pliocene climatic events and hominid evolution, in *Evolutionary History of the "Robust" Australopithecines*, F. E. Grine, ed., Aldine De Gruyter, New York, pp. 405-426.

Webb, S. D. (1984). Ten million years of mammal extinctions in North America, in *Quaternary Extinctions: A Prehistoric Revolution*, P. S. Martin and R. G. Klein, eds., University of Arizona Press, Tucson.

Wesselman, H. B. (1985). Fossil micromammals as indicators of climatic change about 2.4 Myr ago in the Omo Valley, Ethiopia, *South African Journal of Science 81*, 260-261.

Weyl, P. K. (1968). The role of the oceans in climatic change: A theory of the ice ages, *Meteorological Monograph 12*, 37-62.

Wolfe, J. A. (1979). Temperature parameters of humid to mesic forests of eastern Asia and relation to forests of other regions of the Northern Hemisphere and Australasia, *U.S. Geological Survey Professional Paper 1106*, 1-37.

Woodruff, F., and S. M. Savin (1989). Miocene deepwater oceanography in the Mediterranean and associated record of climatic change, *Paleoceanography 4*, 87-140.

Zachariasse, W. J., L. Gudjonsson, F. J. Hilgen, C. G. Langereis, L. J. Lourens, P. J. J. M. Verhallen, and J. D. A. Zijderveld (1990). Late Gauss to early Matuyama invasions of *Neogloboquadrina atlantica* in the Mediterranean and associated record of climatic change, *Paleoceanography 5*, 239-252.

Zimmerman, H. B. (1984). Lithostratigraphy and clay mineralogy of the western margin of the Pockall Plateau and the Hatton sediment drift, *Initial Reports Deep Sea Drilling Project 81*, 683-694.

8

The Response of Hierarchially Structured Ecosystems to Long-Term Climatic Change: A Case Study Using Tropical Peat Swamps of Pennsylvanian Age

WILLIAM A. DIMICHELE
National Museum of Natural History, Smithsonian Institution

TOM L. PHILLIPS
University of Illinois

ABSTRACT

Carboniferous coal-forming swamps are an excellent system in which to evaluate the effects of regional to global climatic changes on ecosystem structure and dynamics. Stressful physical conditions restrict the access of most species, creating semiclosed conditions in which the signal-to-noise ratio should be high. We examine patterns of change in coal-swamp systems during the Pennsylvanian at three levels: landscapes, habitats within landscapes, and species within habitats. The timing and extent of turnover at these three levels suggest a hierarchial organization, in the sense that patterns at one level emerge from interactions among elements at a lower level, and can have subsequent reciprocal effects on the dynamics at that lower level.

Changes in the species composition of coals, and in the dominance-diversity structure they define, occur continuously throughout the Pennsylvanian. However, there are distinct breakpoints that allow us to recognize five basic organizational themes. Each of these breakpoints corresponds to an independently inferred change in regional or global climate. Examination of species-level turnover patterns reveals highest values at the landscape-level breakpoints, suggesting at first that climate change may be affecting species turnover, which then scales upward directly into landscape changes. Estimation of the relationship between species turnover and habitat patterns, however, suggests an intermediate level of organization. Species turnover throughout the Westphalian occurs mostly within habitats and on strongly ecomorphic themes; species of the same or closely related and morphologically similar genera tend to replicate each other through time. It is the proportion of habitats that changes at the landscape-breakpoint boundaries, and habitats contain the ecomorphic elements that give the landscape its apparent dominance-diversity structure.

At the Westphalian-Stephanian boundary, high levels of extinction eliminated sufficient numbers of species that Westphalian ecomorphic patterns were destroyed and a new set of species-habitat dynamics was established. In combination with replacement patterns during the Westphalian, the extinction suggests that biotic interactions among species assist in creating a "fabric" or multiniche system that helps constrain the ecomorphic nature of species replacements, whether such replacements occur through evolution within or migration into the system. Loss or decline of a species creates a vacant niche, whose limits are partially defined by the remaining biota. Major disruption of this system by extrinsically induced extinction permits the system to reestablish the interaction fabric based on the biologies of the new suite of species.

INTRODUCTION

One of the most important tenets of conventional wisdom currently dominant in plant community ecology is the concept of individualism. This is a highly reductionist principle that attributes all apparent structure in plant communities to the properties of individual organisms; associations of species are a consequence only of their similar tolerances to physical conditions and similar resource requirements. Such a view allows no "emergent" properties, characteristics of ecosystems that result from the interactions of coexisting species populations (avatars in the terminology of Damuth, 1985). Some plant ecologists have turned to the fossil record, almost exclusively that of Quaternary pollen, to test the predictions of this hypothesis when dynamics are viewed over a longer time scale (e.g., Delcourt and Delcourt, 1987; Overpeck et al., 1992) and have found much congruence. Consequently, the pollen record of the past 10,000 yr has come to stand as proxy for the past 420 million years (m.y.) of terrestrial plant history. What about the rest of the record?

Most of the terrestrial fossil record is less complete than that of the Quaternary. This is not to imply that it cannot be sampled at the same scale. It can, as studies of spores and pollen from very ancient sediments indicate (e.g., Smith and Butterworth, 1967; Mahaffy, 1985; Farley, 1989; Eble and Grady, 1993). The major difference lies in our understanding of contemporaneous, potentially causative changes in climate and paleogeography, resolved much more coarsely as one looks deeper into time. Yet the older record preserves patterns of change over a much broader span of time intervals than 10,000 to 100,000 yr.

Where studied in detail, the past does not reveal a pattern of ever-changing community structure and composition. Rather, remarkable periods of persistence of species associations and ecomorphic patterns are found, in which species turnover may occur but structure and dynamics remain largely unchanged for a few to many millions of years. Examples do not abound because little detailed research on community paleoecological patterns has been pursued outside the Late Carboniferous (e.g.; Scott, 1978; Phillips and DiMichele, 1981; Gastaldo, 1987; Raymond, 1988) or the early Tertiary (e.g., Wing, 1988, Farley, 1989). However, these vastly different ecosystems show remarkably similar patterns and suggest that the Quaternary record cannot be used reliably as a blanket model of terrestrial plant community dynamics through time.

In this chapter we focus on changes in the taxonomic composition and structural properties of late Paleozoic plant communities from wetland habitats, in particular those of peat-forming swamps. Peat-forming communities of the Late Carboniferous (Pennsylvanian) provide some of the best opportunities to characterize and analyze an ancient ecosystem. Each of the five major plant groups (lycopsids, ferns, sphenopsids, pteridosperms, and cordaites) is highly distinctive morphologically and anatomically, the generic and species diversities are quite low, and taxa were largely pantropical in distribution within the wetlands. Particularly important in resolving the plants and their intraswamp habitats are coal balls, which are essentially *in situ* accumulations of plant litter preserved relatively uncompacted in concretions within coal seams. Coal balls occur in many coals, mostly in the upper Carboniferous. Coal balls, miospores, and compression fossils have been studied from across the ancient Euramerican paleocontinent largely as a consequence of their exposure during coal mining and because of their relevance to coal geology as well as paleobotany. Similarly, studies of depositional patterns, and interpretations of regional and global climate have engendered considerable debate (e.g., Wanless et al., 1969; Horne et al., 1978; Phillips and Cecil, 1985; Heckel, 1986; Klein and Willard, 1989; Cecil, 1990). Despite disagreements, the data, and varying conclusions drawn from them, have yielded greater insight than for any pre-Tertiary systems.

The implications of this research may go far beyond the late Paleozoic. Paleoecology provides a baseline against which we can examine, for greater generality, our concepts formulated almost entirely from extant ecosystems. The fossil record is our only access to long-term patterns and their implications for questions of community persistence, stability, and response to major extrinsic perturbations. By examining the response of ancient ecosystems to major disruptions we hope to gain insights into the ways in which plant communities of today may behave—not the relatively trivial specifics of change, but the general mechanistic basis underlying change. We need to learn how to recognize responses of ancient ecosystems that are unique products of times and circumstances, as opposed to inherent qualities that reflect principles of organization that transcend evolutionary diversification of the biota. When comparing modern plant communities to any from the Late Carboniferous we must recognize that the levels of Carboniferous species diversity were at least three orders of magnitude lower than those at present, that there are difficulties in recognizing those environments that are most stable and geologically persistent, particularly in extant systems, and that differences in temporal scale can confound comparisons of ecological patterns.

LATE CARBONIFEROUS WETLANDS

The Carboniferous was the first and most extensive coal age. The pantropical wetlands of the Late Devonian and Early Carboniferous from which the coal-forming forests evolved were the Earth's primeval forests (Cleal, 1987). Most of the major lineages of plants that dominated these environments originated during the Late Devonian and earliest Carboniferous. In taxonomic terms, all the generally recognized major class-level groups of vascular plants (save the flowering plants) and many of the major ordinal-level groups originated during this narrow window of time. In the establishment of all basic life history and architectural patterns in vascular plants, the Late Devonian and earliest Carboniferous parallels the Cambrian radiation of marine invertebrates.

The vascular plant radiation involved a strong association of class- and ordinal-level clades with specific ecological conditions. In effect, the basic ecologies and habitats of the landscape were strongly partitioned along taxonomic lines. This pattern of ecological partitioning along taxonomic and life history lines was to distinguish the Carboniferous from all subsequent times. The landscape partitioning of the Carboniferous began to break down over an extended time interval, from the end of the Late Carboniferous through the Early Permian. During this time there were major extinctions in the wetlands (Phillips *et al.*, 1974, 1985; Pfefferkorn and Thomson, 1982), accompanied by expansion of seed plants into vacated resource space (Knoll, 1984). The result, by the Late Permian, was dominance of the landscape by one major life history, the seed habit, which ushered in the pattern still prevalent today.

The lowlands were partitioned, in the broadest sense, between seed plants and lycopsids as the dominant tree groups. Seed plants occupied mainly the better-drained environments; lycopsids grew in swampy, semiflooded habitats. These lineages had very different life histories and morphologies, and the ecosystems they dominated had widely differing dynamics (Phillips and DiMichele, 1992). Today the seed plants, particularly flowering plants, dominate nearly all terrestrial ecosystems; despite the diversity of dynamics encompassed by extant plants and ecosystems, they do not include some of the structure and dynamical properties that were part of the Carboniferous ecological spectrum, the lycopsid-dominated swamps, for example.

Eco-Taxonomic Patterns

Peat Swamps

Peat-forming coal swamps had organic substrates, suggesting chelation of mineral nutrients, typically low pH, and periodic flooding. Such environments had high levels of abiotic stress, a result of the combination of low nutrient levels, anaerobic conditions within water-saturated rooting zones, water-table fluctuations, and regular disturbance. The high abiotic stress (sensu Grime, 1979; DiMichele *et al.*, 1987) appears to have selected strongly against most species, leading to a specialized flora. The tolerant species tended to be parts of peat swamp-centered evolutionary lineages, such as many of the lycopsid genera (DiMichele and Phillips, 1985), or specialized offshoots that became somewhat isolated from main line of the clade, such as cordaitean gymnosperms and some of the marattialean ferns (Rothwell and Warner, 1984; Costanza, 1985; Trivett and Rothwell, 1985; Lesnikowska, 1989).

Late Carboniferous peat swamps were not a single monotonous habitat, as depicted in most museum dioramas. Rather they encompassed a range of subhabitats controlled largely by flooding regime and influenced by disturbances such as incursions of mineral matter, brackish water, and wildfires. Some swamps or parts of swamps may have been domed (ombrotrophic), as in modern areas of Sumatra and Borneo (Cecil *et al.*, 1985; Esterle and Ferm, 1986; Kvale and Archer, 1990; Staub and Esterle, 1992; Eble and Grady, 1993). Others were clearly planar (rheotrophic) swamps, subject to flooding and introduction of clastics in floodwaters (Eggert, 1982; Grady and Eble, 1990; Calder, 1993). Both kinds of swamp, which may have coexisted within a single peat body, were sufficiently variable in physical conditions to support a diversity of habitats and the species specific to them.

Knowledge of the plant composition and paleoecology of peat swamps is provided by coal balls and miospore studies. Coal balls (Figure 8.1) are concretions of the original peat fabric petrified (permineralized) before the peat was coalified. They occur within coal seams, generally as calcium carbonate and occasionally as siliceous concretions. The plants in coal balls are anatomically preserved, often in exquisite detail, and range in size from tiny fragments to large axes. Coal balls occur in more than 65 upper Carboniferous coal seams, but may be confined largely to planar peats (Cecil et al., 1985).

Miospores are approximate indicators of the flora of coal forests. They can be recovered from most coals that have not been subject to intense metamorphosis because of the decay-resistant chemical composition of their outer walls. Correlation of spores with source plants (e.g., Courvoisier and Phillips, 1975; Willard, 1989) provides a baseline for inferring paleoecological patterns from palynological data. In this chapter we focus on the macrofossil (coal-ball) data base we have constructed, although palynological data will be referenced where appropriate.

Clastic Wetlands

The wetlands surrounding peat swamps were a typically diverse array of lowland settings including marshes, clastic (muddy) swamps, levees, splays, and channel margins, all part of larger flood basins. Most wetlands preserved in the fossil record were periodically flooded or had wet substrates (Scott, 1978; Gastaldo, 1987). The dominant plants included pteridosperms (seed ferns), ferns, and other groups that were locally important, such as sphenopsids (horsetails) or lycopsids. Such lowland areas encompassed a much wider array of habitats than did coal swamps. Consequently, there was much more potential for species migrations, habitat gradients, and generally broadly transitional boundaries (ecotones) between habitats, resulting in considerable floristic variation on basic themes.

Among the wetland habitats were swamps with large clastic influxes. These are often preserved as dark, organic shales. Many such swamps were lycopsid-dominated, like the peat swamps, and provide us with unusual information on spatial relationships of the plants, tree sizes, and gross morphology (Wnuk, 1985; DiMichele and Gastaldo, 1986; DeMaris, 1987; Wnuk and Pfefferkorn, 1987), important in the estimation of areal variability of peat-swamp assemblages (e.g., DiMichele and Nelson, 1989). In other cases, pteridosperm dominance of some clastic swamps and wet mineral substrates occurred in the Late Devonian and throughout the Carboniferous.

Plants of the clastic wetlands are generally preserved as compressions and impressions in relatively inorganic sediments, such as shales and sandstones, which provide external morphological features. Coal balls, on the other hand, preserve anatomical structure. Consequently, there are constraints in comparing taxonomic identifications from the two fossil forms. To the extent that identifying characteristics can be compared, the plants of peat deposits, and to a lesser degree those of flooded mineral-rich substrates, mostly represent different species from those in

FIGURE 8.1 Coal balls *in situ*; Herrin (No.6) Coal, Illinois. Large, light-colored masses are calcium carbonate permineralized coal balls, within the darker coal.

the moist lowlands (DiMichele et al., 1986, 1991). This is a consequence of the edaphic barriers created by the physical character of swamps. Overlap is greatest among medullosan pteridosperms, many of which apparently colonized those parts of peat swamps subject to flood and fire disturbance, areas with the greatest similarity to parts of the clastic wetlands.

COAL SWAMPS AS HIERARCHICALLY ORGANIZED SYSTEMS

Coal swamps were an integral part of Carboniferous lowland environments and have become the epitome of the "coal age" as generally presented to nonspecialists. In fact, coal swamps represent a special subset of the total range of environments existing during that time. Although this subset is more diverse and heterogeneous than dioramas depict, it constitutes a distinctive basis for making ecological comparisons through Late Carboniferous time. Peat-forming environments are a recognizable environmental subset, providing a taphonomically comparable basis for pattern analysis. In addition, the system was semiclosed to species introduction because of the edaphic constraints imposed on plants by the usually saturated peat substrate. This limitation increases the resolution of ecological patterns through time. The effort of dozens of paleobotanists, for more than a century, has provided the taxonomic and morphological baseline for us to generate a large quantitative data base from approximately 40 coals in western Europe and the United States (Figure 8.2).

Organization of Coal-Swamp Ecosystems

The organization of coal-swamp ecosystems is complex, and can be examined at a variety of spatial and temporal scales (Figure 8.3). Each of these scales reveals different elements of the overall organizational hierarchy. In most instances the relationships of organizational patterns and dynamics at different levels are more clearly seen when a larger view of the system is taken.

Resolution at 10^0- to 10^4-yr Time Scales:
Habitats and Species Assemblages

Coal-ball occurrences vary from scattered specimens to distinct zones or massive aggregates, in situ within coal seams (Figure 8.1). Although many deposits are locally extensive, distinct zones of coal balls can be followed laterally no more than a few meters in most instances. Such "zones" of coal balls can be treated as the litter layer of a single forest stand, preserving a time period of <1 yr (instantaneous events such as fires preserved in fusain layers) to 100 yr. Study of the species composition and

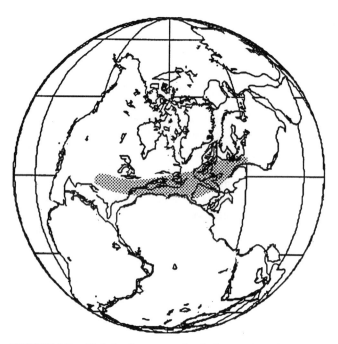

FIGURE 8.2 Global paleogeography during the Late Carboniferous. Coal balls come from the shaded area of the tropical Euramerican subcontinent. Map adapted from 1988 version of Terra Mobilis produced by C. R. Scotese and C. R. Denham.

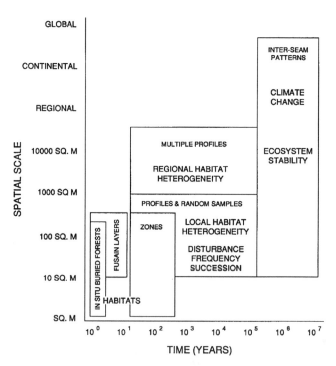

FIGURE 8.3 Spatial and temporal scales of resolution available from coal-ball data analysis. SQ.M.= square meters. Small capitals refer to fossil samples, large, bold capitals refer to inferences made from fossil samples.

relative species abundances in individual zones (Phillips et al., 1977; Phillips and DiMichele, 1981) is the basic unit for further quantitative studies. Inferences can be drawn regarding the vegetational structure of the forest from which the litter sample was drawn, and the dynamics of that forest can be deduced from analysis of the relative proportions, life histories, and habits of the plants.

Aspects of the physical habitat also can be determined or deduced for individual zones, revealing the conditions under which the parent forest grew. Physical evidence of mineral matter in coal, either through ash determinations (Eble and Grady, 1993) or petrography (Johnson, 1979; Esterle and Ferm, 1986; Grady and Eble, 1990), suggests susceptibility of the swamp to floods, that introduce clastics, or peat decay, which may concentrate mineral matter (Cecil et al., 1979; Ruppert et al., 1985). Relative abundances of mineral charcoal (fusain) indicate the presence of fires in the swamp and can point to particular plant groups or tissues that are most frequently preserved as charcoal, or to patterns of preservation that suggest fire intensity (charred outer rinds on tree bark) (Phillips and DiMichele, 1981; DiMichele and Phillips, 1988). The proximity of a coal-ball zone to the underclay, or to thick mineral partings in the coal bed, indicates significant changes in edaphic conditions to or from those characteristic of the peat substrate. In many cases, the mineral partings represent disturbances by clastic-laden floodwaters. Specific plant groups or assemblages have been found to have a disproportionate association with these indicators.

Taphonomic criteria, that is, the characteristics of the preservation of the coal-ball peat, also provide insights into the physical character of the habitat associated with a particular zonal assemblage (Scott and Rex, 1985; Raymond, 1987; Phillips and DiMichele, 1989). Peat exposure can be gauged from the degree to which the peat is rotted and the plant remains poorly preserved. Exposure is also indicated by detritivore activity, such as frass-filled borings in wood, periderm, or soft tissues, or extensive fecal pellet deposits left by surface feeders. The framework-to-matrix ratio of the peat is an indicator of the degree to which fine, particulate material has been flushed from the peat fabric by water through-flow, which may also have affected peat acidity.

Finally, habitat inferences can be drawn from the biotic character of the assemblage itself, through ecomorphic analysis. For example, the relative proportion of ground cover and free-sporing life histories (the "fern" life cycle) can suggest the frequency or length of surface water cover. Relative proportions of opportunistic/invasive life histories, in combination with high species richness and high ecomorphic diversity of an assemblage, can point to recent disturbance and subsequent colonization. The proportions of monocarpic (one reproduction at the end of the lifetime) versus polycarpic (multiple reproductions during lifetime) lycopsid life histories suggest the relative extent to which the environment was stable and "predictable."

These basic patterns are the fundamental building blocks of our higher-order inferences about coal-swamp structure and temporal dynamics. We extend them through time and space to construct a larger, more complex picture.

Resolution at the 10^3- to 10^5-yr Time Scale: Landscape Patterns

Coal-ball occurrences in multiple layers, or profiles, provide access to changes in vegetation, vegetational dynamics, and habitats through time. The time interval may vary from 10^3 to 10^5 yr, depending on how thick the coal seam is and how much of it was preserved as coal-ball peat.

Dominance and diversity patterns on a site through time can be quantified through profile analysis. In many cases, patterns of vegetational change are associated with physical markers that can be correlated between coal-ball masses within a single seam, such as clastic partings, fusain layers, or distinctive petrographic composition. The evidence we have accumulated to date suggests little directional succession at a whole-seam level, consistent with a planar-peat origin for most coal-ball bearing coals, and no discernible evolutionary change in the morphology of the component species within the time necessary to accumulate a 1- to 2-m thick coal seam.

Landscape patterns within a single coal swamp can be reconstructed from profile data, using single or multiple profiles, or from random sample data. Within any one profile the succession of zones can be taken to represent an atemporal record or sample of the kinds of habitats and assemblages that existed within the swamp. Ordination (see Digby and Kempton, 1987) of individual zones, or of individual coal balls in the case of random samples, provides a base map of landscape-level variability within the coal, including the relative proportions of habitats, dominance-diversity patterns, and the relative similarity of assemblages from different subenvironments. The patterns at the landscape level, including the relative proportions of habitats, plant assemblages, life histories, and community dynamics, are characteristic of a coal or group of coals. These provide a "fingerprint" of a particular coal swamp, and more broadly of a time and geographic region.

Resolution at the 10^5- to 10^7-yr Time Scale: Interseam Patterns

It is on the longest time scales that evolutionary change can be detected, and on which the effects of climatic or other extrinsic factors on swamp community dynamics and structure are seen most clearly. Long-term trends in landscape, habitat, and species composition must be examined among coals. Each can be evaluated independently, or the relationships among them can be investigated. For example: Is the timing of change similar at all levels of observation? Are there relationships among levels of organization that are detectable only on the longer time scales? Do events or changes at one level appear to dictate or constrain those at other levels?

Climatic change is often brought about by geological phenomena that develop over millions of years, such as the movement of continental plates, the elevation of mountain ranges, or the waxing and waning of ice sheets and associated eustatic sea-level fluctuations. It is extremely difficult to correlate such events, which have their own considerable margins of error, with specific changes in community composition. Broadly based correlations can be developed, however, and there is great likelihood that widespread biotic changes are the consequence of events of global or at least regional scale. Thus, the long-term analysis of swamp patterns is a scale at which such phenomena can be examined. Certainly, glacial eustatic and climatic effects occur at time scales approximately the same as those over which we believe coal seams accumulated. However, the peat accumulation may have required a relatively narrow range of conditions, such that coal seams occur between the "events," which may mark their beginnings and ends. Consequently, it is only on the interseam level of observation that we can make a case for the climatic changes that affected the dynamics observable at lower levels.

Coal-Swamp Species and Ecomorphs

Well over 100 whole-plant species have been identified in coal balls from the upper Carboniferous of Europe and United States, across a time span of 12 to 15 m.y. Most of these can be assigned broadly to several ecomorphic groups, based on their habit, reproductive morphology, and life history attributes (Figure 8.4). The ecomorphic groups, particularly trees, conform fairly strictly to taxonomic groups or clades, that is, to groups of closely related taxa, which is the basic pattern of the Carboniferous. Much of what we know of evolution based on studies of modern plants and animals suggests that this should not be the case; ecological overlap between groups of plants with different ancestries is the rule in modern forests. It was not an expected result when we began to look for ecomorphic patterns among coal-swamp plants. The explanation for this pattern may lie partly in the ecological structure of the Carboniferous world, including the origin of that pattern.

Habits of the major ecomorphic tree groups are illustrated in Figure 8.4. Ecomorphic determination is based mainly on critical life history attributes: timing of reproduction; approximate reproductive output; morphology of, and energy investment in, disseminules; and dispersibility of disseminules. Habit and vegetative morphology also are important: size at maturity; adult shape; growth habit

FIGURE 8.4 Major tree ecomorphs in Late Carboniferous coal swamps. *Left to right*: Mature monocarpic lycopsid about 20 m in height; juvenile lycopsid; polycarpic lycopsid (these came in a wide range of sizes); medullosan pteridosperm (seed fern); small *Psaronius* typical of Westphalian swamps; large *Psaronius* typical of the latest Westphalian and Stephanian; cordaite reconstructed as possible mangrove; scrambling cordaite; large arboreous cordaite; sphenopsids illustrating clonal habit.

(tree, vine, shrub, ground cover); degree of xeromorphy; and expense of construction. The latter two are comparative, and hence relative, depending on the spectrum of morphologies present in Carboniferous swamps.

Major Intraswamp Habitats

Habitats within a peat-forming swamp are all highly modified from extraswamp analogues by the presence of the predominantly organic substrate. The edaphic qualities of this substrate are variable, and these variations define many of the intraswamp habitats.

Flooded habitats are recognized mainly on biotic or ecomorphic criteria: low species richness, little ground cover, few free-sporing plants (which need an exposed substrate to complete their life cycle), and dominance by plants with specialized semiaquatic or flood-tolerant morphologies. Such habitats also are relatively low in charcoal. Lycopsids were the most common elements of these environments in the Westphalian.

Peat-to-clastic ecotonal habitats cover a wide range of conditions, recognizable mainly on physical criteria, although they encompass plants with corroborative life history strategies. There are several variants. Peat-to-mud transitional environments are associated with underclays, clastic partings, or usually high ash and mineral matter in coal; they are populated in the Westphalian by a small arboreous lycopsid (*Paralycopodites*), and often by medullosan seed ferns. Fire-prone habitats are associated with elevated levels of fusain and often with increased clastic material; the most common components of these environments are medullosan pteridosperms, sphenopsids, and in some cases, small, scrambling cordaitean gymnosperms. Habitats with long periods of exposure and presumed drying of the peat surface are characterized by heavily rotted and rerooted peats, often with evidence of extensive invertebrate burrowing; such environments are associated with larger cordaites, some sigillarian lycopsids, and medullosan pteridosperms.

Cryptic, or irregular disturbance, habitats are recognizable by physical and ecomorphic attributes. They generally have little fusain or mineral matter. Evidence from an unusual buried forest deposit (Wnuk and Pfefferkorn, 1987), drawn from species compositional similarities, suggests irregular floods. In some swamps, storms associated with the influx of marine waters may have been a major disturbance agent, suggested by multiple coal-ball and coal layers containing marine invertebrates. Such habitats are generally dominated by polycarpic, long-lived trees, lycopsids in the Westphalian, and possibly tree ferns in the Stephanian. Species richness is intermediate, growth architectures often are very variable among the subdominants, and a groundcover component is generally important.

PATTERNS OF CHANGE IN COAL-SWAMP COMMUNITIES DURING THE PENNSYLVANIAN PERIOD

Objectives

In this section we focus on the temporal patterns of change in coal swamps. This pattern is examined first at the landscape level—changes that are the easiest to describe and thus to relate to larger questions of environmental influence. The timing and extent of landscape-level changes is compared with patterns in habitat and species composition of successive swamps. We examine the relative timing and extent of change in these elements, and the relationship between species turnover and habitat persistence. We then argue that these relationships suggest a hierarchical structure in which biotic factors influence patterns of species replacement.

Changes at the Landscape Level

Change in the relative abundance of the major plant groups comprising swamp communities is the major, and simplest, indicator of community change and has been discussed elsewhere (Phillips and Peppers, 1984; Phillips *et al.*, 1985). Because the major plant groups (lycopsids, ferns, pteridosperms, sphenopsids, cordaites) are broadly distinct in habitat preference, changes in their relative abundances also reflect changes in the physical characteristics of swamps. Figure 8.5 summarizes the changes by geographic region, with the general pattern summarized in the right-hand column.

Biomass distribution is spread among enough major tree groups, and the patterns are sufficiently distinct, that some important generalities can be resolved at this level. We use mostly western European chronostratigraphic terminology because that of the United States varies widely among geographic regions.

1. Lycopsids, of several ecomorphic forms, dominate most coal swamps for the 9 m.y. (using the Hess and Lippolt, 1986, time scale) of the Westphalian (late early and middle Pennsylvanian). Major extinctions in North America occurred near the Westphalian-Stephanian (middle-upper Pennsylvanian) boundary, eliminating most of the lepidodendrids and removing the lycopsids from a position of ecological dominance in swamps (Phillips *et al.*, 1974).

2. Cordaitean gymnosperms are the major subdominants or dominants during the midportion of this time interval, from the Westphalian B to the early Westphalian D. Two distinct phases are represented. During the Westphalian B and C, cordaitean taxa that produced *Mitrospermum*-type seeds (ovules) were the most common

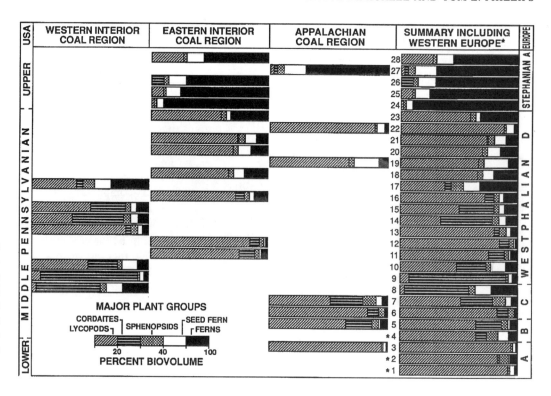

FIGURE 8.5 Patterns of change in the abundance of the major plant groups in 28 selected upper Carboniferous coals. Summary at right includes coals from western Europe (labeled with an asterisk).

forms. During the late Westphalian C and early Westphalian D the most abundant cordaites produced *Cardiocarpus*-type seeds; these may have had a more shrubby or scrambling habit than earlier coal-swamp forms (Cridland, 1964; Costanza, 1985).

3. Tree ferns replace lycopsids as the ecological dominants of the Stephanian (Late Pennsylvanian). Tree fern abundances begin to rise during the Westphalian C, and especially during the Westphalian D. However, systematic studies suggest that largely different suites of *Psaronius* species and ecomorphs occupied Westphalian and Stephanian swamps (Lesnikowska, 1989).

4. Pteridosperms (seed ferns) fluctuate in abundance during the Westphalian, but attain uniformly higher and less variable abundances in the Stephanian. Studies of seed morphology (Taylor, 1965) suggest substantial species turnover at the Westphalian-Stephanian boundary.

5. Sphenopsids are present in coal swamps throughout the Late Carboniferous, but generally at low levels. Palynology suggests local times and regions of abundance (Peppers, 1979).

The existence of breakpoints, marked by minor to major extinctions of swamp-centered lineages, is one of the most conspicuous aspects of this data summary. These breakpoints also are associated with landscape-level changes in dominance-diversity patterns (e.g., the rise and decline of cordaites, and the dramatic expansion of tree ferns).

Palynological analyses (Peppers, 1979; Phillips et al., 1985) reveal the same basic patterns as the coarser megafossil data and point particularly to the Westphalian A-B boundary and the Westphalian-Stephanian boundary as times of major species turnover (Peppers, 1979; Phillips et al., 1974). The patterns at this level led us to suggest a typological classification of coal swamps some time ago (DiMichele and Phillips, 1981). This classification was based on the premise that there were periods of little change in dominance and diversity from one coal swamp to the next in stratigraphic sequence—periods of persistence—punctuated by much shorter time intervals of abrupt vegetational change. Our original formulation was based on a smaller data base, but the basic pattern has remained as further data have accrued.

In order to evaluate this pattern more formally we have used some statistical methods to examine the data for floristic breakpoints (Figure 8.6). Jaccard coefficients and sign tests were used to assess similarity based on species presence or absence; through inspection we examined the data for points of low similarity. Although these are weak tests, used on the weakest manifestation of the data (presence-absence of species), they strongly corroborate the breakpoints at the Westphalian A-B boundary and at the Westphalian/Stephanian boundary. They support less strongly the presence of breakpoints at other boundaries where proportions of dominant tree taxa shift but are associated with few extinctions.

RESPONSE OF ECOSYSTEMS TO LONG-TERM CLIMATIC CHANGE

		COAL BED	SIMILARITY (JACCARD)	SIGN TEST SIGNIFICANCE
PENNSYLVANIAN	STEPHANIAN	CALHOUN		
		DUQUESNE	.72	
		FRIENDSVILLE	.83	
		BRISTOL HILL	.78	.03
		----------	.50	.02
	WESTPHALIAN	BAKER		
	D	L. FREEPORT	.84	.02
		HERRIN	.96	
		M. KITTANING	.82	
		SPRINGFIELD	.92	
		IRON POST	.88	
		SUMMUM	.94	
		----------	.92	
		BEVIER	.94	
		FLEMING	.81	
		SECOR	.86	

		MURPHYSBORO	.91	
		BUFFALOVILLE	.98	
		URBANDALE		
		----------	.65	.01
	C	ROCK SPRINGS	.58	.05

		HAMLIN	.92	
	B	PATHFORK		
		----------	.50	
	A	BOUXHARMONT	.68	.02
		UNION		

FIGURE 8.6 Quantitative assessment of landscape-level patterns of change through the Westphalian and Stephanian. Dotted lines mark breakpoints in coal bed average vegetational composition; breakpoints were determined by inspection and statistical analysis. Jaccard similarity was calculated for adjacent coal. Sign tests are listed where significant differences were found between adjacent coals.

A more intuitive, or subtle, way to analyze landscape-level vegetational patterns is to use ordinations of zones or coals balls to examine the structure within individual coals. Patterns within successive coals then can be compared. Ordinations cluster the stands, be those stands individual coal balls or zones of coal balls from profiles, based on their quantitative taxonomic similarity. In two or three dimensions they serve as base maps on which additional information can be superimposed: for example, life history occurrences, diversity, physical attributes associated with assemblages, and ecomorphic characteristics. Ordinations thus serve as a transition between landscape-level and habitat-level patterns in that they illustrate the biotic and physical variability of the landscape in detail.

Because of the nature of this review, only selected coals can be illustrated to show the basic patterns of vegetational, landscape-level changes through time (Figure 8.7). The nature of landscape changes detected in ordination again corroborates the pattern detected at the level of average coal seam composition: change in swamp organization at the Westphalian A-B boundary, minor changes between the Westphalian B and early Westphalian D, substantial restructuring during the early Westphalian D, and total overhaul of swamp community organization near the Westphalian-Stephanian (middle-upper Pennsylvanian) boundary.

Changes in the Habitat Composition of Landscapes

Ordination patterns reveal how the species assemblages of landscapes changed during the Late Carboniferous. During this time there is a close correlation between physical attributes of swamps and species assemblages, particularly in the Westphalian. Because physical attributes of swamps, such as fusain or mineral matter, can be mapped onto ordinations, as can taphonomic, life history, and structural patterns, it is possible to translate ordinations based on taxa into maps of habitat diversity. This permits us to infer patterns of habitat change through time.

Westphalian A swamps had complex habitat organization, harboring a variety of taxa with different physical preferences. The Union Seam of England (Figure 8.7) had four basic assemblages—each, we believe, characteristic of a distinct subset of swamp environments: assemblages dominated by monocarpic lycopsids (*Lepidophloios harcourtii* or *Lepidodendron hickii*), growing in areas with long periods of standing water sufficient to reduce the abundance and diversity of ground cover and free-sporing plants; ecotonal assemblages (*Paralycopodites* and medullosan seed ferns) enriched in fusain, and ecomorphically distinct (no data exist on their position within the seam); cryptic disturbance habitats dominated by polycarpic lycopsids (*Diaphorodendron vasculare* and *Sigillaria* spp.), with a diversity of growth architectures, including abundant ground cover; *Lyginopteris* assemblages for which the physical attributes are unclear, but which overlap to some extent with *Diaphorodendron vasculare*-dominated assemblages and may have been part of the broader cryptic disturbance set of environments. Other Namurian and Westphalian A swamps offer variations on these themes (Holmes and Fairon-Demaret, 1984; Bertram, 1989).

The transition from the Westphalian A to the Westphalian B was marked by changes in the dominance-diversity composition of swamps. The patterns that appeared persisted through the Westphalian B and C, and into the early

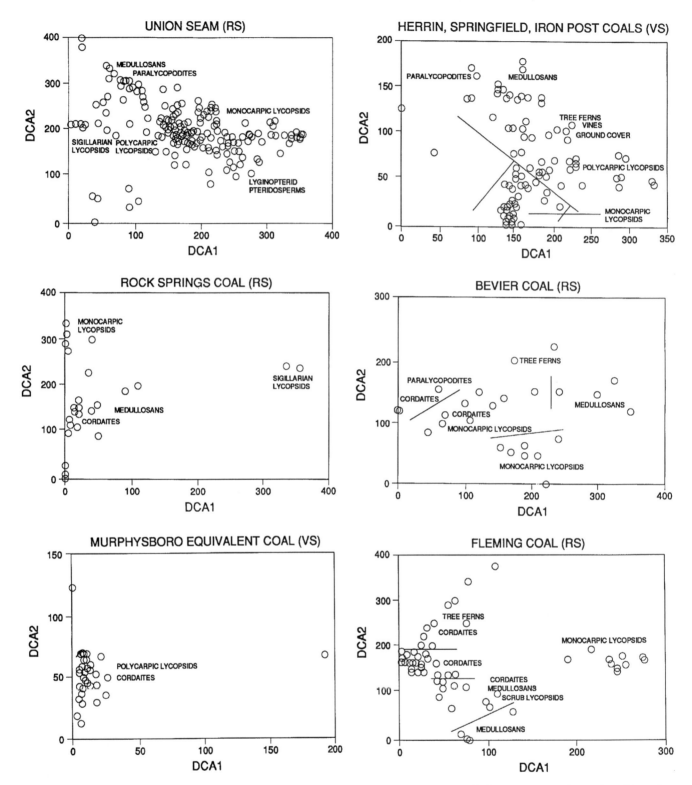

FIGURE 8.7 Detrended correspondence analysis ordinations of six Westphalian swamps. Oldest to youngest in stratigraphic order: Union seam (Westphalian A); Rock Springs coal (Westphalian C); Murphysboro equivalent coal (early Westphalian D); Fleming coal (early Westphalian D); Bevier coal (early Westphalian D); combined ordination of Iron Post, Springfield, and Herrin coals (late Westphalian D). VS = based on vertical profiles of coal balls; each point represents one zone from a profile. RS = based on random samples of profiles; each point represents one coal ball.

Westphalian D. The most notable changes were the extinction of *Lyginopteris* and the increase in the abundance of cordaitean gymnosperms. The cause of the extinction of *Lyginopteris* is not clear; the genus disappeared both in coal swamps and in surrounding clastic lowland settings of Euramerica.

Post-Westphalian A swamps not only lacked *Lyginopteris* but differed in other major aspects from swamps of the Westphalian A. In general, there was a decline in the habitat diversity of any one coal swamp, with the habitats present representing a subset of those in the Westphalian A, complemented by the addition of cordaitean-dominated assemblages. Cordaites appeared initially in ecotonal assemblages characterized by heavily rotted peats and abundant fusain, sometimes associated with medullosans, ultimately becoming a locally dominant component of some swamps (Phillips et al., 1985). Later in the interval, in the late Westphalian C and early Westphalian D, opportunistic cordaites and tree ferns began to expand in importance. The tree ferns transcended habitats, appearing in ecotonal assemblages in numbers following one of the minor breakpoints, becoming interstitial opportunists by the early Westphalian D, and eventually occurring in all but persistently flooded assemblages during the Westphalian.

Despite the breakup of Westphalian A patterns of landscape organization, the basic habitats of the earlier time are recognizable in Westphalian B to early Westphalian D swamps. The persistent habitats, recognizable by the physical, taphonomic, and structural attributes (as discussed earlier), retained their ecomorphic character, although the component species were largely different. Any one of the several possible habitats may have been either absent from a given coal, or present in such low frequency that it was not sampled, including those characterized by cordaitean dominance. Various combination of habitats thus appear throughout the interval (Figure 8.8).

Late Westphalian D coal swamps represent a return to the basic habitat organization characteristic of the Westphalian A: flooded habitats dominated by monocarpic lycopsids, ecotonal habitats dominated by the lycopsid *Paralycopodites* and medullosans, and cryptic disturbance habitats dominated by polycarpic lycopsids with structurally complex vegetation. Cordaiteans disappeared as an ecologically significant component, associated with the extinctions of several numerically important species. The reassembly of the Westphalian A type of swamp habitat organization is modified by the persistence and further expansion of tree ferns as interstitial opportunists. Identifiably opportunistic species equivalent to the tree ferns were rare in Westphalian A swamps. *Psaronius* expansion in coal swamps correlates with a similar expansion in lowland wetlands in general (Pfefferkorn and Thomson, 1982).

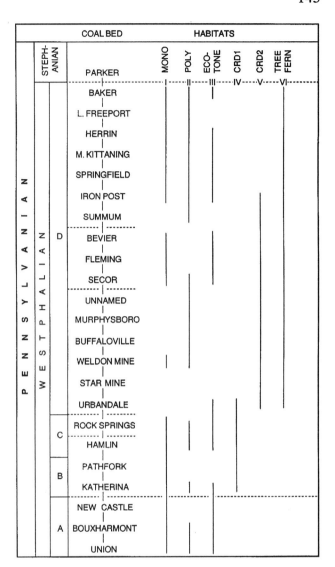

FIGURE 8.8 Patterns of change in physically determined habitats through time. Habitat characteristics discussed in text. Dotted lines are breakpoints determined by inspection and statistical analysis. MONO = habitats dominated by monocarpic lycopsids; POLY = habitats dominated by polycarpic lycopsids; ECOTONE = habitats dominated by medullosans and/or *Paralycopodites*; CRD1/CRD2 = variants on ecotonal habitats enriched in or dominated by cordaitean gymnosperms; TREE FERN = designates the occurrence of significant amounts of tree ferns as interstitial elements in other associations.

The habitats of the late Westphalian D each have an ecomorphic character that had persisted throughout the 9 m.y. of the Westphalian. This character persisted through landscape-level breakpoint boundaries, despite species-level changes within the habitats. In relatively few instances, the same species can be identified within these habitats over the entire 9-m.y. period. Species replacements within habitats were largely on ecomorphic themes, especially for trees. The pattern suggests that landscape-level patterns of persistence and change are driven fundamentally by changes in the proportions of the basic habitats represented, not by species-level turnover, which occurred largely *within* habitats during the Westphalian.

Major extinctions of coal-swamp plants occurred during the Westphalian-Stephanian transition. These were first recognized in miospore studies as the loss of the major *Lycospora*-producing lepidodendrids (Phillips et al., 1974). Recent systematic studies of the tree ferns (Lesnikowska, 1989) revealed a nearly complete extinction of species in this group in North America at the end of the Westphalian; Stephanian swamps presumably were recolonized from surrounding lowlands, where major extinctions probably also occurred, given that there is limited overlap between Westphalian and Stephanian species (e.g., Corsin, 1951; Remy and Remy, 1977). Examination of the literature on pteridosperms (e.g., Taylor, 1965; Rothwell, 1981; Pigg, 1987) and small ferns (e.g., Phillips, 1974) suggests substantial species level turnover in the coal-swamp members of these groups as well.

In North America this was the first Pennsylvanian extinction to eliminate whole ecomorphic groups of trees entirely. It was these trees that defined the biotic or ecomorphic aspect of most intraswamp habitats. They also defined the biotic limits to habitat space, that is, the nature of the biotic partitioning of the habitat resources. Thus, with this extinction, the fabric of Westphalian swamp communities—the persistent ecomorphic organization of several specific physical habitats—was destroyed.

Stephanian swamps were reorganized into a different set of norms of reaction between the plants and the physical environment. We do not fully understand these at present, in part because vertical profile analysis of Stephanian coal swamps has begun only recently. The swamps did become heavily tree fern dominated and thus generally more like the surrounding clastic lowlands. Tree fern-dominated areas may have had a great deal of local spatial heterogeneity, with cordaites, sphenopsids, and pteridosperms liberally intermixed (Willard and Phillips, 1993). The lycopsids *Sigillaria*, a tree, and *Chaloneria*, a robust herb, were locally abundant in parts of many of these swamps (DiMichele et al., 1979). The degree to which any of the Stephanian habitats overlap with those of the Westphalian has yet to be determined; however, preliminary evidence suggests different patterns of resource partitioning.

Changes in the Species-Level Composition of Habitats

The species composition of coal swamps underwent considerable turnover during the Late Carboniferous. In analyzing the distribution and extent of that turnover through time, our objective was to relate it to changes in structure at the other levels of organization and to evaluate the locus of change with respect to habitats. In this analysis stratigraphic range data are used, which we believe is justified because most species are swamp-centered and therefore had potential access to peat swamp ecosystems throughout their existence. However, our sampling may miss some of the smaller or rarer taxa. There are numerous problems with these data: inadequate sampling, inadequate taxonomy for many groups, low sampling density in the Westphalian B and C, and regional patterns that obscure the overall pattern. Nonetheless, the basic message is consistent with the pattern seen at the landscape and habitat levels. We expect to be able to refine the data as our studies proceed.

The data are drawn from our own profile and random sample analyses in order to avoid problems of inflation in the numbers of taxa known from particularly well studied coals, and to provide a uniform basis in taxonomic usage and sampling design. Thus, even though a species is known from a particular coal, if it did not appear in profile or random sample studies it was listed as absent in the raw data compilation; a species will appear as present in the data used for the turnover calculation, however, if the coal from which it is known falls between first and last occurrences in our samples. Species turnover was calculated as

$$\frac{O_{t_{0+1}} + E_{t_0}}{\frac{1}{2}(n_{t_0} + n_{t_{0+1}})}$$

Where E_{t_0} = species disappearing after a coal at time t_0; $O_{t_{0+1}}$ = species first appearing in the next coal at time t_{0+1}; n_{t_0} = number of species at time 0; and $n_{t_{0+1}}$ = number of species in next coal in sequence.

Turnover pattern is illustrated in Figure 8.9, with landscape-level breakpoint boundaries shown as dotted lines. Note that the highest turnovers occur around the major breakpoints, the Westphalian A-B boundary and the Westphalian-Stephanian boundary. Turnover within any one of the five major interbreakpoint intervals is much lower. The Westphalian D has notably higher diversity than the Westphalian A-C or the Stephanian.

Analysis of species replacements by habitat suggests that species replacement patterns are not random across the landscape during the Westphalian. Species with similar ecomorphic characteristics may have complementary stratigraphic distributions, or newly appearing species may replace older ones as ecological dominants. These species tend to replace one another within a given type of biotic assemblage and habitat. As a result the ecomorphic character of the basic habitats is maintained throughout the Westphalian and, presumably, will prove to be maintained on a different set of themes during the Stephanian. Al-

		COAL BED	SIMILARITY (JACCARD)	FIRST	LAST	TURNOVER %	SPECIES
PENNSYLVANIAN	STEPHANIAN	CALHOUN		3	-		31
			.72			14.9	
		DUQUESNE		5	7		36
			.83			7.6	
		FRIENDSVILLE		5	0		30
			.78			12.3	
		BRISTOL HILL		5	2		27
			.50			31.8	
	WESTPHALIAN	BAKER		0	17		39
			.84			8.2	
	D	L. FREEPORT		0	7		46
			.96			2.1	
		HERRIN		7	2		48
			.82			9.9	
		M. KITTANING		0	2		43
			.92			5.4	
		SPRINGFIELD		3	4		47
			.88			5.4	
		IRON POST		1	2		45
			.94			3.3	
		SUMMUM		2	2		46
			.92			4.4	
		BEVIER		2	2		46
			.94			3.3	
		FLEMING		3	1		45
			.81			10.6	
		SECOR		5	7		49
			.86			6.3	
		MURPHYSBORO		4	1		46
			.91			4.6	
		BUFFALOVILLE		1	0		42
			.98			1.2	
		URBANDALE		13	0		41
			.65			21.1	
	C	ROCK SPRINGS		11	2		30
			.58			26.9	
		HAMLIN		0	3		22
			.92			4.4	
	B	PATHFORK		6	2		24
			.50			33.4	
	A	BOUXHARMONT		9	12		30
			.68			17.3	
		UNION		-	1		22

FIGURE 8.9 Patterns of turnover of species between coal swamps; first occurrences and last occurrences of species, and total species refer to individual coals. Similarity (Jaccard) and turnover percentage are comparisons between successive coals.

though species turnover tends to occur in greatest numbers at the breakpoints, where landscape patterns also change, the replacements that occur at these times are on ecomorphic themes, except at the Westphalian-Stephanian boundary.

At the Westphalian-Stephanian boundary the basic pattern changes. Many of the dominant tree ecomorphs are eliminated in the swamp extinctions. As a consequence, the dynamics of species interactions alone appear to dictate the initial patterns of swamp reorganization in the early Stephanian. Great variation in the suite of dominant species is detected palynologically (Phillips *et al.*, 1974, 1985) among the coals that occur successively after the extinction. Ultimately, the system appears to reequilibrate, with fern dominance and pteridosperm subdominance becoming the rule for the remainder of the Stephanian. These patterns suggest that following the disruption of Westphalian ecosystem structure, coal-swamp dynamics became more stochastic for a relatively short period, until a new system of interactions was established.

CLIMATE CHANGE AND CAUSATION

Evidence for Climatic Variability

The morphological characteristics of tropical Euramerican floras indicate a humid, warm climate, lacking seasonality for much of the time: large, evergreen leaves; trunks and leaves with only minor armoring; woods lacking growth rings; trees and shrubs with structure suggesting rapid growth. The characteristics of Late Carboniferous plants are magnified as climate indicators by plants of the younger Permian Period in Euramerica, many of which are much more xeromorphic in character and suggest growth under periodic moisture limitation. One of the most extreme examples of xeromorphy in the Permian is the Hermit Shale flora of Arizona (White, 1929), which contains only seed plants, many of which are armored with spines and have thick, highly sclerotic leaves; the flora is associated with clear sedimentological indicators of seasonal rainfall, and may have inhabited a dry coastal area on the margin of the craton (Stevens and Stone, 1988). Such botanical evidence has led paleobotanists (e.g., White, 1933) to suggest climatic factors as the driving force behind floristic change.

Over the past ten years there has been a reawakening of interest in late Paleozoic climate prompted by studies of paleogeography (Scotese et al., 1979; Ziegler et al., 1981), and the resulting models of climatic dynamics (Parrish, 1982; Parrish et al., 1989). Most of the focus has been on climatic fluctuations during the time of Pennsylvanian coal formation. The most explicit of these have focused on the availability of moisture in the lowland tropics. Inferences have been based on the stratigraphic patterns of several indicators: coal resource abundance (Phillips and Peppers, 1984), coal sulfur and ash (Cecil et al., 1985), coal underclay mineralogy (Dulong and Cecil, 1989), chemical characteristics of rocks associated with coal-bearing strata (Cecil et al., 1985; Cecil, 1990), and the abundances of environmentally sensitive fossil plants in coal (Winston and Stanton, 1989).

Additional evidence for climatic changes during the Pennsylvanian and Permian comes from smaller-scale studies of rocks with depositional histories indicating alternation of wet and dry conditions on a regional level. Examples are underclay mineralogies indicative of long periods of subaerial exposure within coal-bearing sequences (Prather, 1985; Spears and Sezgin, 1985), complex paleosols indicating marine regression and increasingly drier climate (Prather, 1985; Goebel et al., 1989), alternating wetter and drier intervals in ancient dune deposits (Driese, 1985), and mixed sequences including both fluvial and eolian deposits (Johnson, 1987, 1989a,b). Deposits that indicate alternation of wet and dry conditions are generally attributed to changes in base level and associated changes in regional rainfall patterns. Eustasy, linked to South Polar glaciation, has been cited in most instances as the proximal cause of climatic variations (e.g., Wanless and Shepard, 1936; Heckel, 1986, 1989; Rust et al., 1987; Rust and Gibling, 1990; and virtually all of the above citations in this paragraph), and there is considerable evidence of Gondwanan glaciation in the late Paleozoic (Veevers and Powell, 1987). Regional and global climate may have been affected by a variety of additional factors related to regional tectonics. The effects of plate collisions on crustal deformation (Klein and Willard, 1989), changes in circulation patterns associated with the uplift of mountain ranges (Parrish, 1982), and the movement of continents and climatic belts relative to each other (Ziegler, 1990) all complicate climatic patterns.

The principal difficulty with most climatic scenarios is correlation. It is extremely difficult, some would say impossible, to correlate identifiable glacial deposits with identifiable changes in base level half the world away. We sympathize with such concerns and recognize that local, autocyclic depositional factors will overprint and often confound climatic interpretation; how does one differentiate dry climate from locally well-drained conditions based on the limited exposures usually available? Nonetheless, it is clear that one cannot turn to a "default" climate. It is equally clear that climate has a major, if not *the* major, impact on many aspects of sedimentation and the distribution of biotas. Ziegler (1990) demonstrates this well in his summary of an enormous amount of physical and paleontological data on Permian climatic and biogeographic patterns. He notes that sharp biogeographic boundaries, often attributed by paleontologists to physiographic barriers, can be the result of subtle climatic variability, a phenomenon well marked in the modern world and in the postglacial (Holocene) migrational patterns of plants (e.g., Delcourt and Delcourt, 1987).

Relationships of Climatic Patterns to Vegetational Patterns

The landscape-level breakpoint boundaries detected in Late Carboniferous coal swamps correspond closely to times of inferred climatic change during the Westphalian and Stephanian (DiMichele et al., 1986). In eastern North America, comparison of three Pennsylvanian climatic scenarios with points of vegetational reorganization is illustrated in Figure 8.10. The correspondence is remarkably close, and the data bases are independent, suggesting a correlation-causation relationship. The differences in the inferred climatic patterns are not as important as the points at which departures from a norm are detected. It is these departures, or major inflections in the climate curves, that appear to dictate the times of vegetational change.

The vegetational changes at the landscape and species levels appear to scale approximately to the magnitude of inferred climatic variability. The largest inferred climatic changes were near the Westphalian A-B boundary and the Westphalian-Stephanian boundary, also the times of the greatest vegetational changes. However, habitat-level changes offer a different perspective. Habitats persist through several climatic excursions during the Westphalian, retaining ecomorphic attributes and basic generic composition. It is only during the major climatic changes at the end of the Westphalian that habitat structure crumbles. This suggests that there are aspects of the structure of at least *some* ecosystems that do not follow climatic patterns in a linear manner. Rather, thresholds may exist, and once exceeded, a breakdown in organization or a reduction in the number of hierarchical levels within the system may occur rapidly.

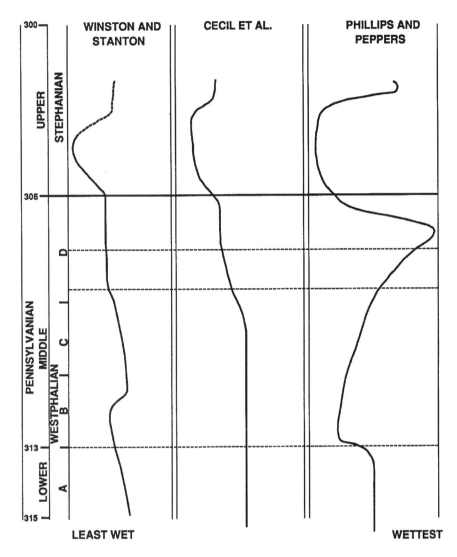

FIGURE 8.10 Comparison of three climate curve diagrams. The Winston and Stanton (1989) curve is based on abundances in Appalachian basin coals of plants thought to be climatically diagnostic; the Cecil *et al.* (1985) curve is based on geochemical characteristics of coals and associated rocks in the Appalachian basin; and the Phillips and Peppers (1984) curve is a smoothed logarithmic curve plotting the distribution of coal resources in North America.

Examination of compression-impression floras from the Pennsylvanian and Permian suggests a similar pattern. There are no climate curves for the Stephanian and Early Permian. However, analyses of Pfefferkorn and Thomson (1982), Ziegler (1990), and DiMichele and Aronson (1992) indicate discontinuous vegetational changes, ultimately resulting in a tropical vegetation with highly xeromorphic aspect. It appears that lowland-wetland vegetation persisted as a unit into the Early Permian and was replaced rather than displaced by what came to be the "Mesophytic" flora (Knoll, 1984; DiMichele and Aronson, 1992). The evidence of increasingly seasonally dry climate into the Early Permian is consistent with both the vegetational and the morphological patterns. The compression floras also suggest a level of organization above that of the landscapes observed in coal swamps, at which taxonomic composition is highly conserved. It may be at this level that the ultimate effect of evolutionary ancestry is manifested.

SUMMARY AND IMPLICATIONS

The relationships of turnover at the landscape, habitat and species levels in coal swamps are summarized in Figure 8.11. Turnovers in landscape organization—breakpoints—appear to be a consequence of changes in the proportions of several major types of intraswamp habitats. Turnover in the species composition of swamps correlates with turnover at the landscape level. Consequently, the reshuffling of habitat proportions may reflect periods of intraswamp disruption and dislocation, leading to increased, but usually not catastrophic, levels of species extinction. Species replacements across breakpoints are on strongly ecomorphic themes and, therefore, occur largely within the confines of habitats. For this reason, species turnover does not appear to be the major underlying mechanism of change at the landscape level. Intervals of very high extinction result in the loss of framework ecomorphic species. The result is the breakdown of intrahabitat spe-

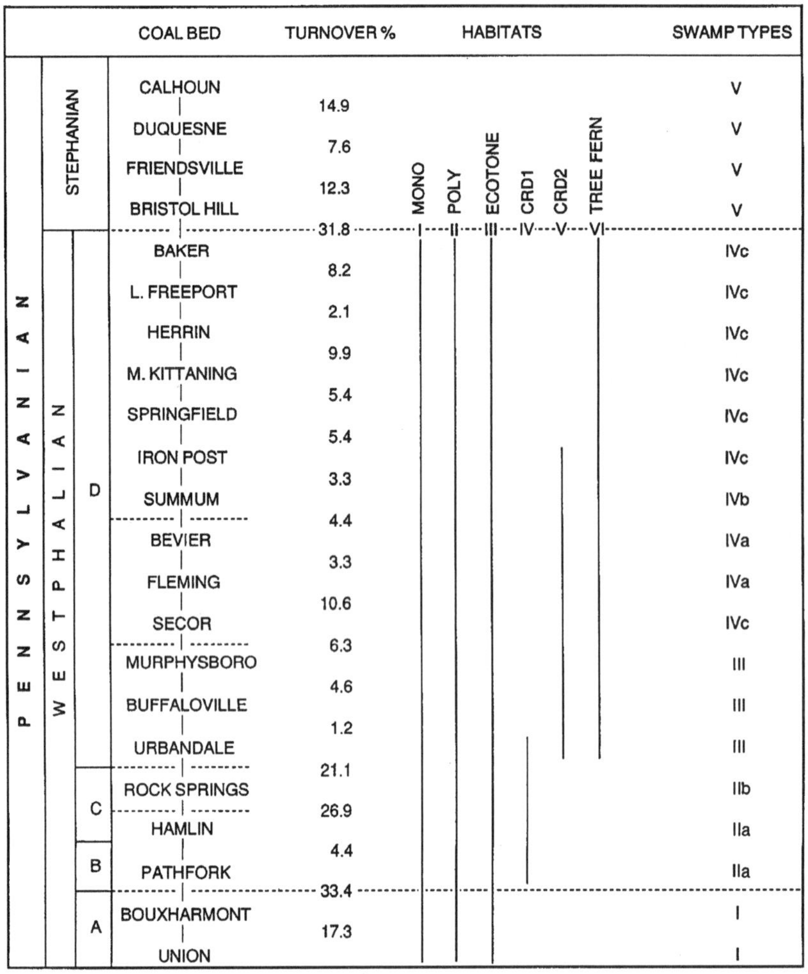

		COAL BED	TURNOVER %	HABITATS							SWAMP TYPES
				MONO	POLY	ECOTONE	CRD1	CRD2	TREE FERN		
	STEPHANIAN	CALHOUN	14.9								V
		DUQUESNE	7.6								V
		FRIENDSVILLE	12.3								V
		BRISTOL HILL									V
			31.8	I	II	III	IV	V	VI		
P E N N S Y L V A N I A N	W E S T P H A L I A N	BAKER	8.2								IVc
		L. FREEPORT	2.1								IVc
		HERRIN	9.9								IVc
	D	M. KITTANING	5.4								IVc
		SPRINGFIELD	5.4								IVc
		IRON POST	3.3								IVc
		SUMMUM	4.4								IVb
		BEVIER	3.3								IVa
		FLEMING	10.6								IVa
		SECOR	6.3								IVc
		MURPHYSBORO	4.6								III
		BUFFALOVILLE	1.2								III
		URBANDALE	21.1								III
	C	ROCK SPRINGS	26.9								IIb
		HAMLIN	4.4								IIa
	B	PATHFORK									IIa
			33.4								
	A	BOUXHARMONT	17.3								I
		UNION									I

FIGURE 8.11 Comparison of species-, habitat-, and landscape-level changes in coal swamps during the upper Carboniferous. Habitat distributions are listed in a "range-through" fashion, based on first and last occurrence. Swamp type is a typological characterization of average quantitative landscape composition on a coal bed average basis.

cies replacement patterns, followed by ecosystem reorganization along different lines. During the Pennsylvanian, such major periods of extinction were rare, permitting the biotic-abiotic linkage to persist for millions of years.

Hierarchical Organization

These patterns are consistent with an interpretation of coal-swamp plant communities as hierarchically organized. Species have characteristic ecological amplitudes, which are more likely to be shared with closely related than distantly related species, at least in the less diverse, preangiosperm world. In a given habitat, the entire plexus of species defines a biotic network of evolved interactions. Loss of a few species from this network (extinction or extirpation) can be accommodated because the system has sufficient biotic linkages to be buffered; this notion runs counter to the findings of ecosystem models, where the greater the number of linkages, the lower is the stability of the system. Released resources and severed patterns of interaction are most likely to be utilized by species with similar morphological attributes. Particularly during the Carboniferous, in which ecosystems are strongly partitioned taxonomically, this means by a species related closely to the earlier occupant, and with similar growth form and life history. Thus, the ecomorphic-biotic structure of a habitat, given sufficient time to evolve, may strongly constrain the nature and dynamics of species replacement.

Breakdown of the biotic habitat structure will occur during catastrophic extinction, as near the Westphalian-Stephanian boundary. The result is that the biotic fabric collapses and no longer can constrain the selection of replacement species. At such times the system may go into a lottery-like period of species interactions, perhaps controlled largely by interspecific competition (admittedly, this is almost impossible to document in the fossil record). The new system may reequilibrate and establish a new set of biotic limits to species replacement dynamics.

The coal-swamp floras of the Stephanian bear close resemblance to Westphalian floras of the clastic wetlands (e.g., Pfefferkorn and Thomson, 1982). This suggests that many of the species or clades in Stephanian coal swamps may have had a prior history of cohabitation in the tropical wetlands, which may have reduced the levels of interspecific competition necessary to establish a stable system of biotic interactions within swamps. The ultimate dominance of tree ferns, with subdominant pteridosperms, is a pattern shared by Stephanian coal swamps and the clastic wetlands. We do not know if the same species occupied both kinds of habitats, but it seems unlikely, because of the edaphic constraints imposed by peat substrates. If these plants were capable of tolerating disturbances, they may have established a very different set of equilibrium interactions than those of Westphalian swamps.

Ecosystem Persistence

In a system with organization like that of Pennsylvanian coal swamps, ecosystem persistence is a real property, empirically measurable. The biotic interactions within a system tend to conserve its ecomorphic structure and dynamics, even if there is species turnover. This generality may be true only during times when physical or extrinsic conditions remain relatively uniform for millions of years. It may also require sufficient time for a system of interactions among component species to evolve.

The concept of persistence may need to be evaluated at hierarchical levels above species composition. Although species associations can persist for millions of years, as our data suggest, these patterns may not reveal the major structural attributes of the ecosystem. To view all ecosystems as organized strictly as happenstance associations of species with similar resource requirements may reflect a bias that has grown out of studying short time intervals available to neoecology, complemented by the somewhat longer patterns of Recent palynology. The climatic fluctuations of the last 10,000 to 100,000 yr are not typical of all of Earth history (moreover, the Northern Hemisphere is not a good proxy for the tropics). The dogma of "individualism" should be reevaluated for generality.

Long-Term Species Replacement Dynamics: Evolutionary Implications

Species with opportunistic life histories may have a significant advantage during intervals of ecosystem disruption. During a postextinction lottery, high reproductive output, high dispersibility, tolerance of a diversity of physical conditions, ability to grow in disturbed areas, and a tendency to spawn peripheral isolates (due to dispersal capacities), all favor invasive opportunists. During the Late Carboniferous this is seen in the marked expansion of tree ferns and scrambling cordaites following breakpoint boundaries (Phillips and Peppers, 1984) and by the massive expansion of tree ferns following the terminal Westphalian extinctions. Tree ferns were cheaply constructed; had massive reproductive output, causing vast overrepresentation in the miospore record (Willard, 1993); produced small and widely dispersed isospores capable of founding a population from one spore; and underwent a major radiation in coal swamps and clastic lowlands during the late Westphalian and Stephanian. Similar patterns have been detected at the Cretaceous-Tertiary boundary, the now famous "fern spike" (Tschudy *et al.*, 1984), and the great radiation of angiosperms in the early Tertiary may have been made possible in part by their fundamentally opportunistic biologies during a period of ecosystem disruption (Wolfe and Upchurch, 1986; Wing and Tiffney, 1987).

A secondary dynamic to result from this process is the evolution of larger, longer-lived, presumably more competitive species from opportunistic ancestors. It is the larger forms that ultimately dominate subsequent ecosystems. The tree ferns of the Stephanian again exhibit this clearly, and hints of similar patterns are found among the pteridosperms, lycopsids, and sphenopsids. Most Westphalian marattialeans of coal swamps were not truly large trees; Lesnikowska (1989) has reconstructed scramblers and small trees in which virtually all measurable indicators of size are smaller than descendant forms in the Stephanian, in some instances up to an order of magnitude smaller. Gigantism in Stephanian coal-swamp plants has been noted for sigillarians, which our studies suggest were much larger than intraswamp sigillarians of the Westphalian. Galtier and Phillips (1985) and Willard and Phillips (1993) note record large sizes for medullosan and sphenopsid stems in Stephanian peat deposits. Large sizes point to longer-lived plants in Stephanian coal swamps, and taxonomy points to a new suite of species. Together, these observations suggest that Stephanian descendant forms, which had become ecosystem dominants, were growing in more long-persistent habitats than were Westphalian forms, which had been largely subdominant to lycopsids or ecologically restricted by them during the Westphalian.

The fabric created by the biotic interactions among organisms plays a large role in dictating evolutionary dynamics. Different kinds of opportunities for establishment of divergent phenotypes appear to exist during times of environmental stability than during times of instability and disruption. This is strongly suggested by species turnover, both its magnitude and the ecomorphic nature of species replacement. We suggest that during periods of environmental stability the targets of opportunity for divergent forms are strictly defined by both biotic and abiotic factors. Occupied niches are for the most part not available to new phenotypes, creating a very fine-meshed selective filter. In contrast, during times of environmental change, and some degree of consequent disruption of the ecological status quo, the role of the existing biotic structure is reduced. This may be a threshold response (i.e., the biotic fabric exists or it does not). If the fabric is severely disrupted, as evidently happened during the early Stephanian within peat swamps, the selective filter becomes coarse meshed as numerous opportunities are created for the survival of divergent forms. It is during such times that the role of interspecific competition would be expected to be most powerful as an active agent of selection.

This scenario integrates environmental change with the evolving lineages. It is in effect the model presented by Valentine (1980) for the origin of putative higher taxa. Higher taxa may certainly have a greater chance of survival during times of ecosystem disruption than during periods of long-term stability; however, the model applies to speciation in general. The patterns of the fossil record may require little more than an understanding that ecosystems have not been in constant flux for the past 600 m.y. Recognition that *opportunities* for survival of divergent phenotypes have varied enormously through time may help reconcile mechanisms based on studies of living organisms with patterns in the fossil record.

REFERENCES

Bertram, U. (1989). Untersuchungun an coal balls aus dem Namur A von Ostrau Unter Spezieller Berucksichtigung der Gattungen *Heterangium*, *Lyginopteris*, und *Microspermopteris*, *Palaeontographica 214B*, 125-244.

Calder, J. (1993). The evolution of a ground water-influenced (Westphalian B) peat-forming ecosystem in a piedmont setting: The No. 3 seam, Springhill Coal Field, Cumberland Basin, Nova Scotia, *Geological Society of America, Special Paper*.

Cecil, C. B. (1990). Paleoclimate controls on stratigraphic repetition of chemical and siliciclastic rocks, *Geology 18*, 533-536.

Cecil, C. B., J. C. Renton, R. W. Stanton, and F. T. Dulong (1979). Some geologic factors controlling mineral matter in coal, in *Carboniferous Coal Short Course Guidebook*, A. C. Donaldson, M. W. Presley, and J. J. Renton, eds., West Virginia Geological and Economic Survey Bulletin B-37-1, pp. 224-239.

Cecil, C. B., R. W. Stanton, S. G. Neuzil, F. T. Dulong, L. F. Ruppert, and B. S. Pierce (1985). Paleoclimate controls on late Paleozoic sedimentation and peat formation in the central Appalachian Basin (U.S.A.), *International Journal of Coal Geology 5*, 195-230.

Cleal, C. J. (1987). This is the forest primaeval, *Nature 326*, 828.

Corsin, P. (1951). *Flore Fossile du Bassin de la Sarre et de la Lorraine*, 4ᵐᵉ Fascicule, Pecopteridees. Etudes des Gites Mineraux de la France, 370 pp.

Costanza, S. H. (1985). *Pennsylvanianioxylon* of middle and upper Pennsylvanian coals from the Illinois Basin, and its comparison with *Mesoxylon*, *Palaeontographica 197B*, 81-121.

Courvoisier, J. M, and T. L. Phillips (1975). Correlation of spores from Pennsylvanian coal-ball fructifications with dispersed spores, *Micropaleontology 21*, 45-59.

Cridland, A. A. (1964). *Amyelon* in American coal balls, *Palaeontology 7*, 186-209.

Damuth, J. (1985). Selection among "species": A formulation in terms of natural functional units, *Evolution 39*, 1132-1146.

Delcourt, P. A., and H. R. Delcourt (1987). *Long-Term Forest Dynamics of the Temperate Zone*, Springer-Verlag, New York, 439 pp.

Digby, P. G. N., and R. A. Kempton (1987). *Multivariate Analysis of Ecological Communities*, Chapman and Hall, London, 206 pp.

DiMichele, W. A., and R. B. Aronson (1992). The Pennsylvanian-Permian vegetational transition: A terrestrial analogue to the onshore-offshore hypothesis, *Evolution 46*, 807-824.

DiMichele, W. A., and P. J. DeMaris (1987). Structure and dynamics of a Pennsylvanian-age *Lepidodendron* forest: Colonizers of a disturbed swamp habitat in the Herrin (No. 6) Coal of Illinois, *PALAIOS 2*, 146-157.

DiMichele, W. A., and W. J. Nelson (1989). Small-scale spatial heterogeneity in Pennsylvanian-age vegetation from the roof shale of the Springfield Coal (Illinois Basin), *Palaios 4*, 276-280.

DiMichele, W. A., and T. L. Phillips (1981). Stratigraphic-geographic patterns of change in Pennsylvanian coal-swamp vegetation, *Botanical Society of America, Miscellaneous Series Publication 160*, 44 (Abstract).

DiMichele, W. A., and T. L. Phillips (1985). Arborescent lycopod reproduction and paleoecology in a coal-swamp environment of late middle Pennsylvanian age (Herrin Coal, Illinois, U.S.A.), *Review of Palaeobotany and Palynology 44*, 1-26.

DiMichele, W. A., and T. L. Phillips (1988). Paleoecology of the middle Pennsylvanian-age Herrin coal swamp (Illinois) near a contemporaneous river system, the Walshville paleochannel, *Review of Palaeobotany and Palynology 56*, 151-176.

DiMichele, W. A., J. F. Mahaffy, and T. L. Phillips (1979). Lycopods of Pennsylvanian age coals: *Polysporia*, *Canadian Journal of Botany 57*, 1740-1753.

DiMichele, W. A., T. L. Phillips, and R. G. Olmstead (1987). Opportunistic evolution: Abiotic environmental stress and the fossil record of plants, *Review of Palaeobotany and Palynology 50*, 151-178.

DiMichele, W. A., T. L. Phillips, and D. A. Willard (1986). Morphology and paleoecology of Pennsylvanian coal-swamp plants, *University of Tennessee, Department of Geological Sciences, Studies in Geology 15*, 97-114.

DiMichele, W. A., T. L. Phillips, and G. E. McBrinn (1991). Quantitative analysis and paleoecology of the Secor coal and roof shale floras (Pennsylvanian-age, Oklahoma), *Palaios 6*, 390-409.

Driese, S. G. (1985). Interdune pond carbonates, Weber Sandstone (Pennsylvanian-Permian), northern Utah and Colorado, *Journal of Sedimentary Petrology 55*, 187-195.

Dulong, F. T., and C. B. Cecil (1989). Stratigraphic variation in bulk sample mineralogy of Pennsylvanian underclays from the Central Appalachian Basin, *Carboniferous Geology of the Eastern United States*, 28th International Geological Congress, Field Trip Guidebook T143, pp. 112-118.

Eble, C. F., and W. C. Grady (1993). Paleoecological interpretation of two middle Pennsylvanian coal beds in the central Appalachian Basin, *Geological Society of America, Special Paper*.

Eggert, D. L. (1982). A fluvial channel contemporaneous with deposition of the Springfield Coal Member (V), Petersburg Formation, Northern Warrick County, Indiana, *Indiana Geological Survey Special Report 28*, 20 pp.

Esterle, J. S., and J. C. Ferm (1986). Relationship between petrographic and chemical properties and coal seam geometry, Hance Seam, Breathitt Formation, southeastern Kentucky, *International Journal of Coal Geology 6*, 199-214.

Farley, M. B. (1989). Palynological facies fossils in nonmarine environments in the Paleogene of the Bighorn Basin, *Palaios 4*, 565-573.

Galtier, J., and T. L. Phillips (1985). Swamp vegetation from Grand Croix (Stephanian) and Autun (Autunian), France and comparisons with coal-ball peats of the Illinois Basin, *C.R. 9th International Congress of Carboniferous Stratigraphy and Geology 4*, 13-24.

Gastaldo, R. A. (1986). Implications on the paleoecology of autochthonous lycopods in clastic sedimentary environments of the early Pennsylvanian of Alabama, *Palaeogeography, Palaeoclimatology, Palaeoecology 53*, 191-212.

Gastaldo, R. A. (1987). Confirmation of Carboniferous clastic swamp communities, *Nature 326*, 869-871.

Goebel, K. A., E. A. Bettis III, and P. H. Heckel (1989). Upper Pennsylvanian paleosol in Stranger Shale and underlying Iatan Limestone, southwestern Iowa, *Journal of Sedimentary Petrology 59*, 224-232.

Grady, W. C., and C. F. Eble (1990). Relationships among macerals, miospores and paleoecology in a column of the Redstone coal (upper Pennsylvanian) from north-central West Virginia (U.S.A.), *International Journal of Coal Geology 15*, 1-26.

Grime, J. P. (1979). *Plant Strategies and Vegetation Processes*, John Wiley & Sons, New York, 222 pp.

Heckel, P. H. (1986). Sea-level curve for Pennsylvanian eustatic marine transgressive-regressive depositional cycles along midcontinent outcrop belt, North America, *Geology 14*, 330-334.

Heckel, P. H. (1989). Updated middle-upper Pennsylvanian eustatic sea level curve for midcontinent North America and preliminary biostratigraphic characterization, *Compte Rendu, XI International Congress of Carboniferous Stratigraphy and Geology, Beijing 4*, 160-185.

Hess, J. C., and H. J. Lippolt (1986). $^{40}Ar/^{39}Ar$ ages of tonstein and tuff sanidines: New calibration points for the improvement of the upper Carboniferous time scale, *Isotope Geoscience 59*, 143-154.

Holmes, J. C., and M. Fairon-Demaret (1984). A new look at the flora of the Bouxharmont coal balls from Belgium, *Annals de la Societe Geologique de Belgique 107*, 73-87.

Horne, J. C., J. C. Ferm, F. T. Caruccio, and B. P. Baganz (1978). Depositional models in coal exploration and mine planning in Appalachian region, *American Association of Petroleum Geologists Bulletin 62*, 2379-2411.

Johnson, P. R. (1979). Petrology and environments of deposition of the Herrin (No. 6) Coal Member, Carbondale Formation, at the Old Ben Coal Company Mine No. 24, Franklin County, Illinois, M.S. thesis, University of Illinois, Urbana-Champaign, 169 pp.

Johnson, S. Y. (1987). Sedimentology and paleogeographic significance of six fluvial sand bodies in the Maroon Formation, Eagle Basin, northwest Colorado, *U.S. Geological Survey Bulletin 1787-A*, 1-18.

Johnson, S. Y. (1989a). Significance of loessites in the Maroon Formation (middle Pennsylvanian to lower Permian), Eagle Basin, northwest Colorado, *Journal of Sedimentary Petrology 59*, 782-791.

Johnson, S. Y. (1989b). The Frying Pan Member of the Maroon Formation, Eagle Basin, northwest Colorado, *U.S. Geological Survey Bulletin 1787-I*, 1-11.

Klein, G. deV., and D. A. Willard (1989). Origin of the Pennsylvanian coal-bearing cyclothems of North America, *Geology 17*, 152-155.

Knoll, A. H. (1984). Patterns of extinction in the fossil record of vascular plants, in *Extinctions*, M. Nitecki, ed., University of Chicago Press., Chicago, pp. 21-68.

Kvale, E. P., and A. W. Archer (1990). Tidal deposits associated with low-sulfur coals, Brazil Fm. (lower Pennsylvanian), Indiana, *Journal of Sedimentary Petrology 60*, 563-574.

Lesnikowska, A. D. (1989). Anatomically preserved Marattiales from coal swamps of the Desmoinesian and Missourian of the midcontinent United States: Systematics, ecology and evolution. Ph.D. thesis, University of Illinois, Urbana-Champaign, 227 pp.

Mahaffy, J. F. (1985). Profile patterns of coal and peat palynology in the Herrin (No. 6) Coal Member, Carbondale Formation, middle Pennsylvanian of southern Illinois, *Ninth International Congress of Carboniferous Stratigraphy and Geology, C.R. 5*, 25-34.

Overpeck, J. T., R. S. Webb, and T. Webb III (1992). Mapping eastern North American vegetation change of the past 18 ka: No-analogs and the future, *Geology 20*, 1071-1074.

Parrish, J. M., J. T. Parrish, and A. M. Ziegler (1989). Permian-Triassic paleogeography and paleoclimatology and implications for therapsid distributions, in *The Biology and Ecology of Mammal-Like Reptiles*, J. Roth, C. Roth, and N. Hotton III, eds., Smithsonian Institution Press, Washington, D.C.

Parrish, J. T. (1982). Upwelling and petroleum source beds, with reference to the Paleozoic, *American Association of Petroleum Geologists Bulletin 66*, 750-774.

Peppers, R. A. (1979). Development of coal-forming floras during the early part of the Pennsylvanian in the Illinois Basin, *Ninth International Congress of Carboniferous Stratigraphy and Geology, Guidebook to Field Trip 9*, part 2, 8-14.

Pfefferkorn, H. W., and M. C. Thomson (1982). Changes in dominance patterns in upper Carboniferous plant-fossil assemblages, *Geology 10*, 641-644.

Phillips, T. L. (1974). Evolution of vegetative morphology in coenopterid ferns, *Annals of the Missouri Botanical Garden 61*, 427-461.

Phillips, T. L., and C. B. Cecil (1985). Paleoclimatic controls on coal resources of the Pennsylvanian System of North America: Introduction and overview of contributions, *International Journal of Coal Geology 5*, 1-6.

Phillips, T. L., and W. A. DiMichele (1981). Paleoecology of middle Pennsylvanian age coal swamps in southern Illinois/ Herrin Coal Member at Sahara Mine No. 6, in *Paleobotany, Paleoecology, and Evolution*, Volume 1, K. J. Niklas, ed., Praeger Press, New York, pp. 205-255.

Phillips, T. L., and W. A. DiMichele (1989). From plants to coal: Peat taphonomy of upper Carboniferous coals, *28th International Geological Congress Abstracts 2*, 605 (Abstract).

Phillips, T. L., and W. A. DiMichele (1992). Comparative ecology and life-history biology of arborescent lycopods in Late Carboniferous swamps of Euramerica, *Annals of the Missouri Botanical Garden 79*, 560-588.

Phillips, T. L., and R. A. Peppers (1984). Changing patterns of Pennsylvanian coal-swamp vegetation and implications of climatic control on coal occurrence, *International Journal of Coal Geology 3*, 205-255.

Phillips, T. L., R. A. Peppers, M. J. Avcin, and P. F. Laughnan (1974). Fossil plants and coal: Patterns of change in Pennsylvanian coal swamps of the Illinois Basin, *Science 184*, 1367-1369.

Phillips, T. L., A. B. Kunz, and D. J. Mickish (1977). Paleobotany of permineralized peat (coal balls) from the Herrin (No. 6) Coal Member of the Illinois Basin, *Geological Society of America, Microform Publication 7*, 18-49.

Phillips, T. L., R. A. Peppers, and W. A. DiMichele (1985). Stratigraphic and interregional changes in Pennsylvanian coal-swamp vegetation: Environmental inferences, *International Journal of Coal Geology 5*, 43-109.

Pigg, K. B. (1987). Paleozoic seed ferns: *Heterangium kentuckyensis* sp. nov., from the upper Carboniferous of North America, *American Journal of Botany 74*, 1184-1204.

Prather, B. E. (1985). An upper Pennsylvanian desert paleosol in the D-zone of the Lansing-Kansas City Groups, Hitchcock County, Nebraska, *Journal of Sedimentary Petrology 55*, 213-221.

Raymond, A. (1987). Interpreting ancient swamp communities: Can we see the forest in the peat? *Review of Palaeobotany and Palynology 52*, 217-231.

Raymond, A. (1988). Paleoecology of a coal-ball deposit from the middle Pennsylvanian of Iowa dominated by cordaitalean gymnosperms, *Review of Palaeobotany and Palynology 53*, 233-250.

Remy, W., and R. Remy (1977). *Die Floren des Erdaltertums*, Verlag Gluckauf GMBH, Essen, 468 pp.

Rothwell, G. W. (1981). The Callistophytales (Pteridospermopsida): Reproductively sophisticated gymnosperms, *Review of Palaeobotany and Palynology 32*, 103-121.

Rothwell, G. W., and S. Warner (1984). *Cordaixylon dumusum* n. sp. (Cordaitales) I. Vegetative structures, *Botanical Gazette 145*, 275-291.

Ruppert, L. F., C. B. Cecil, R. W. Stanton, and R. P. Christian (1985). Authigenic quartz in the upper Freeport coal bed, west-central Pennsylvania, *Journal of Sedimentary Petrology 55*, 334-339.

Rust, B. R., and M. R. Gibling (1990). Braidplain evolution in the Pennsylvanian South Bar Formation, Sydney Basin, Nova Scotia, Canada, *Journal of Sedimentary Petrology 60*, 59-72.

Rust, B. R., M. R. Gibling, M. A. Best, S. J. Dilles, and A. G. Masson (1987). A sedimentological overview of the coal-bearing Morien Group (Pennsylvanian), Sydney Basin, Nova Scotia, Canada, *Canadian Journal of Earth Sciences 24*, 1869-1885.

Scotese, C. R., R. K. Bambach, C. Barton, R. Van der Voo, and A. M. Ziegler (1979). Paleozoic base maps, *Journal of Geology 87*, 217-277.

Scott, A. C. (1978). Sedimentological and ecological control of Westphalian B plant assemblages from West Yorkshire, *Proceedings of the Yorkshire Geological Society 41*, 461-508.

Scott, A. C., and G. Rex (1985). The formation and significance of Carboniferous coal balls, *Philosophical Transactions of the Royal Society of London B 311*, 123-137.

Smith, A. V. H., and M. A. Butterworth (1967). Miospores in the coal seams of the Carboniferous of Great Britain, *Palaeontological Society, Special Paper No. 1*.

Spears, D. A., and H. I. Sezgin (1985). Mineralogy and geochemistry of the Subcrenatum Marine Band and associated coal-bearing sediments, Langsett, South Yorkshire, *Journal of Sedimentary Petrology 55*, 570-578.

Staub, J. R., and J. S. Esterle (1992). Evidence for a tidally influenced upper Carboniferous ombrogeneous mire system: Upper bench, Beckley bed (Westphalian A), southern West Virginia, *Journal of Sedimentary Petrology 62*, 411-428.

Stevens, C. H., and P. Stone (1988). Early Permian thrust faults in east-central California, *Geological Society of America Bulletin 100*, 552-562.

Taylor, T. N. (1965). Paleozoic seed studies: A monograph of the genus *Pachytesta*, *Palaeontographica 117B*, 1-46.

Trivett, M. L., and G. W. Rothwell (1985). Morphology, systematics and paleoecology of Paleozoic fossil plants: *Mesoxylon priapi*, sp. nov. (Cordaitales), *Systematic Botany 10*, 205-233.

Tschudy, R. H., C. L. Pillmore, C. J. Orth, J. S. Gilmore, and J. D. Knight (1984). Disruption of the terrestrial plant ecosystem at the Cretaceous-Tertiary boundary, western interior, *Science 225*, 1030-1032.

Valentine, J. W. (1980). Determinants of diversity in higher taxonomic categories, *Paleobiology 6*, 444-450.

Veevers, J. J., and C. M. Powell (1987). Late Paleozoic glacial episodes in Gondwanaland reflected by transgressive-regressive depositional sequences in Euramerica, *Geological Society of America Bulletin 98*, 475-487.

Wanless, H. R., and F. P. Shepard (1936). Sea level and climatic changes related to late Paleozoic cycles, *Geological Society of America Bulletin 47*, 1177-1206.

Wanless, H. R., J. R. Baroffio, and P. C. Trescott (1969). Conditions of deposition of Pennsylvanian coal beds, *Geological Society of America, Special Paper 114*, 105-142.

White, D. (1929). *Flora of the Hermit Shale, Grand Canyon, Arizona*, Carnegie Institute of Washington, Publication No. 405, 221 pp.

White, D. (1933). Some features of the Early Permian flora of North America, *16th International Geological Congress 1*, 679-689.

Willard, D. A. (1989). Source plants for Carboniferous microspores: *Lycospora* from permineralized *Lepidostrobus*, *American Journal of Botany 76*, 820-827.

Willard, D. A. (1993). Vegetational patterns in the Springfield Coal (Middle Pennsylvanian; Illinois Basin): Comparison of miospore and coal-ball records, in *Modern and Ancient Coal-Forming Environments*, J. C. Cobb and C. B. Cecil, eds., Geological Society of America Special Paper 286, Boulder, Colo., pp. 139-152.

Willard, D. A., and T. L. Phillips (1993). Paleobotany and palynology of the Bristol Hill Coal Member (Bond Formation) and Friendsville Coal Member (Mattoon Formation) of the Illinois Basin (upper Pennsylvania), *Palaios 8*, 574-586.

Wing, S. L. (1988). Taxon-free paleoecological analysis of Eocene megafloras from Wyoming, *American Journal of Botany 75*(6-part 2), 120 (Abstract).

Wing, S. L., and B. H. Tiffney (1987). The reciprocal interaction of angiosperm evolution and tetrapod herbivory, *Review of Palaeobotany and Palynology 50*, 179-210.

Winston, R. B., and R. W. Stanton (1989). Plants, coal, and climate in the Pennsylvanian of the central Appalachians, in *Carboniferous Geology of the Eastern United States*, 28th International Geological Congress, Field Trip Guidebook T143, pp. 118-126.

Wnuk, C. (1985). The ontogeny and paleoecology of *Lepidodendron rimosum* and *Lepidodendron bretonense* trees from the middle Pennsylvanian of the Bernice Basin, Sullivan County, Pennsylvania, *Palaeontographica 195B*, 153-181.

Wnuk, C., and H. W. Pfefferkorn (1987). A Pennsylvanian-age terrestrial storm deposit: Using plant fossils to characterize the history and process of sediment accumulation, *Journal of Sedimentary Petrology 57*, 212-221.

Wolfe, J. A., and G. R. Upchurch (1986). Vegetation, climatic and floral changes at the Cretaceous-Tertiary boundary, *Nature 324*, 148-152.

Ziegler, A. M. (1990). Phytogeographic patterns and continental configurations during the Permian Period, in *Palaeozoic Palaeogeography and Biogeography*, W. S. McKerrow and C. R. Scotese, eds., Geological Society of London Memoir, pp. 363-379.

Ziegler, A. M., R. K. Bambach, J. T. Parrish, S. F. Barrett, E. H. Gierlowski, W. C. Parker, A. Raymond, and J. J. Sepkoski (1981). Paleozoic biogeography and climatology, in *Paleobotany, Paleoecology and Evolution*, Volume 2, K.J. Niklas, ed., Praeger, New York, pp. 231-266.

The Late Cretaceous and Cenozoic History of Vegetation and Climate at Northern and Southern High Latitudes: A Comparison

ROSEMARY A. ASKIN
University of California, Riverside

ROBERT A. SPICER
Oxford University

ABSTRACT

The Late Cretaceous and Cenozoic high-latitude land vegetation bequeathed a sensitive paleobotanical and palynological record of regional and global environmental change. Foliar physiognomy provides the most reliable indicators. This record is frequently available for northern localities, augmented by wood and palynomorph data. Southern data are provided mainly by palynomorphs, with some foliar, cuticular, and wood information. The northern high-latitude vegetation was mainly deciduous, whereas evergreen taxa locally predominate in the south. Major northern clades were all derived from lower latitudes; in contrast, Antarctica was a center of evolutionary innovation and dispersal. Differences in northern and southern vegetation are a function of continental configurations, interrelated with continentality (winter-summer temperature range), seasonality, moisture/aridity regimes, sea-level cycles, and overprinted by biotic stress or selective mechanisms. Vast land areas encircled the North Pole (to within 85°N), enhancing climatically driven northward and southward migrations, whereas an Antarctic continent continuously occupied the South Polar latitudes, had relatively restricted dispersal corridors, and became increasingly isolated as the other Gondwana fragments spread northward.

INTRODUCTION

Fossils of plants that lived at high latitudes provide a sensitive and unparalleled record of the complex interplay of global climatic change and polar conditions through time. High-latitude plants, presently portrayed by extant polar desert, tundra, and taiga floras, require adaptations for stringent "icehouse" conditions. An icehouse world is, however, infrequently encountered in earth history. More typical "greenhouse" conditions necessitate other strategies for plant survival (Spicer, 1989a; Spicer and Chapman, 1990). In the Mesozoic and Cenozoic greenhouse world, forests thrived near the poles despite the seasonal stress of polar light cycles (the near congruity of rotational and magnetic poles is assumed here).

Light and particularly temperature are the principal controlling factors for polar vegetation, whereas precipitation is generally the prime factor at lower latitudes (Ziegler, 1990). Realistic quantitative estimates of these physical conditions can be ascertained from fossil floras. These estimates are crucial data for climatic modeling and for understanding and predicting global climate change. Environmental parameters (mean annual temperature, MAT; mean annual temperature range, MAR; coldest month mean temperature, CMM) that can be determined from fossil floras, and their validity and limitations are reviewed by Spicer and Parrish (1990a). Parameters derived from fossil floras reflect vegetational response to the environment and are obtained by careful analysis and interpretation of leaf physiognomy (size and margin morphology, Bailey and Sinnott, 1915; Wolfe, 1971, 1979, 1985; Wolfe and Upchurch, 1987) and overall floral composition (Wolfe, 1979). Such information is available for northern high-latitude floras, especially from upper Cretaceous to Eocene strata from northern Alaska. In southern high latitudes, fossil leaf assemblages are less common, and environmental interpretations are not yet well developed. Ongoing work on leaf floras of southern basins (e.g., Daniel *et al.*, 1990; Parrish *et al.*, 1991) should help correct this imbalance in data types, but in the meantime much of the southern data are derived, by necessity, from palynomorphs.

Useful qualitative information can be obtained from wood anatomy, tree-ring data, and palynomorphs. Palynomorph assemblages contribute a broad-brush view of the regional vegetation or, if locally derived, can provide a more detailed picture. Generalized climatic conditions can be inferred from palynomorph and leaf fossils by analogy with presumed modern counterparts (nearest living relatives), but this method can be risky because it assumes evolutionary stasis. For conservative taxa (conifers, ferns), such conclusions may be reasonably reliable.

The northern and southern high-latitude (~60 to 90°) vegetational history from the middle Cretaceous through the Cenozoic is presented in this overview, along with its suggested relationship to global change, in particular climatic change.

Vegetational changes evolved along different pathways in the northern and southern regions, although basic physiologic constraints of polar conditions are similar. Physiognomic parallels at both poles (e.g., highly dissected ginkgo leaves, the broad-leaved conifers *Podozamites* and *Agathis*-type, *Sphenopteris*-like ferns) illustrate this latter point.

PALEOGEOGRAPHIC FRAMEWORK

The difference in continental configurations between the northern and southern polar regions is the overriding cause of differences in the evolution of their respective floras. These differences are illustrated (Figures 9.1 and 9.2) in the paleogeographic reconstructions of Smith *et al.* (1981). During the Late Cretaceous and Cenozoic, vast land areas encircled the North Pole (to within 85°N), facilitating climatically driven northward and southward floral migrations, whereas an Antarctic continent continuously occupied the South Polar position, had relatively restricted dispersal corridors, and became increasingly isolated as other Gondwana fragments spread northward.

SUMMARY OF HIGH-LATITUDE VEGETATIONAL CHANGES

Significant botanical events, and vegetational types and trends for northern and southern high latitudes are outlined in Figures 9.3 to 9.6, plotted alongside "global change" information. These charts are based on studies and fossil localities cited below and in Figures 9.1 and 9.2.

Northern Cretaceous

Albian-Cenomanian and Arrival of Angiosperms

In the middle Cretaceous, land areas extended northward to 75°N. In these latitudes, prior to the arrival of angiosperms near the end of the Albian, forests were conifer dominated with *Podozamites*, *Arthrotaxopsis*, and *Elatocladus* being the most common foliage (Smiley, 1966, 1967, 1969a,b; Samylina, 1973, 1974; Spicer and Parrish, 1986, 1990a; Spicer, 1987). Needle-leaved conifers were common. Ginkgophytes (*Ginkgo*, *Sphenobaiera* or *Sphenarion*, *Ginkgoites*) were diverse, but restricted to river margins (*Ginkgo*-like forms) or back levees (*Sphenobaiera*), and cycads were relatively common but spatially restricted. Ferns (e.g., *Onychiopsis*, *Sphenopteris*-like forms) and *Equisetites* were early colonizers and common as ground cover. Tree productivity and water availability were high, and temperatures were typical of cool temperate regimes (Spicer and Parrish, 1986; Parrish and Spicer, 1988a). All vegetation was deciduous, could enter dormancy, or could overwinter as underground organs or seeds (Spicer and Parrish, 1986). Palynological evidence shows that bryophytes, lycopods, and fungi were prevalent (May and Shane, 1985; Spicer *et al.*, 1988), particularly in mire environments that gave rise to extensive coals (Youtcheff *et al.*, 1987; Grant *et al.*, 1988).

Angiosperms that produced tricolpate pollen reached the northern high latitudes, including the Canadian Arctic, during the Albian (Jarzen and Norris, 1975; Singh, 1975; Scott and Smiley, 1979). In northern Alaska (Spicer, 1987) and Siberia (Samylina, 1974; Lebedev, 1978), leaf floras indicate that in the Cenomanian, platanoid angiosperms (e.g., "*Platanus*," *Protophyllum*, *Pseudoprotophyllum*,

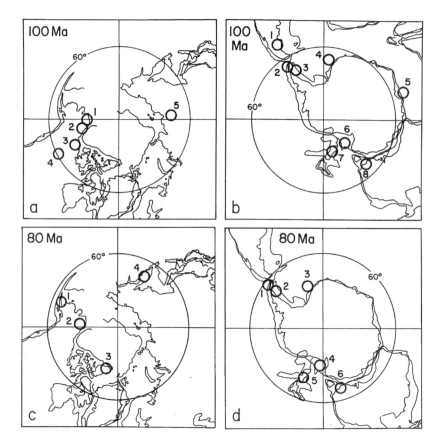

FIGURE 9.1 Cretaceous high-latitude plant fossil localities on 100 and 80 Ma paleocontinental reconstructions (polar Lambert equal-area projections) of Smith et al. (1981):

(a) Northern Albian-Cenomanian localities, 1: Smiley (1966, 1967, 1969a,b), Scott and Smiley (1979); 2: May and Shane (1985), Spicer and Parrish (1986, 1990a,b), Spicer (1987), Parrish and Spicer (1988a,b), Grant et al. (1988), Youtcheff et al. (1987); 3: Jarzen and Norris (1975); 4: Singh (1975); 5: Samylina (1973, 1974), Lebedev (1978).

(b) Southern Aptian, Albian-Cenomanian localities, 1: Volkheimer and Salas (1975), Archangelsky (1980), Romero and Archangelsky (1986); 2: Rees and Smellie (1989), Rees (1990), Chapman and Smellie (1992); 3: Dettmann and Thomson (1987), Baldoni and Medina (1989); 4: Truswell (1983), Truswell and Anderson (1985); 5: Truswell (1990); 6: Truswell (1983); 7: Couper (1960), Raine (1984); 8: Douglas and Williams (1982), Dettmann (1986a), Taylor and Hickey (1990), Parrish et al. (1991), Dettmann et al. (1992).

(c) Northern Turonian to Maastrichtian localities, 1: Hollick (1930), Spicer (1983); 2: Parrish et al. (1987), Frederiksen et al. (1988), Parrish and Spicer (1988a), Frederiksen (1989), Spicer and Parrish (1990a,b); 3: Hickey et al. (1983); 4: Krassilov (1979).

(d) Southern Turonian to Maastrichtian localities, 1: Birkenmajer and Zastawniak (1989); 2: Cranwell (1969), Baldoni and Barreda (1986), Francis (1986), Dettmann and Thomson (1987), Askin (1988a,b, 1989, 1990a,b), Dettmann and Jarzen (1988), Baldoni and Medina (1989), Dettmann (1989), Jarzen and Dettmann (1990), Askin et al. (1991); 3: Truswell (1983), Truswell and Anderson (1985); 4: Truswell (1983); 5: Couper (1960), Mildenhall (1980), Raine (1984, 1988), Daniel et al. (1990); 6: Stover and Partridge (1973), Martin (1977), Dettmann and Jarzen (1988), Dettmann (1989), Dettmann et al. (1992).

Pseudoaspidiophyllum, *Crednaria*) locally dominated riparian habitats, where they successfully replaced *Ginkgo* and *Ginkgo*-like plants.

By the late Cenomanian, angiosperm diversity had risen to more than 60 leaf forms in Alaska (Spicer and Parrish, 1990a). Pollen diversity in Alaska has yet to be fully evaluated. The vegetation was still conifer dominated, but needle-leaved conifers were less common. Angiosperm leaf sizes were large, and leaf physiognomy suggests a wet regime with MATs of 10°C (Parrish and Spicer, 1988a). Tree rings show little intra-annual variation (few false rings). There is some inter-annual variation, possibly a result of fluctuations in water availability (Parrish and Spicer, 1988b), although overall water stress was lacking, based on the high productivity and large cell size. Latewood was very limited, suggesting rapid onset of dark-induced dormancy (Spicer and Parrish, 1990b). There are no periglacial sediments known, or any features indicative of sea ice.

Turonian-Coniacian-Santonian

The Turonian was characterized by a major global sea-level highstand, reducing the nonmarine sedimentary record.

By the Coniacian, needle-leaved conifers and *Podozamites* had disappeared and taxodiaceous foliage was common. Platanoid angiosperms were still dominant along river and lake margins, and had begun to penetrate forests. Cycads were rare or absent, *Ginkgo* diversity was much reduced, and *Equisetites* and ferns formed the main ground cover. Leaf margin analysis of a small number of specimens imply an MAT of 13°C at about 78°N (Parrish and Spicer, 1988a), and conditions were still wet although coals are thinner and less numerous. Angiosperm diversity was high, but possibly less than in the Cenomanian. All taxa were deciduous or capable of winter dormancy.

Santonian nonmarine sediments are rare and not yet sampled for plant fossils.

Campanian-Maastrichtian

The North Slope of Alaska was at 85°N, and overall diversity had dropped in the megafloral record. Woody angiosperm diversity was much reduced, and conifers were represented by only two foliage and five wood taxa (Spicer and Parrish, 1990a). *Equisetites* and ferns continued as the main ground cover, and *Ginkgo* and cycads had disappeared. Fires were common.

The palynomorph record is extensive and indicates angiosperm predominance in the Maastrichtian, with a high turnover and spatial heterogeneity of taxa (Frederiksen *et al.*, 1988; Frederiksen, 1989). Plants that produced the pollen are believed to be mostly herbaceous and probably annuals. A "weedy" strategy would have been favored by short growing seasons and fire disturbance. Some woody Betulaceae/Ulmaceae may have been present (Parrish *et al.*, 1987; Frederiksen, 1989); however tree taxa were dominated by the limited variety of conifers.

Tree rings reflect lower productivity and more inter- and intra-annual variation. Latewood to earlywood ratios are higher, suggesting thermal limitations on spring or summer growth rather than sudden dark-induced dormancy (Spicer and Parrish, 1990b). By the Maastrichtian, vegetation was more open, with smaller trees. The environment was not as wet, with periodic drying. Vegetational physiognomy suggests MAT of 2.5 to 5°C and long, cold, dark winters with CMM probably no lower than −11°C (Parrish *et al.*, 1987). Large wood tracheid cross sections, and lack of periglacial sediments, imply no severe drought or freezing (Spicer and Parrish, 1990b).

On the Alaskan Peninsula, conditions were much warmer than on the North Slope. Leaf forms were more advanced, migration from the south continued, and *Nilssonia* (cycad) and diminuitive *Ginkgo dawsonii* survived into the Maastrichtian (Hollick, 1930; Spicer, 1983).

Floras in both the northern (Spicer, 1989b) and the southern (Askin, 1988b, 1990a; Raine, 1988) high latitudes apparently did not suffer any major ecological trauma at the Cretaceous-Tertiary boundary.

Southern Cretaceous

Albian-Cenomanian and Early Angiosperms

By the middle Cretaceous, angiosperms were well established in the southern high latitudes. At least 8 an-

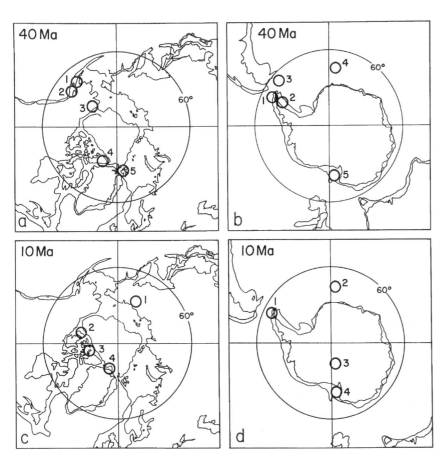

FIGURE 9.2 Cenozoic high-latitude plant fossil localities on 40 and 10 Ma paleocontinental reconstructions (polar Lambert equal-area projections) of Smith *et al.* (1981):

(a) Northern Paleocene-Eocene localities, 1: Wolfe (1980); 2: Wolfe (1966), Wolfe and Poore (1982); 3: Hickey *et al.* (1983), Parrish *et al.* (1987), Frederiksen *et al.* (1988), Spicer and Parrish (1990a); 4: Francis and McMillan (1987); 5: Schweitzer (1974).

(b) Southern Paleocene-Eocene localities, 1: Stuchlik (1981), Lyra (1986), Torres and LeMoigne (1988), Birkenmajer and Zastawniak (1989), Torres and Meon (1990); 2: Dusen (1908), Fleming and Askin (1982), Zamaloa *et al.* (1987), Askin (1988a,b, 1990b), Case (1988), Askin *et al.* (1991); 3. Mohr, 1990; 4: Kennett and Barker (1990), Mohr (1990); 5: Truswell (1983).

(c) Northern Oligocene to Pliocene localities, 1: Sher *et al.* (1979); 2: Hills *et al.* (1974); 3: Hills and Matthews (1974), Kuc *et al.* (1983); 4: Funder *et al.* (1985).

(d) Southern Oligocene to Pliocene localities, 1: Palma-Heldt (1987), Birkenmajer and Zastawniak (1989); 2: Mohr (1990); 3: Askin and Markgraf (1986), Carlquist (1987), Webb *et al.* (1987), Harwood (1988); 4: Kemp (1975), Kemp and Barrett (1975), Truswell (1983, 1990), Hill (1989), Mildenhall (1989).

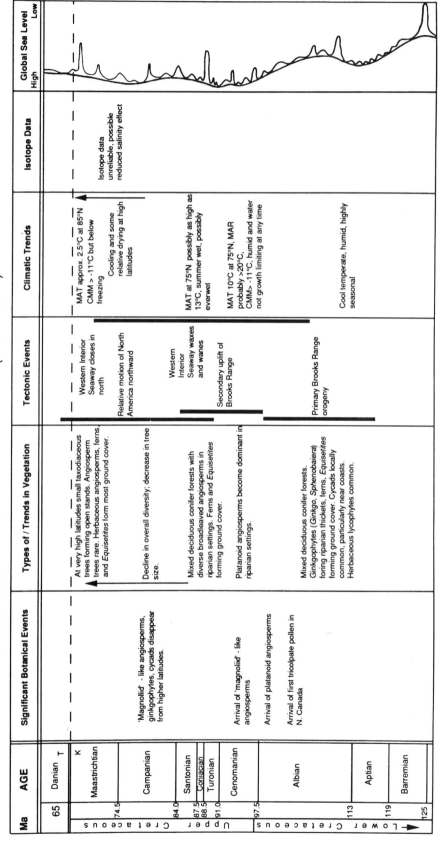

FIGURE 9.3 Vegetational trends and global change for the Cretaceous northern high latitudes. Sources of data are in the text. Sea-level curve for Figures 9.3 to 9.6 from Haq et al. (1987); time scale for Figures 9.3 and 9.4 from Harland et al. (1990).

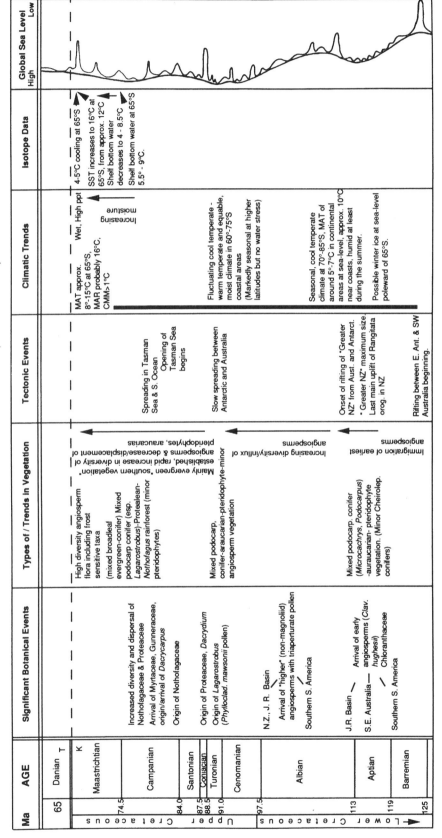

FIGURE 9.4 Vegetational trends and global change for the Cretaceous southern high latitudes. Isotope data for Figures 9.4 and 9.6 from Barrera et al. (1987), Stott and Kennett (1990), and Stott et al. (1990).

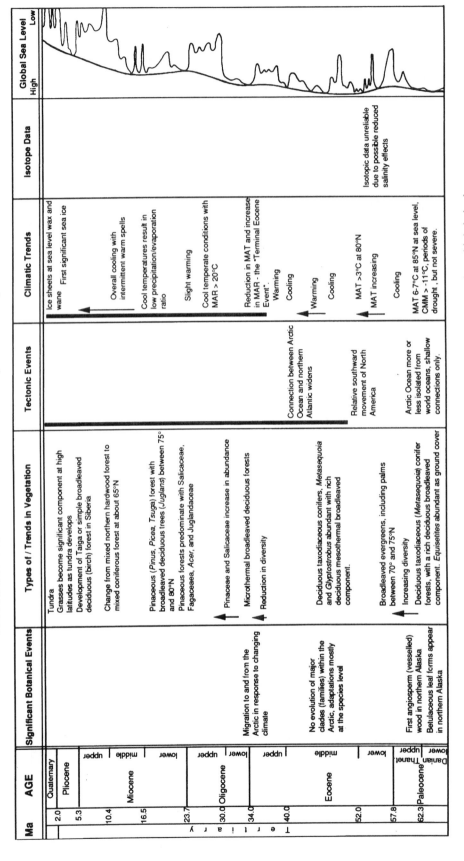

FIGURE 9.5 Vegetational trends and global change for the Cenozoic northern high latitudes. Time scale for Figures 9.5 and 9.6 from Berggren *et al.* (1985) and Harland *et al.* (1990).

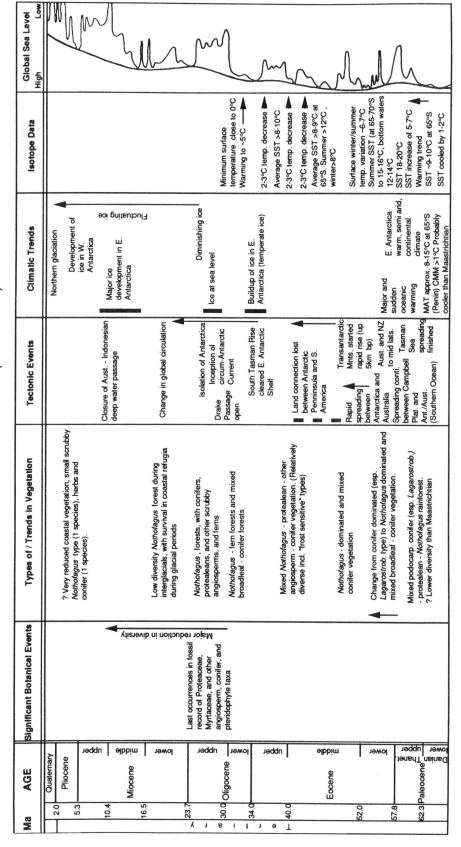

FIGURE 9.6 Vegetational trends and global change for the Cenozoic southern high latitudes.

giosperm pollen taxa were present near the tip of the Antarctic Peninsula (about 66°S) by the Albian-Cenomanian (Dettmann and Thomson, 1987; Baldoni and Medina, 1989). There, and slightly further north in southern South America (Volkheimer and Salas, 1975; Archangelsky, 1980; Romero and Archangelsky, 1986), palynomorphs and leaves represent monocotyledonous and several dicotyledonous families, including "higher" (nonmagnoliid) forms that produced triaperturate pollen. Vegetation was predominantly conifer forest with podocarps (*Podocarpus* and *Microcachrys*-type) and araucarians, pteridosperms, and diverse fern understory and ground cover.

For a small riparian collection (seven leaf taxa) of mainly microphyllous and deciduous angiosperms from Livingston Island, South Shetland Islands (59 to 65°S), a MAT of 13 to 20°C was suggested (Rees and Smellie, 1989; Rees, 1990), although values are tentative at best for samples with fewer than 20 species (Wolfe, 1971). Associated foliage includes various conifers and cycads. Several types of fern foliage and *Equisetites* may have formed ground cover, as in northern high latitudes. Well-defined growth rings in conifer wood attests to strong seasonality, but the lack of rings in angiosperm woods (Chapman and Smellie, 1992) is difficult to explain.

Angiosperms had penetrated close to 80°S by the Albian-Cenomanian in New Zealand (in Motuan and Ngaterian Stages; Raine, 1984). They have an older record in southeastern Australia (Aptian Koonwarra beds, ~60°S) based on pollen (Dettmann, 1986a) and foliage (with flowers) of a small, rhizomatous, perennial angiosperm (Taylor and Hickey, 1990).

Douglas and Williams (1982) interpreted the Albian-Cenomanian climate in southern Australia as wet warm temperate, with moderate seasonality, probably seasonally dry, and no widespread winter freezing. Cooler temperatures are indicated by Parrish *et al.* (1991) based on re-evaluation of the Albian vegetation, though not as cold as suggested by oxygen isotope results (Gregory *et al.*, 1989; MAT less than 0°C). Parrish *et al.* (1991) noted that the Albian floras include both deciduous, thin-cuticled leaves, commonly occurring as mats (e.g., *Phyllopteroides*), plus more abundant microphyllous conifers and small-leaved bennettitaleans with thick cuticles and specialized stomata for reducing water loss. The latter could have retained their leaves throughout the winter if temperatures were low enough to slow metabolic processes significantly. In their rift valley continental setting, these floras apparently experienced greater MAR than those in coastal sites, and thus with colder winters could contain evergreen taxa.

In contrast, coastal Cenomanian floras from South Island, New Zealand (Daniel *et al.*, 1990) are physiognomically more comparable with coeval floras of Alaska. Many major groups, including angiosperms, conifers, ferns, and cycadophytes, are broad leaved, have mostly thin cuticles, and appear deciduous.

On the coast of East Antarctica, in Prydz Bay, Albian palynomorph assemblages in Ocean Drilling Program (ODP) site 119 drillhole samples (Truswell, 1990) record podocarpaceous conifer vegetation with ferns (especially schizaeaceous types) and various hepatics (liverworts). Fungal material is common. Recycled palynomorphs in Recent muds of the Weddell Sea (Truswell, 1983; Truswell and Anderson, 1985) indicate this vegetational type probably extended around much of the East Antarctic coastal margin.

Albian conifer wood from James Ross Island (~66°S) has uniform rings, absence of false rings, and large earlywood cell size, which indicate little water stress and no evidence for frost (Francis, 1986). Narrow latewood indicates sudden dark-induced dormancy, similar to coeval northern high-latitude wood.

All the southern high-latitude fossil occurrences fall on the marginal areas of Gondwanaland fragments. Palynomorph assemblages indicate the predominance of temperate, podocarp-araucarian-fern forest vegetation, with lycopod and fern moorland vegetation in some areas (Douglas and Williams, 1982; Dettmann, 1986a,b; Dettmann and Thomson, 1987; Truswell, 1990; Dettman *et al.*, 1992). The nature of the vegetation in the more inland, continental craton of East Antarctica remains unknown.

Turonian-Coniacian-Santonian

Palynological data spanning the Turonian-Coniacian-Santonian interval is available from James Ross Island (Baldoni and Medina, 1989) and from the Australasian high-latitude sector. The podocarp-araucarian-fern forest association continued through this interval in southern high latitudes. In the Turonian-Coniacian, the abundance and diversity of angiosperms remained low (Dettmann, 1989), while cryptogams were still an important part of the vegetation. This was, however, an important transitional time when the established southern podocarpaceous conifer forest vegetation was diversifying, and new angiosperm families that henceforth typified southern vegetation started to appear. The southern high-latitude region was a locus of evolutionary innovation from the Turonian to the end of the Cretaceous. Among the conifers, *Lagarostrobus* originated in the Turonian (Dettmann, 1989). This important forest tree dominated southern high-latitude forests from the Santonian through the Paleocene, and produced pollen (*Phyllocladidites mawsonii*) identical to that of the Huon pine (*L. franklinii*), now restricted to wet, cool temperate, maritime western margins of Tasmania. *Dacrydium*-type conifers originated in the Coniacian (Dettmann, 1989). Based on the angiosperm pollen record, *Ilex* (Aquifoliaceae) originated in southern high latitudes. Its earliest fossil record was in the Turonian of the Otway Basin, southeastern Australia (Martin, 1977; Dettmann, 1989). Proteaceae, now a major southern family, also evolved at about this time (Dettmann, 1989).

Campanian-Maastrichtian

Evolutionary innovation in southern floras reached its zenith during the Campanian. Evidence is provided by a rich palynomorph record from James Ross, Seymour, and adjacent islands near the tip of the Antarctic Peninsula (e.g., Dettmann and Thomson, 1987; Dettmann and Jarzen, 1988; Askin, 1989, 1990b), from New Zealand (e.g., Couper, 1960; Mildenhall, 1980; Raine, 1984), and from the Otway and Gippsland Basins of southeastern Australia (e.g., Stover and Partridge, 1973; Dettmann and Jarzen, 1988; Dettman *et al.*, 1992). The Nothofagaceae (we accept the family status, discussion and references cited in Dettmann *et al.*, 1990) originated in southern high latitudes in the early Campanian, with ancestral pollen types widespread from Australasia to the Antarctic Peninsula. *N. brassii, fusca,* and *menziesii* types differentiated soon after (Dettmann *et al.*, 1990), with the southern South America-Antarctic Peninsula area the center of *Nothofagus* diversification. In the Proteaceae, several types of both Proteoideae and Grevilleoideae occur in Campanian rocks, with earlier arrival of ancestral Proteaceae from northern Gondwana (Dettmann, 1989). The Antarctic Peninsula and New Zealand-southeastern Australia areas were both early Campanian diversification sites for various proteaceaeous groups (*Beauprea* type, *Macadamia* type, *Gevuina-Hicksbeachia* type, *Knightia* type, *Xylomelum* type; Dettmann and Jarzen, 1988; Pocknall and Crosbie, 1988; Dettmann, 1989). *Gunnera* (Gunneraceae) immigrated southward from northern Gondwana via the Antarctic Peninsula in the Campanian (Jarzen and Dettmann, 1990). Members of Myrtaceae may have entered via this route, while Winteraceae appeared in the New Zealand-southeastern Australian area (Dettmann, 1989). Loranthaceae appeared near the end of the Campanian in the Antarctic Peninsula area (Askin, 1989). The conifer *Dacrycarpus* also appeared in the high southern latitudes during the Campanian (Dettmann, 1989).

Diversification continued through the Maastrichtian, with many endemic angiosperm pollen taxa of unknown botanical affinities appearing in the fossil record. This may represent parallel (to northern) development of a herbaceous, "weedy" strategy during the general cooling trend. The latest Maastrichtian on Seymour Island (~66°S) included a short warm interval characterized by members of Bombacaceae, Olacaceae (*Anacolosa* type), and Sapindaceae (Cupanieae tribe) (Askin, 1989).

A rich Campanian leaf assemblage from King George Island (Zamek locality), South Shetland Islands, suggests a broad-leaved forest community, including evergreen types (with thick, coriaceous leaves) and deciduous *Nothofagus* growing in a subhumid mesothermal climate (Birkenmajer and Zastawniak, 1989). Abundant ferns, especially gleicheniaceous types (Cao, 1989), grew on adjacent moist lowland areas at the end of the Cretaceous.

Conifer and angiosperm wood from the James Ross Basin have wide, uniform rings and low latewood-to-earlywood ratios, indicating high productivity, sudden dark-induced dormancy, and no water stress (Francis, 1986). Rainfall of 1000 to 2000 mm/yr was suggested. From the late Maastrichtian (and Early Paleocene) on Seymour Island, dispersed plant cuticle from evergreen foliage of araucarians, Cupressaceae, and podocarp conifers, plus numerous angiosperm taxa, including Myrtaceae and Lauraceae, indicates a wet climate, MAT approximately 8 to 15°C, MAR probably <16°C, and CCM >1°C (G. R. Upchurch, University of Colorado, personal communication, 1990).

Podocarpaceous conifer (especially *Lagarostrobus*)-*Nothofagus*-Proteaceae forest grew throughout much of the southern high latitudes. Presumed paleoclimates were humid, warm-cool temperate (Cranwell, 1969; Mildenhall, 1980; Francis, 1986; Dettmann and Thomson, 1987; Dettmann and Jarzen, 1988; Askin, 1989, 1990a; Truswell, 1991).

Northern Cenozoic

Paleocene

After a slow start, woody angiosperms became the vegetation dominants, with about 20 leaf taxa in the fossil record. During the Paleocene, angiosperm wood (with vessels and other angiospermous features) occurred on the North Slope of Alaska. Betulaceous-Ulmaceous forms became established, and also migrating from lower latitudes were members of Palmae, Fagaceae, and Juglandaceae (Wolfe, 1966, 1980; Wolfe and Wahrhaftig, 1970; Spicer *et al.*, 1987). "*Metasequoia*" was the dominant conifer. Leaf assemblages include rare Cupressaceous foliage and the first record of a rosaceous leaf on the Alaskan North Slope (J. A. Wolfe, U.S. Geological Survey, personal communication, 1989). The MAT was 7°C, and the climate periodically dry, although coals were extensive, albeit thin (less than 1-m thick). The palynomorph record has yet to be examined in detail.

Eocene

The fossil record is good in Canada and southwest Alaska, though poor in northern Alaska. On Ellesmere Island, at 66°N, the vegetation included high-productivity taxodiaceous forests. Floras across this region, and in Svalbard and Greenland, suggest a possible thermal maximum for the past 100 m.y. (Heer, 1868; Schweitzer, 1974; Wolfe and Poore, 1982; Francis and McMillan, 1987).

During warm intervals of the Middle and Late Eocene, the vegetation of southern Alaska included multistratal rain forest dominated by Palmae, Lauraceae, Menispermaceae, Fagaceae, Theaceae, and other mesothermal families. Further north, the Alaska Eocene vegetation was dominantly broad-leaved deciduous (e.g., Meshik floras). Important elements were Taxodiaceae, "*Cercidiphyllum*," "*Cocculus*," *Fagus*, *Quercus*, and Juglandaceae (Wolfe, 1966, 1980, 1985; Spicer *et al.*, 1987). These taxa were all derived from lower-latitude ancestors.

Oligocene-Miocene-Pliocene

Overall diversity decreased into the Early Oligocene, although Pinaceae and Salicaceae increased in diversity and ecological importance. The vegetation included cool temperate, angiosperm-dominated, broad-leaved deciduous forest. Thick coals (at Healey, Alaska) interspersed with very thick conglomerates suggest very high and evenly distributed rainfall (R.A.S., personal observation).

Cooling produced a general pattern of southward migrations of particular lineages of closely related species. *Salix*, which originated in lower latitudes, is diverse in Early Miocene and even more diverse in the later Miocene of the Kenai Group, contrasting with its lower diversity in coeval midlatitude assemblages (Wolfe, 1985). Early Middle Miocene floras on Banks Island (70°-80°N) include *Pinus*, *Picea*, *Tsuga*, and *Juglans* (Hills *et al.*, 1974).

Grasses first became a major component of the pollen flora at about the Miocene-Pliocene transition, and taiga and tundra began to develop on a wide scale (Wolfe, 1985). Taiga appears to develop first in Siberia at about this time (Baranova *et al.*, 1968; Sher *et al.*, 1979) and provides evidence for periglacial conditions.

Southern Cenozoic

The present ice cover on Antarctica greatly hampers the retrieval of information on Cenozoic (and older) vegetation and climates. Much of our knowledge is derived from careful interpretation of recycled palynomorphs in recent seafloor muds around the coast of Antarctica, with sediment provenance inferred from tracing ice stream paths (Truswell, 1983; Truswell and Drewry, 1984). Only a generalized view of vegetation and climate could be derived from these data, in part because there was no refined age control for many of the (endemic) palynomorphs. This situation should improve as more Antarctic and Subantarctic sediments are drilled and as precise stratigraphic ranges of distinctive palynomorph species are established from these drillholes, and from the rare outcrop localities such as Seymour Island.

Paleocene

On the Antarctic Peninsula, the southern high-latitude lower Paleocene vegetation apparently had decreased diversity (Askin, 1988a). Cuticular evidence suggests that conditions may have been slightly cooler than in the Maastrichtian; however, the composition of dispersed cuticle assemblages and the presence of frost-sensitive epiphyllous fungi preclude severe frost, at least in coastal areas (G. R. Upchurch, University of Colorado, personal communication, 1990). Tree-ring data from conifer and angiosperm wood also suggest cooler conditions through the upper Maastrichtian into the Paleocene (Francis, 1986, 1991). Podocarpaceous conifer-*Nothofagus*-Proteaceae forest floras continued through the Paleocene. *Lagarostrobus* remained an important component, and its predominance in Paleocene coals in Australia (Stover and Partridge, 1973; Martin, 1984; Truswell, 1990) and Seymour Island (Fleming and Askin, 1982) highlight its preference for wet habitats. In marginally high latitudes, Cunoniaceae-dominated vine forests with conifers covered the central Australian landscape (Sluiter, 1990), and further south, *Nothofagus* was becoming more prominent. By the Paleocene, the New Zealand-Campbell Plateau block had separated and was drifting into lower latitudes.

Leaf floras from King George Island with *Nothofagus*, Myrtaceae, various other angiosperms, and ferns, represent a temperate broad-leaved forest with MAT of 10 to 12°C and rainfall of 1000 to 4000 mm/yr (Birkenmajer and Zastawniak, 1989). Seymour Island Paleocene leaf floras are dominated by fern foliage, with microphyllous *Nothofagus*, other angiosperms (some notophyllous), and conifer foliage, and probably reflect a cool temperate climate (Dusen, 1908; Case, 1988). Evidence for warm, humid conditions in East Antarctica (adjacent to Maud Rise) in the earliest Paleocene is provided by ODP 113 material, with further evidence from subsequent Paleocene lateritic and possible aeolian sediments for warm, semiarid, continental climate in inland East Antarctica (Kennett and Barker, 1990).

Eocene

The Eocene is marked by diverse floras and a general shift from conifer-dominated forest to *Nothofagus*-dominated vegetation in the coastal high latitudes. Warmth-loving (frost-sensitive) taxa are present through much of the Eocene, and a warm, moist climate is indicated. Information is available from leaf floras (Dusen, 1908; Case, 1988) and palynomorphs (Zamaloa *et al.*, 1987; Askin *et* al., 1991) from the Seymour Island area; recycled palynomorphs from around the Antarctic coast (Truswell, 1983); leaves (e.g., Birkenmajer and Zastawniak, 1989; Li and Shen, 1989), wood (Torres and LeMoigne, 1988), and palynomorphs (Stucklik, 1981; Lyra, 1986; Torres and Meon, 1990) in the South Shetland Islands; and palynomorphs in Middle Eocene ODP samples in the South Orkney region (Mohr, 1990), close to 60°S. Terrestrial dispersal corridors that linked Antarctica with the other austral land masses were severed by the Late Eocene, keeping Antarctic plant taxa from emigrating northward, and preventing active immigration to Antarctica.

Oligocene-Miocene-Pliocene

The Oligocene epoch, during which major tectonic and climate changes of lasting effect took place, saw a drastic reduction in vegetational diversity. For Early Oligocene *Nothofagus*-fern and fern bush floras on King George Island, Birkenmajer and Zastawniak (1989) suggest MAT of 11.7 to 15°C and rainfall between 1220 and 3225 mm. Also on King George Island, leaf floras of Oligocene-Miocene age suggest *Nothofagus*-podocarp conifer forest communities with possible MAT of 5 to 8°C and rainfall of 600 to 4300 mm (Birkenmajer and Zastawniak, 1989).

In the Maud Rise area (ODP 113 material), late Early Oligocene palynomorph assemblages contain no terrestrially derived spores or pollen (Mohr, 1990), implying possible ice cover or greatly reduced vegetation on adjacent East Antarctica.

In the Ross Sea area (close to 70°S paleolatitude), Late Oligocene material (from Deep Sea Drilling Project (DSDP) site 270) indicates that a *Nothofagus*-dominated cold temperate forest, with Proteaceae, conifer (*Podocarpus*), Myrtaceae, and rare ferns still existed, despite evidence for sea-level glaciation. Also in this area, other drillholes penetrating Oligocene sediment (MSSTS-1, CIROS-1) recovered evidence of mixed *Nothofagus*-conifer forests, with other angiosperms also present (results summarized in Truswell, 1990). Mildenhall (1989, CIROS-1) suggested that the coastal fringes of the Ross Sea supported a *Nothofagus* forest with podocarps, Proteaceae, and other shrubby angiosperms. Upper Oligocene sediments in CIROS-1 contain a leaf (Hill, 1989) that resembles *Nothofagus gunnii*, an extant deciduous alpine Tasmanian species. Clay mineralogy, together with the fossil evidence, is consistent with a cool or cold temperate environment, and temperate rather than polar ice is indicated (Barrett et al., 1989).

There is at present no good information on Miocene Antarctic vegetation; however some elements of the earlier flora must have survived in sheltered coastal refugia during major glacial advances. Wood and pollen remains, albeit reflecting a stunted, very depauperate flora, have been found near the head of the Beardmore Glacier in Pliocene glacial deposits now at 1800- to 1990-m elevations (Webb et al., 1987; Harwood, 1988). The wood is from a shrubby *Nothofagus* similar to extant species in Tasmania (*N. gunnii*) and southernmost South America (Carlquist, 1987), and the palynomorph assemblage contains only a few taxa, among them *Nothofagus* (*fusca* group), a podocarp and pollen resembling herbaceous Labiateae or Polygonaceae (Askin and Markgraf, 1986).

COMPARISON OF NORTHERN AND SOUTHERN HIGH-LATITUDE VEGETATION

Cretaceous Evolutionary Trends

Southern Cretaceous floras are notable for their high-latitude origins of major southern clades. In contrast, ancestral forms for all major northern clades (e.g., families) occur first at lower latitudes (Spicer et al., 1987). This contrast is due to differences in continental configurations and resulting differences in biotic stress.

The early (Cenomanian) diversity of angiosperm, and particularly platanoid leaf forms, in the northern latitudes was at the generic or lower level. Diversity at the family level was low. These trends are a function of evolutionary novelty, coupled with hybridization and polyploidy, giving rise to morphological intergradation (Spicer, 1986). Vast land areas surrounded the North Pole through the Late Cretaceous to Cenozoic, facilitating migrations to and from the polar latitudes. Thus, new taxa were exposed to broad selection pressures, suppressing development of polar specialists.

In the high southern latitudes, Antarctica remained in the polar position and was connected to lower-latitude land areas by relatively restricted dispersal corridors. The Antarctic plant communities were thus buffered to a certain extent from the competitive stress of invading taxa. Changing climatic conditions in this relatively isolated region would favor origin of specialists adapted to the changing environment, and low competitive stress would allow these specialists to become established. Major evolutionary innovation started in the Santonian and peaked in the Campanian-Maastrichtian. The increased rate of evolution coincides approximately with the end of the "Cretaceous quiet zone" (Harland et al., 1982) at the Santonian-Campanian transition and, more importantly, occurs during the Late Cretaceous cooling.

Northern and Southern Floras: Deciduous Versus Evergreen

In fossil floras, evergreen habit is recognized from the thick coriaceous nature of leaves and diagnostic (thick) dispersed plant cuticle fragments. Deciduousness can often be inferred from fossil leaf mats representing autumnal fall, dehiscence structures on conifer short shoots, etc. It is not reliable to extrapolate past leaf retention-shedding from habits of modern analogues, because at least some plants (e.g., *Sequoia*, which was possibly ancestrally deciduous; Spicer, 1987) may not be obligate evergreen or deciduous taxa, but capable of either habit over time, depending on ambient light related to latitude.

Microphyllous-xeromorphic (e.g., small-leaved, needle-leaved) plants can retain their leaves and survive winter cold. In extreme cold, evergreenness is not detrimental because metabolic processes are essentially shut down. In warmer greenhouse winters, however, leaf shedding was suggested as a better strategy to conserve stored metabolites (Spicer, 1987; Spicer and Chapman, 1990).

Northern Cretaceous high-latitude conifers and angiosperms were almost entirely deciduous, and in the Late Cretaceous the geographic boundary between deciduous (poleward) and evergreen floras coincided approximately with the paleo-Arctic Circle (66°N—limit of 24-hr light-dark regime; Wolfe, 1985). Deciduousness still typifies present mid- to high-latitude northern vegetation, along with an evergreen mixed coniferous zone.

Deciduousness in the north evolved at low to midlatitudes in response to seasonally arid conditions (like modern savanna). Large leaves (for higher productivity) could be produced only if they were shed in the dry season. In conifers, inefficient vascular systems necessitated small leaves, even under relatively favorable conditions, although many conifers were also deciduous. For plant taxa that subsequently grew at high latitudes under a polar light regime, deciduousness was fortuitously advantageous.

Extant southern floras are typically evergreen, but it is uncertain what habit prevailed in Cretaceous southern high latitudes. Evergreenness can be advantageous in polar latitudes because less energy is required to produce photosynthetic organs each spring, and a quick start to the short growing season is possible (Spicer and Chapman, 1990). This is true, however, only if freeze damage can be avoided in the winter or, at the other extreme, if warm winter temperatures do not lead to metabolite depletion during respiration. The southern middle to late Cretaceous floras, close to the paleo-Antarctic Circle, included some deciduous taxa. At the end of the Cretaceous, and continuing into the Paleocene, the Seymour area vegetation was evergreen. At those paleolatitudes (60 to 65°S, and this applies to parts of the Antarctic margin in the Paleogene) a prolonged dark period would not occur and an evergreen habit with mild winters would not have been a problem. Thus, warm, moist coastal areas of Antarctica could have supported an evergreen forest. (In northern latitudes, evergreens extended to 70 to 75°N during the Eocene.) There is, however, no fossil evidence to determine which habit predominated in inland Antarctica. There, deciduousness would have been beneficial and perhaps some southern plant taxa assumed a deciduous habit when growing under polar light regimes during greenhouse warmth. However, the greater continentality experienced in the south appears to have given rise to winter temperatures that were low enough to favor evergreenness as a successful strategy (*sensu* Spicer and Chapman, 1990). Habit of ancestral *Nothofagus* that evolved under these conditions remains unknown (Dettmann *et al.* (1990), and extant *Nothofagus* includes both deciduous (in Patagonia-Tierra del Fuego and Tasmania) and evergreen species.

Cenozoic Vegetational Changes

In both hemispheres, after a cooler earliest Paleocene, climatic warming into the Eocene was reflected by concomitant changes in high-latitude vegetation. Late Paleocene and Eocene floras exhibit increased diversity and complexity and, in both northern and southern high latitudes, include typically warmer-climate (mesothermal) taxa and communities. In the Cenozoic, global climates ultimately were controlled by tectonism that occurred in southern mid- to high latitudes, culminating in the isolation of Antarctica, inception of the Antarctic Circumpolar Current, development of the modern cryospheric ocean, buildup of Antarctic ice, and deterioration into icehouse conditions (references in Kennett and Barker, 1990). Vegetational changes in the north and south correspond to tectonic events and climatic changes (Figures 9.5 and 9.6). Whereas broad dispersal corridors surrounding the North Pole allowed climatically driven northward and southward migrations, post-Eocene Antarctica was completely isolated, resulting in stepwise extinction of taxa with each climatic deterioration. Important sequential cooling steps identified in the Cenozoic southern oceans (summarized by Kennett and Barker, 1990) occurred in the Middle Eocene, near the Eocene-Oligocene boundary, in the middle Oligocene, the Middle Miocene, the early Late Miocene, the latest Miocene, and the Late Pliocene, with some intervening, short-lived warming trends.

Southern high-latitude floras suffered the loss of many taxa (including mesothermal types) during and at the end of the Eocene. Subsequent loss of most remaining taxa occurred through the Oligocene and Miocene, with an extremely depauperate flora surviving to the Pliocene in coastal refugia. Along ice-free margins, present-day Antarctica supports a cryptogamic flora (e.g., mosses, lichens) with only two vascular plant species found in sheltered areas of the Antarctic Peninsula.

A similar diversity decline occurred in northern high latitudes during and at the end of the Eocene with the southward migration of many taxa and permanent disappearance from high latitudes of mesothermal forms. Superimposed on the poleward and equatorward migrations with warming or cooling trends were significant vegetational changes such as the increase of Pinaceae and Salicaceae in high latitudes and development of an evergreen, mixed conifer forest community (including *Pinus*, *Picea*, and *Tsuga*). During the final cooling stages, the taiga forest community developed and, subsequently, the tundra.

Possible Future Changes

Geological evidence such as that presented here shows that at times of global warmth, plant productivity and carbon sequestering are high in both Arctic and Antarctic regions. Should there be a future increase in mean global temperature, tundra will be replaced by conifer forest, which in continental interiors may well retain an evergreen component. However in coastal settings, where winter temperatures are likely to be ameliorated by the ocean, deciduous conifers and angiosperms are likely to fare better. This has profound implications for future forestry. Understanding the structure and dynamics of polar forests, by using the past natural systems as models, underscores the value of polar paleobotanical studies and provides a blueprint for engineered high-latitude ecosystems designed for carbon capture.

ACKNOWLEDGMENTS

This review is largely based on a presentation at a symposium, arranged by the Geophysics Study Committee, at the 1990 annual meeting of Geological Society of America. We appreciate the helpful comments of our reviewers, Drs. Charles J. Smiley and Thomas N. Taylor, and Stephen R. Jacobson for an earlier review. R.A.A gratefully acknowledges support of a National Science Foundation grant (DPP 9019378), and R.A.S. the support of a NERC grant (GR3/7939).

REFERENCES

Archangelsky, S. (1980). Palynology of the lower Cretaceous in Argentina, *Proceedings of the Fourth International Palynological Conference*, v. 2, Lucknow (1976-77), pp. 425-428.

Askin, R. A. (1988a). The Campanian to Paleocene palynological succession of Seymour and adjacent islands, northeastern Antarctic Peninsula, in *Geology and Paleontology of Seymour Island, Antarctic Peninsula*, R. M. Feldmann and M. O. Woodburne, eds., Geological Society of America Memoir 169, pp. 131-153.

Askin, R. A. (1988b). The palynological record across the Cretaceous/Tertiary transition on Seymour Island, Antarctica, in *Geology and Paleontology of Seymour Island, Antarctic Peninsula*, R. M. Feldmann and M. O. Woodburne, eds., Geological Society of America Memoir 169, pp. 155-162.

Askin, R. A. (1989). Endemism and heterochroneity in the Late Cretaceous (Campanian) to Paleocene palynofloras of Seymour Island, Antarctica: Implications for origins, dispersal and palaeoclimates of southern floras, in *Origins and Evolution of the Antarctic Biota*, J. A. Crame, ed., Geological Society Special Publication 147, pp. 107-119.

Askin, R. A. (1990a). Cryptogam spores from the upper Campanian and Maastrichtian of Seymour Island, Antarctica, *Micropaleontology 36*(2), 141-156.

Askin, R. A. (1990b). Campanian to Paleocene spore and pollen assemblages of Seymour Island, Antarctica, *Review of Palaeobotany and Palynology 65*, 105-113.

Askin, R. A., and V. Markgraf (1986). Palynomorphs from the Sirius Formation, Dominion Range, Antarctica, *Antarctic Journal of the United States 21*(5), 34-35.

Askin, R. A., D. H. Elliot, J. D. Stilwell, and W. J. Zinsmeister (1991). Stratigraphy and paleontology of Cockburn Island, Antarctic Peninsula, *South American Journal of Earth Science 4*, 99-117.

Bailey, I. W., and E. W. Sinnot (1915). A botanical index of Cretaceous and Tertiary climates, *Science 41*, 831-834.

Baldoni, A. M., and V. Barreda (1986). Estudio palinologico de las Formaciones Lopez de Bertodano y Sobral, Isla Vicecomodoro Marambio, Antartida, *Boletim IG-USP, Ser. Cient. 17*, University of Sao Paulo, 89-98.

Baldoni, A. M., and F. Medina (1989). Fauna y microflora del Cretacico, en bahia Brandy, isla James Ross, Antartida, *Ser. Cient. INACH 39*, 43-58.

Baranova, I. P., S. F. Biske, V. F. Goncharov, I. A. Kulkova, and A. S. Titkov (1968). Cenozoic of the Northeast USSR [in Russian], *Tr. Akad. Nauk. SSSR Sib. Fil. Inst. Geol. and Geofiz. 38*, 125 pp.

Barrera, E., B. T. Huber, S. M. Savin, and P. N. Webb (1987). Antarctic marine temperatures: Late Campanian through Early Paleocene, *Paleoceanography 2*, 21-47.

Barrett, P. J., M. J. Hambery, D. M. Harwood, A. R. Pyne, and P. N. Webb (1989). Synthesis, in *Antarctic Cenozoic History from the CIROS-1 Drillhole, McMurdo Sound*, P. J. Barrett, ed., D.S.I.R. Bulletin 245, pp. 241-251.

Berggren, W. A., D. V. Kent, J. J. Flynn, and J. A. Van Couvering (1985). Cenozoic geochronology, *Geological Society of America Bulletin 96*, 1407-1418.

Birkenmajer, K., and E. Zastawniak (1989). Late Cretaceous-early Tertiary floras of King George Island, West Antarctica: Their stratigraphic distribution and palaeoclimatic significance, in *Origins and Evolution of the Antarctic Biota*, J. A. Crame, ed., Geological Society Special Publication 147, pp. 227-240.

Cao Liu (1989). Late Cretaceous sporopollen flora from Half Three Point on Fildes Peninsula of King George Island, Antarctica, in *Proceedings, International Symposium on Antarctic Research*, China Ocean Press, Tianjin, pp. 151-156.

Carlquist, S. (1987). Pliocene *Nothofagus* wood from the Transantarctic Mountains, *Aliso 11*(4), 571-583.

Case, J. A. (1988). Paleogene floras from Seymour Island, Antarctic Peninsula, in *Geology and Paleontology of Seymour*

Island, Antarctic Peninsula, R. M. Feldmann and M. O. Woodburne, eds., Geological Society of America Memoir 169, pp. 523-530.

Chapman, J. L., and J. L. Smellie (1992). Cretaceous fossil wood and palynomorphs from Williams Point, Livingston Island, Antarctic Peninsula, *Review of Palaeobotany and Palynology 74*, 163-192.

Couper, R. A. (1960). New Zealand Mesozoic and Cainozoic plant microfossils, *New Zealand Geological Survey Paleontology Bulletin 32*, 87 pp.

Cranwell, L. M. (1969). Palynological intimations of some pre-Oligocene Antarctic climates, in *Palaeoecology of Africa*, van Zinderen Bakker, ed., S. Balkema, Cape Town, pp. 1-19.

Daniel, I. L., J. D. Lovis, and M. B. Reay (1990). A brief introductory report on the mid-Cretaceous megaflora of the Clarence Valley, New Zealand, *Proceedings of the 3rd IOP Conference*, Melbourne, 1988, 27-29.

Dettmann, M. E. (1986a). Early Cretaceous palynoflora of subsurface strata correlative with the Koonwarra Fossil Bed, Victoria, in *Plants and Invertebrates from the Lower Cretaceous Koonwarra Fossil Beds, South Gippsland, Victoria*, P. A. Jell and J. Roberts, eds., Association of Australasian Palaeontologists Memoir 3, pp. 79-110.

Dettmann, M. E. (1986b). Significance of the Cretaceous-Tertiary spore genus *Cyatheacidites* in tracing the origin and migration of *Lophosoria* (Filicopsida), *Special Papers in Palaeontology*, No. 35, pp. 63-94.

Dettmann, M. E. (1989) Antarctica: Cretaceous cradle of austral temperate rainforests? in *Origins and Evolution of the Antarctic Biota*, J. A. Crame, ed., Geological Society Special Publication 147, pp. 89-105.

Dettmann, M. E., and D. M. Jarzen (1988). Angiosperm pollen from uppermost Cretaceous strata of southeastern Australia and the Antarctic Peninsula, *Association of Australasian Palaeontologists Memoir 5*, 217-237.

Dettmann, M. E., and M. R. A. Thomson (1987). Cretaceous palynomorphs from the James Ross Island area, Antarctica—A pilot study, *British Antarctic Survey Bulletin 77*, 13-59.

Dettmann, M. E., D. T. Pocknall, E. J. Romero, and M. del C. Zamaloa (1990). *Nothofagidites* Erdtman ex Potonie, 1960; A catalogue of species with notes on the paleogeographic distribution of *Nothofagus* Bl. (Southern Beech), *New Zealand Geological Survey Paleontology Bulletin 60*, 79 pp.

Dettmann, M. E., R. E. Molnar, J. G. Douglas, D. Burger, C. Fielding, H. T. Clifford, J. Francis, P. Jell, T. Rich, M. Wade, P. V. Rich, N. Pledge, A. Kemp, and A. Rozefelds (1992). Australian Cretaceous terrestrial faunas and floras: Biostratigraphic and biogeographic implications, *Cretaceous Research 13*, 207-262.

Douglas, J. G., and G. E. Williams (1982). Southern polar forests: The Early Cretaceous floras of Victoria and their palaeoclimatic significance, *Palaeogeography, Palaeoclimatology, Palaeoecology 39*, 171-185.

Dusen, P. (1908). Uber die tertiare Flora der Seymour Insel, *Wissenschaft. Ergeb schwed. Sudpolarexped. 1901-1903 Bd3(3)*, 127 pp.

Fleming, R. F., and R. A. Askin (1982). An early Tertiary coal bed on Seymour Island, Antarctic Peninsula, *Antarctic Journal of the United States 17(5)*, 67.

Francis, J. E. (1986). Growth rings in Cretaceous and Tertiary wood from Antarctica and their palaeoclimatic implications, *Palaeontology 29*, 665-684.

Francis, J. E. (1991). Palaeoclimatic significance of Cretaceous-Early Tertiary fossil forests of the Antarctic Peninsula, in *Geological Evolution of Antarctica*, M. R. A. Thomson et al., eds., Cambridge University Press, Cambridge, pp. 623-627.

Francis, J. E., and N. J. McMillan (1987). Fossil forests of the Far North, *Geos 16*, 6-9.

Frederiksen, N. O. (1989). Changes in floral diversities, floral turnover rates, and climates in Campanian and Maastrichtian time, North Slope of Alaska, *Cretaceous Research 10*, 249-266.

Frederiksen, N. O., T. A. Ager, and L. E. Edwards (1988). Palynology of Maastrichtian and Paleocene rocks, lower Colville River region, North Slope of Alaska, *Canadian Journal of Earth Science 25*, 512-527.

Funder, S., N. Abrahamsen, O. Bennile, and R. W. Feyling-Nanssen (1985). Forested Arctic: Evidence from North Greenland, *Geology 13*, 542-546.

Grant, P. R., R. A. Spicer, and J. T. Parrish (1988). Palynofacies of Northern Alaskan Cretaceous coals, *7th International Palynological Congress, Abstracts*, Brisbane, 60.

Gregory, R. T., C. B. Douthitt, I. R. Duddy, P. Rich, and T. H. Rich (1989). Oxygen isotopic composition of carbonate concretions from the lower Cretaceous of Victoria, Australia: Implications for the evolution of meteoric waters on the Australian continent in a paleopolar environment, *Earth and Planetary Science Letters 92*, 27-42.

Haq, B. U., J. Hardenbol, and P. R. Vail (1987). Chronology of fluctuating sea levels since the Triassic, *Science 235*, 1156-1167.

Harland, W. B., A. V. Cox, P. G. Lleyellyn, C. A. G. Pickton, A. G. Smith, and R. Walters (1982). *A Geologic Time Scale*, Cambridge University Press, Cambridge, 131 pp.

Harland, W. B., R. L. Armstrong, A. V. Cox, L. E. Craig, A. G. Smith, and D. G. Smith (1990). *A Geologic Time Scale 1989*, Cambridge University Press, Cambridge, 263 pp.

Harwood, D. M. (1988). Upper Cretaceous and lower Paleocene diatom and silicoflagellate biostratigraphy of Seymour Island, eastern Antarctic Peninsula, in *Geology and Paleontology of Seymour Island, Antarctic Peninsula*, R. M. Feldmann and M. O. Woodburne, eds., Geological Society of America Memoir 169, pp. 55-129.

Heer, O. (1868). Die fossile Flora der Polarlander, *Flora Fossilis Arctica 1*, 192 pp.

Hickey, L. J., R. M. West, M. R. Dawson, and D. K. Choi (1983). Arctic terrestrial biota: Palaeomagnetic evidence for age disparity with mid northern latitudes during the Late Cretaceous and early Tertiary, *Science 222*, 1153-1156.

Hill, R. S. (1989). Fossil leaf, in *Antarctic Cenozoic History from the CIROS-1 Drillhole, McMurdo Sound*, P. J. Barrett, ed., D.S.I.R. Bulletin 245, pp. 143-144.

Hills, L. V., and J. V. Matthews (1974). A preliminary list of fossil plants from the Beaufort Formation, Meighen Island, District of Franklin, *Geological Survey of Canada Paper 74-1B*, 224-226.

Hills, L. V., J. E. Klovan, and A. R. Sweet (1974). *Juglans eocinerea* n.sp., Beaufort Formation (Tertiary), southwestern Banks Island, Arctic Canada, *Canadian Journal of Botany 52*, 65-90.

Hollick, A. (1930). The upper Cretaceous floras of Alaska, *U.S. Geological Survey Professional Paper 156*, 123 pp.

Jarzen, D. M., and M. E. Dettmann (1990). Taxonomic revision of *Tricolpites reticulatus* Cookson ex Couper, 1953 with notes on the biogeography of *Gunnera*, L, *Pollen et Spores 31*, 97-112.

Jarzen, D. M., and G. Norris (1975). Evolutionary significance and botanical relationships of Cretaceous angiosperm pollen of the Western Canadian Interior, *Geoscience and Man 11*, 47-60.

Kemp, E. M. (1975). Palynology of Leg 28 drillsite, Deep Sea Drilling Project, in *Initial Reports of the Deep Sea Drilling Project 28*, D. E. Hayes, L. A. Frakes, *et al.*, eds., U.S. Government Printing Office, Washington, D.C., pp. 599-621.

Kemp, E. M., and P. J. Barrett (1975). Antarctic glaciation and early Tertiary vegetation, *Nature 258*, 507-508.

Kennett, J. P., and P. F. Barker (1990). Latest Cretaceous to Cenozoic climate and oceanographic developments in the Weddell Sea, Antarctica: An ocean-drilling perspective, in *Proceedings of the Ocean Drilling Program, Scientific Results, 113*, P. F. Barker, J. P. Kennett, *et al.*, eds., Ocean Drilling Program, College Station, Texas, pp. 937-960.

Krassilov, V. A. (1979). *The Cretaceous flora of Sakhalin*, Cambridge University Press, Cambridge, 131 pp., pp. 1-183.

Kuc, M. (1983). Fossil flora of the Beaufort Formation, Meighen Island, NWT-Canada, *Canadian-Polish Research Institute, ser. A, 1*, 1-44

Lebedev, Y. L. (1978). Evolution of Albian-Cenomanian floras of Northeast USSR and the association between their composition and facies conditions, *International Geology Review 19*, 1183-1190.

Li Haomin, and Shen Yanbin (1989). A preliminary study of the Eocene flora from the Fildes Peninsula of King George Island, Antarctica, in *Proceedings, International Symposium on Antarctic Research*, China Ocean Press, Tianjin, pp. 128-135.

Lyra, C. (1986). Palinologia de sedimentos terciários da peninsula Fildes, ilha Rei George (ilhas Shetland do Sul Antártica) e algumas considerações paleoambientais, *An. Acad. Brasil. Cienc. 58*(Supl.), 137-147.

Martin, H. A. (1977). The history of *Ilex* (Aquifoliaceae) with special reference to Australia: Evidence from pollen, *Australian Journal of Botany 25*, 655-673.

Martin, J. A. (1984). The use of quantitative relationships and palaeoecology in stratigraphic palynology of the Murray Basin in New South Wales, *Alcheringa 8*, 253-272.

May, F. E., and J. D. Shane (1985). An analysis of the Umiat delta using palynologic and other data, North Slope, Alaska, *U.S. Geological Survey Bulletin 1614*, 97-120.

Mildenhall, D. C. (1980). New Zealand Late Cretaceous and Cenozoic plant biogeography: A contribution, *Palaeogeography, Palaeoclimatology, Palaeoecology, 31*, 197-233.

Mildenhall, D. C. (1989). Terrestrial palynology, in *Antarctic Cenozoic History from the CIROS-1 Drillhole, McMurdo Sound*, P. J. Barrett, ed., D.S.I.R. Bulletin 245, pp. 119-127.

Mohr, B. A. R. (1990). Early Cretaceous palynomorphs from ODP Sites 692 and 693, the Weddell Sea, Antarctica, in *Proceedings of the Ocean Drilling Program, Science Results, 113*, P. F. Barker, J. P. Kennett, *et al.*, eds, Ocean Drilling Program, College Station, Texas, pp. 449-612.

Palma-Heldt, S. (1987). Estudia palinologico en el terciario de las islas Rey Jorge y Brabante, territorio insular antartico, *Ser. Cient. INACH 36*, 59-71.

Parrish, J. T., and R. A. Spicer (1988a). Late Cretaceous terrestrial vegetation: A near-polar temperature curve, *Geology 16*, 22-25.

Parrish, J. T., and R. A. Spicer (1988b). Middle Cretaceous wood from the Nanushuk Group, central North Slope, Alaska, *Palaeontology 31*, 19-34.

Parrish, J. T., R. A. Spicer, J. G. Douglas, T. H. Rich, and P. Vickers-Rich (1991). Continental climate near the Albian South Pole and comparison with the climate near the North Pole, *Geological Society of America Abstracts with Programs*, A302.

Parrish, M. J., J. T. Parrish, J. H. Hutchison, and R. A. Spicer (1987). Late Cretaceous vertebrate fossils from the North Slope of Alaska and implications for dinosaur ecology, *Palaios 2*, 377-389.

Pocknall, D. T., and Y. M. Crosbie (1988). Pollen morphology of *Beauprea* (Proteaceae): Modern and fossil, *Reviews of Palaeobotany and Palynology 53*, 305-327.

Raine, J. I. (1984). Outline of a palynological zonation of Cretaceous to Paleogene terrestrial sediments in West Coast region, South Island, New Zealand, *New Zealand Geological Survey Report 109*, 82 pp.

Raine, J. I. (1988). The Cretaceous/Cainozoic boundary in New Zealand terrestrial sequences, *7th International Palynological Congress*, Brisbane, Abstract, 137.

Rees, P. M. (1990). Palaeobotanical contributions to the Mesozoic geology of the Northern Antarctic Peninsula region, Ph.D. thesis, University of London, 285 pp.

Rees, P. M., and J. L. Smellie (1989). Cretaceous angiosperms from an allegedly Triassic flora at Williams Point, Livingston Island, South Shetland Islands, *Antarctic Science 1*, 239-248.

Romero, E. J., and S. Archangelsky (1986). Early Cretaceous angiosperm leaves from southern South America, *Science 234*, 1580-1582.

Samylina, V. A. (1973). Correlation of lower Cretaceous continental deposits of Northeast USSR based on palaeobotanical data, *Sovietskaya Geologiya 8*, 42-57.

Samylina, V. A. (1974). Early Cretaceous floras of Northeast USSR (Problems of establishing Cenophytic floras), *Komarovskivc chteniya 27*, Izd-vo Nauka, Leningrad.

Schweitzer, H. J. (1974). Die "tertiaren" Koniferen Spitzbergens, *Palaeontographica 149B*, 1-89.

Scott, R. A., and C. J. Smiley (1979). Some Cretaceous plant megafossils and microfossils from the Nanushuk Group, Northern Alaska. A preliminary report, *U.S. Geological Survey Circular 749*, 89-111.

Sher, A. V., T. N. Kaplina, R. E. Giterman, A. V. Lozhkin, A. A. Arkangelov, S. V. Kiselyov, Y. P. Koutnetsov, E. I. Virina, and V. S. Zashigin (1979). Scientific excursion of problem "Late Cenozoic of the Kolyma Lowland," *Pacific Science Congress 14* (Khabarovsk), *Tour 11 Guidebook*, 115 pp.

Singh, C. (1975). Stratigraphic significance of early angiosperm pollen in the mid-Cretaceous of Alberta, *Geological Association of Alberta Special Paper 13*, 365-389.

Sluiter, I. R. (1990). Early Tertiary vegetation and paleoclimates, Lake Eyre region, northeastern South Australia, in *The Cainozoic of the Australian Region*, P. de Deckker and M. A. J. Williams, eds., Geological Society of Australia Special Publication.

Smiley, C. J. (1966). Cretaceous floras of the Kuk River area, Alaska—Stratigraphic and climatic interpretations, *Geological Society of America Bulletin 77*, 1-14.

Smiley, C. J. (1967). Paleoclimatic interpretations of some Mesozoic floral sequences, *American Association of Petroleum Geologists Bulletin 51*, 849-863.

Smiley, C. J. (1969a). Cretaceous floras of the Chandler-Colville region, Alaska—Stratigraphy and preliminary floristics, *American Association of Petroleum Geologists Bulletin 53*, 482-502.

Smiley, C. J. (1969b). Floral zones and correlations of Cretaceous Kukpowruk and Corwin Formations, northwestern Alaska, *American Association of Petroleum Geologists Bulletin 53*, 2079-2093.

Smith, A. C., A. M. Hurley, and J. C. Briden (1981). *Phanerozoic Paleocontinental World Maps*, Cambridge University Press, Cambridge, 102 pp.

Spicer, R. A. (1983). Plant megafossils from Albian to Paleocene rocks in Alaska, Contract Report to Office of National Petroleum Reserve in Alaska, U.S. Geological Survey.

Spicer, R. A. (1986). Comparative leaf architectural analysis of Cretaceous radiating angiosperms, in *Systematic and Taxonomic Approaches in Palaeobotany*, B. A. Thomas and R. A. Spicer, eds., Spec. Vol. System. Assoc. 31, pp. 221-232.

Spicer, R. A. (1987). The significance of the Cretaceous flora of northern Alaska for the reconstruction of the climate of the Cretaceous, *Geol. Jahrb. Reihe A, 96*, 265-291.

Spicer, R. A. (1989a). Physiological characteristics of land plants in relation to environment through time, *Proceedings of the Royal Society of Edinburgh 80*, 321-329.

Spicer, R. A. (1989b). Plants at the Cretaceous/Tertiary boundary, *Philosophical Transactions of the Royal Society of London B325*, 291-305.

Spicer, R. A., and J. L. Chapman (1990). Climate change and the evolution of high latitude terrestrial vegetation and floras, *Trends in Ecological Evolution 5*, 279-284.

Spicer, R. A., and J. T. Parrish (1986). Paleobotanical evidence for cool North Polar climates in middle Cretaceous (Albian-Cenomanian) time, *Geology 14*, 703-706.

Spicer, R. A., and J. T. Parrish (1990a). Late Cretaceous-early Tertiary palaeoclimates of northern high latitudes: A quantitative view, *Journal of the Geological Society of London 147*, 329-341.

Spicer, R. A., and J. T. Parrish (1990b). Latest Cretaceous woods of the central North Slope, Alaska, *Palaeontology 33*, 225-242.

Spicer, R. A., J. A. Wolfe, and D. J. Nichols (1987). Alaskan Cretaceous-Tertiary floras and Arctic origins, *Paleobiology 13*, 73-83.

Spicer, R. A., J. T. Parrish, and P. R. Grant (1988). Evolution of vegetation and coal-forming environments in the Late Cretaceous of the North Slope of Alaska, *Geological Society of America Abstracts and Program 20*, A29.

Stott, L. D., and J. P. Kennett (1990). The paleoceanographic and paleoclimatic signature of the Cretaceous/Paleogene boundary in the Antarctic: Stable isotope results from ODP Leg 113, in *Proceedings of the Ocean Drilling Program, Science Results, 113*, P. F. Barker, J. P. Kennett, *et al.*, eds, Ocean Drilling Program, College Station, Texas, pp. 829-848.

Stott, L. D., J. P. Kennett, N. J. Shackleton, and R. M. Corfield (1990). The evolution of Antarctic surface waters during the Paleogene: Inferences from the stable isotopic composition of planktonic foraminifers, ODP Leg 113, in *Proceedings of the Ocean Drilling Program, Science Results, 113*, P. F. Barker, J. P. Kennett, *et al.*, eds, Ocean Drilling Program, College Station, Texas, pp. 849-863.

Stover, L. E., and A. D. Partridge (1973). Tertiary and Late Cretaceous spores and pollen from the Gippsland Basin, southeastern Australia, *Royal Society of Victoria Proceedings 85*, 237-286.

Stuchlik, L. (1981). Tertiary pollen spectra from the Ezcurra Inlet Group of Admiralty Bay, King George Island (South Shetland Islands, Antarctica), *Stud. Geol. Polonica 72*, 109-132.

Taylor, D. W., and L. J. Hickey (1990). An Aptian plant with attached leaves and flowers: Implications for angiosperm origin, *Science 247*, 702-704.

Torres, T., and Y. LeMoigne (1988). Maderas fosiles terciarias de la Formacion Caleta Arctowski, isla Rey Jorge, Antartica, *Ser. Cient. INACH 37*, 69-107.

Torres, T., and H. Meon (1990). Estudio palinologico preliminar de cerro Fosil, peninsula Fildes, isla Rey Jorge, Antartica, *Ser. Cient. INACH 40*, 21-39.

Truswell, E. M. (1983). Recycled Cretaceous and Tertiary pollen and spores in Antarctic marine sediments: A catalogue, *Palaeontogr. 186B*, 121-174.

Truswell, E. M. (1990). Cretaceous and Tertiary vegetation of Antarctica: A palynological perspective, in *Antarctic Paleobiology*, T. N. Taylor and E. L. Taylor, eds., Springer-Verlag, New York, pp. 71-88.

Truswell, E. M. (1991). Antarctica: A history of terrestrial vegetation, in *The Geology of Antarctica*, R. J. Tingey, ed., Oxford University Press, New York, pp. 499-528.

Truswell, E. M., and J. B. Anderson (1985). Recycled palynomorphs and the age of sedimentary sequences in the eastern Weddell Sea, *Antarctic Journal of the United States 19*(5), 90-92.

Truswell, E. M., and D. J. Drewry (1984). Distribution and provenance of recycled palynomorphs in surficial sediments of the Ross Sea, Antarctica, *Marine Geology 59*, 187-214.

Upchurch, G. R., and J. A. Wolfe (1987). Mid-Cretaceous to early Tertiary vegetation and climate: Evidence from fossil leaves and woods, in *The Origins of Angiosperms and Their Biological Consequences*, E. M. Friis, W. G. Chaloner, and P. R. Crane, eds., Cambridge University Press, Cambridge, pp. 75-105.

Volkheimer, W., and A. Salas (1975). Die altesten Angiospermen-Palynoflora Argentiniens von der Typuslokalitat der unterkretazischen Huitrin-Folge des Neuquen-Beckens. Mikrofloristische Assoziation und biostratigraphische Bedeutung, *Neues Jahrb. Geol. Palaont., Monat. 7*, 424-436.

Webb, P. N., B. C. McKelvey, D. M. Harwood, M. C. G. Mabin, and J. H. Mercer (1987). Sirius Formation of the Beardmore Glacier region, *Antarctic Journal of the United States 21*, 8-13.

Wolfe, J. A. (1966). Tertiary plants from the Cook Inlet region, Alaska, *U.S. Geological Survey Professional Paper 398-B*, B1-B32.

Wolfe, J. A. (1971). Tertiary climatic fluctuations and methods of analysis of Tertiary floras, *Palaeogeography, Palaeoclimatology, Palaeoecology 9*, 27-57.

Wolfe, J. A. (1979). Temperature parameters of humid to mesic forests of eastern Asia and relations to forests of other regions of the Northern Hemisphere and Australasia, *U.S. Geological Survey Professional Paper 1106*, 1-36.

Wolfe, J. A. (1980). Tertiary climates and floristic relationships at high latitudes in the Northern Hemisphere, *Palaeogeography, Palaeoclimatology, Palaeoeceology 30*, 313-323.

Wolfe, J. A. (1985). Distribution of major vegetational types during the Tertiary, in *The Carbon Cycle and Atmospheric CO_2: Natural Variations Archean to Present*, E. T. Sundquist and W. S. Broecker, eds., American Geophysical Union Monograph 32, pp. 357-375.

Wolfe, J. A., and R. Z. Poore (1982). Tertiary marine and nonmarine climatic trends, in *Climate in Earth History*, Studies in Geophysics, National Academy Press, Washington D.C., pp. 154-158.

Wolfe, J. A., and G. R. Upchurch (1987). North American nonmarine climates and vegetation during the Late Cretaceous, *Palaeogeography, Palaeoclimatology, Palaeoecology 61*, 33-77.

Wolfe, J. A., and C. Wahrhaftig (1970). The Cantwell Formation of the central Alaska Range, *U.S. Geological Survey Professional Paper 1294-A*, A1-A46.

Youtcheff, J. S., P. D. Rao, and J. E. Smith (1987). Variability in two northwest Alaska coal deposits, in *Alaskan North Slope Geology*, I. Taileur and P. Weimer, eds., Society of Economic Paleontologists and Mineralogists (Pacific Section), Bakersfield, Calif., pp. 225-232.

Zamaloa, M. C., E. J. Romero, and L. Stinco (1987). Polen y esporas de la Formacion La Meseta (Eoceno Superior-Oligoceno) de la isla Marambio (Seymour), Antartida, *VII Simp. Argentino Paleobot., Actas Paleobot. Palin.*, Buenos Aires, pp. 199-203.

Ziegler, A. M. (1990). Phytogeographic patterns and continental configurations during the Permian Period, in *Palaeogeography and Biogeography*, W. S. McKerrow and C. R. Scotese, eds., Geological Society Memoir No. 12, pp. 363-379.

10
The Impact of Climatic Changes on the Development of the Australian Flora

DAVID C. CHRISTOPHEL
University of Adelaide

ABSTRACT

Australia in the Tertiary provides an excellent opportunity to examine evolution in an isolated system because for almost 30 million years (m.y.) following the Eocene break with Antarctica, the Australian plate had no major contact with any other. On at least two occasions during that interval, global climatic events were reflected in the fossil plant record. The mid-Eocene cooling is demonstrated by two neighboring floras from either side of the event being dominated by totally different plants. The terminal Eocene cooling is clearly marked by both pollen and megafossil shifts. Thus, through the Tertiary we see a basically greenhouse Eocene Gondwanic flora respond to climatic deterioration and evolve into the sclerophyll and arid communities that dominate the continent today.

UNIQUENESS OF THE AUSTRALIAN SYSTEM

Although the geological and biological uniqueness of Australia is well known and documented, there are certain aspects of that uniqueness that are particularly relevant to the consideration of climate change and the effect on vegetation though the Tertiary Period. The first of these is related to Australia's geographical isolation during much of the Tertiary. It is generally agreed (e.g., Frakes *et al.*, 1987) that from the Late Eocene (from roughly 38 m.y. ago (Ma) to 8 Ma—about 30 m.y.) the Australian Plate was isolated from all other continental plates. This has two important consequences. The first is that climatic changes, and their reflected vegetation patterns, are not significantly masked or diluted by events on other continents. Second, the consequences of change can be followed through time. As an example, a megafossil flora observed for the Early Miocene must have been based on a gene pool present in the Australian Oligocene since no credible external source is available. A corollary is that if a sudden floristic change is observed over a specific time period, the causal factors must be local (e.g., climatic change) because no outside influence can be seriously considered. Therefore, if one thinks of this time interval as an evolutionary experiment, the variables are far more limited than for any other such "experiments" on different continents.

A second important, unique feature of the Australian system is the nature of its plant fossil record. Australia has perhaps the only tropical to subtropical rain forest system in the world that has a well-documented macrofossil record; hence, its evolution can be traced through time. There are three factors contributing to this situation: first, a large number of Australia's macrofossil deposits are preserved as mummified leaves, allowing maximum taxonomic and physiognomic information to be gleaned from them (Christophel, 1981). Second, the isolation of the continent alluded to earlier means that there is a far better chance of actually identifying taxa and communities, and tracing them through time, without having to search for floras of other continents for matches. A consequence is that far greater confidence can be placed on labeling an unidentifiable fossil as extinct because the likelihood of an external taxonomic affinity is much reduced. Finally, a large number of deposits are known from the portion of the Eocene Epoch at or near the time of the early Tertiary plate separation, providing a better than average understanding of the gene pool from which later floristic elements must have been derived. A similar, though somewhat weaker case, can be made for the documentation of some of Australia's less mesic vegetation types, the qualifying feature here being the more recent evolution of these vegetation forms and their components, and hence the greater chance of external influence following Miocene collision with the Sundra Plate (Kemp, 1981).

GEOLOGICAL AND PLATE TECTONIC SETTING

The major events in the physical movements of the Australian Plate during the Tertiary are not contentious. There is general agreement that during the early Paleogene, Australia was attached to Antarctica via its southeastern corner and Tasmania (Figure 10.1) and that, although by mid-Miocene the shelves between components were likely submerged, they were still joined and oceanic circulation over that shelf was minimal. Near the end of the Eocene the rate of northward movement of the Australian Plate increased two- to threefold, and it continued at that rate until the leading, northern edge collided with the island arcs of the Sundra Plate in the Middle- to Late Miocene (Galloway and Kemp, 1981). Tectonic activity was minimal across most of the plate during this isolated rafting period, with the uplift of the eastern highlands likely occurring at an early stage of the Miocene (Ollier, 1986).

MODERN VEGETATION OF AUSTRALIA

In order to assess the impact of Tertiary climatic change on the makeup of the modern Australian flora, it is first necessary to categorize the floristic or vegetational elements in the modern-day flora. A somewhat simplified vegetation map of Australia is shown in Figure 10.2. By distilling that further, it is possible to identify four categories of vegetation: (1) the closed forest systems, (2) the open forest or woodland systems, (3) the heath scrub or mallee systems, and (4) the great arid and semiarid systems that occupy a high percentage of the continental mass. A more thorough treatment of specific vegetation types in Australia may be found in Specht (1981a,b).

In examining the vegetation types one at a time, the first type to be considered is the closed forest system. As may be seen from Figure 10.3A, this system can also

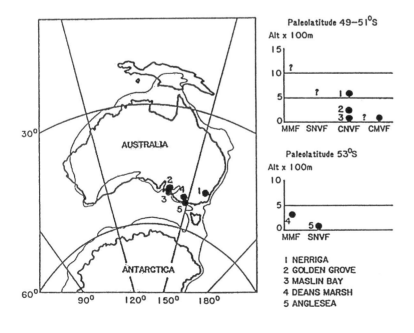

FIGURE 10.1 Reconstruction of Australia in the Eocene showing location and paleolatitude of several Eocene megafossil localities referred to in the text. The estimated altitudes and inferred forest type of each Eocene flora are shown graphically to the right of the map. MMF is microphyll mossy forest, SNVF is simple notophyll vine forest, CNVF is complex notophyll vine forest, and CMVF is complex mesophyll vine forest. (Modified from Christophel and Greenwood, 1989.)

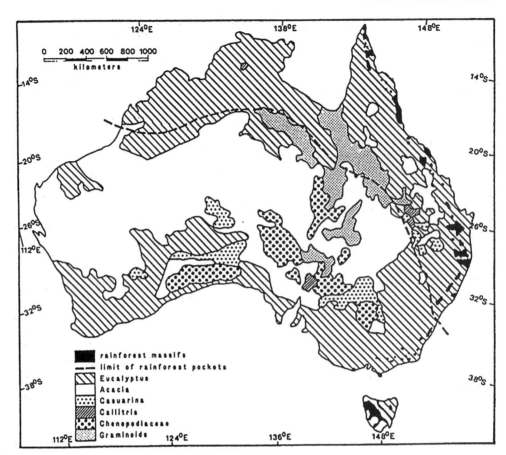

FIGURE 10.2 Simplified vegetation map of Australia. (Modified from Christophel and Greenwood, 1989.)

basically be called a rain forest system. The forest shown in this figure is found in the tropical regions of northern Queensland and is known variously as a Complex Mesophyll Vine Forest (sensu Webb, 1959) or Megathermal Seasonal and Nonseasonal (sensu Nix, 1982). Both authors agree that there is a latitudinal/altitudinal gradient in these closed forests from the tropical in the north to the cool temperate southern beech forests in the south. Although currently covering less than 0.4% of the land mass, the closed forest is particularly important to the evolution of Australian vegetation systems because it contains some of the most ancient plant associations. In general, closed forests may be categorized by high diversity and biomass, a low subcanopy light regime, and constituent plants dominated by Gondwanic taxa.

The second major vegetation type is the open forest or woodland (Figure 10.3B). It is most prevalent in eastern and far southwestern Australia. This community is dominated by *Eucalyptus* species, and has a much lower diversity and biomass accumulation than the closed forest. A far greater amount of light reaches the subcanopy in these forests because of the vertical positioning of *Eucalyptus* leaves in general, and the majority of taxa in this community type are first reported in the Neogene.

The third vegetation system is the heath scrub or mallee vegetation (Figure 10.3C). It is characterized by an unexpectedly high species richness, with a flora of mixed origins but with reasonably low biomass accumulation. A family of shrubs found in this vegetation type is the Ericridaceae, the sister family of the Northern Hemisphere Ericaceae or heath family—hence the labeling of this vegetation system as "heath." The term mallee comes from a growth form of some *Eucalyptus* species as small, multistemmed trees growing from an underground lignotuber (Figure 10.3C). It is interesting to note that some *Eucalyptus* species (e.g., *Eucalyptus baxteri*) can be found growing as either a large tree or a mallee form, depending on the environment in which it is found. The mallee vegetation type is dominated by plants considered to be sclerophyllous—an environmental adaptation that is discussed later.

Finally, the arid and semiarid regions of the continent have a complicated system of vegetation types, of which two are most common. These are represented in Figures 10.3D and 10.3E and are *Acacia* shrublands and chenopod scrub, respectively. Although this vegetation type has exceptionally low biomass and diversity during much of its life, the bi- or triennial rains affecting the region can greatly increase the biomass production and the standing

FIGURE 10.3 Illustrations of major Australian vegetation types shown in Figure 10.2: (A) tropical rain forest near Noah Creek in northern Queensland; (B) *Eucalyptus* woodland near Adelaide, South Australia; (C) mallee or heath scrub in southeastern South Australia (shown is a typical multistemmed mallee form Eucalypt); (D) arid zone vegetation featuring *Acacia* near Alice Springs in Northern Territory; (E) semiarid chenopod scrub featuring blue bush (*Maireana*) and salt bush (*Atriplex*) in northern South Australia; (F) sagebrush habitat from Kansas showing similar vegetation form to Figure 10.3E.

species richness. This vegetation type has a marked resemblance to the sagebrush communities of North America as may be seen by comparing Figures 10.3E and 10.3F, although the taxa filling the various niches are totally unrelated.

It is interesting to note that the highly diverse and successful *Eucalyptus* is absent from the true arid, as it is from the closed forests. The presence and/or relative balance of these four systems can be examined as potential indicators of the climate changes through the Tertiary.

FACTORS AFFECTING VEGETATION AND TAXON-BASED CHANGES

Simplistically, individual green plants can be considered as oxygen-burning food factories, and plant communities as industrial complexes competing for resources and for the general market. The structure of each factory and the balance between individuals (species) can be examined as responses to basic supply parameters. The study of foliar physiognomy has recognized these taxon-independent responses, and early in the twentieth century, Bailey and Sinnott (1916) noted a pattern of response between major climatic factors and specific foliar features. Wolfe (1990), however, demonstrated that univariate comparisons between individual climatic factors and plant responses were likely to lead to oversimplified or even erroneous conclusions.

In Australia, Webb (1959) erected a rain forest classification based on foliar physiognomic features, and Christophel and Greenwood (1988, 1989) demonstrated a predictable relationship between the canopy signatures used by Webb and the signatures of leaf litter. Thus, foliar physiognomy provides a tool for assessing the environment of a plant community that is independent of taxonomic identities of the constituent taxa.

Two climatic factors, available water (usually in the form of precipitation) and temperature (either mean annual, range, or extreme exclusive value), have been most frequently considered as basic to determining physiognomic signatures of floras. Within the Australian system, a third factor, the edaphic feature of soil nutrient availability (particularly phosphate), has been shown to have great importance (Beadle, 1966).

In a particularly important paper, Beadle (1966) concluded that a significant portion of the sclerophyll component of the Australian flora could likely have evolved its characteristic features in response to low nutrients rather than relative aridity. Sclerophyll plants are those with reduced, lignified leaves with short branch internodes and often thickened cuticles. Such features would commonly be thought of as having an adaptive advantage in a drying environment. Acceptance of Beadle's ideas allows a much more realistic mechanism for the floristic changes observed in the Tertiary of Australia. Beadle suggested that many sclerophyllous plants could have evolved in nutrient-poor soils around the margins of Paleogene rain forests. Thus, when climatic deterioration did occur, expansion of existing taxa from those pre-evolved low-nutrient pockets could occur much faster than if all responses to aridification had to be newly speciated. As can be seen later, Beadle's argument makes tenable the sometimes difficult to explain sclerophyllous (often thought xeromorphic) elements that crop up in otherwise mesic Paleogene rain forest plant fossil assemblages. Evidence in modern vegetation of the validity of Beadle's hypothesis comes from the Hawkebury Sandstone region of New South Wales, where in conditions of high rainfall and optimal temperatures a sclerophyll community thrives in the midst of a closed forest system. The only significant environmental difference is the very low nutrient levels of the soils supporting the sclerophyll vegetation.

MAJOR TERTIARY CLIMATIC CHANGES

The scale of climatic change being observed or monitored is related directly to the accuracy of calibration of the tools being used for the monitoring. Thus, when the palynology of late Tertiary or Quaternary deposits is being considered (e.g., Kershaw, 1976, 1981; Kershaw and Sluiter, 1982), it is possible to consider vegetation (and by inference climate) changes on a scale of thousands of years. As yet, however, it has not been possible to calibrate the Australian plant megafossil record to such accuracy.

It is still possible, however, to correlate some of the major climatic changes through the Tertiary with specific megafossil floras. If we examine a generalized chart of oxygen isotope curves for the Tertiary (Figure 10.4), some of the major climatic changes that affected many parts of the globe can be seen. Two are of particular interest. Although the Early Eocene is considered to be the most recent time at which a nearly greenhouse Earth was achieved, there was a significant cooling event at the beginning of the Middle Eocene. McGowran (1986, 1989) correlated this in Australia with an approximately 8-m.y. period during which almost no floral or faunal fossil record exists. This cool period was followed by a rapid rewarming in the late Middle and Late Eocene. However, there was a marked terminal Eocene event resulting in a rapid cooling. This Oligocene cooling is correlated in Australia with the initiation of the circum-Antarctic currents between Australia and Antarctica, and also with the glaciation of Antarctica. This cooling continued to the Middle Miocene, at which point there was a brief return to a warming cycle, followed by a cooling that has continued to the present. The two specific events to be considered relative to the Australian

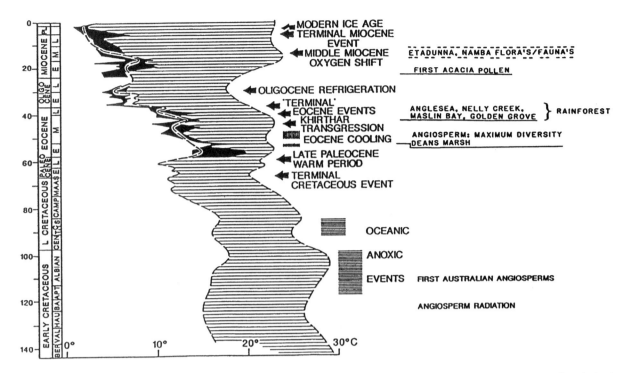

FIGURE 10.4 Oxygen isotope curves calibrated tentatively against a temperature scale. Shaded envelope covers low-latitude Pacific values from surface (to the right) planktonic foraminifera and from bottom (to the left) bethonic foraminifera. The overall trend toward cooling, the reversals, and the progressive differentiation between surface and bottom waters are all more important than the actual temperature values. Black envelope: high southern latitude oceanic profiles; again, surface to the right and bottom to the left. Shading and arrows: events and intervals of global significance. Right: paleobiological record of events of relevance to Australia. (Figure modified from Frakes et al., 1987.)

megafossil record are therefore (1) the Middle Miocene fluctuation and (2) the terminal Eocene event.

PLANT MEGAFOSSIL EVIDENCE FOR CLIMATIC CHANGE

The first area for which to consider plant megafossil evidence is the cooling cycle within the Middle Eocene. As is seen in Figure 10.5, there are a large number of known Eocene localities in southern Australia. Four of them—Maslin's Bay (Christophel and Blackburn, 1978), Golden Grove (Christophel and Greenwood, 1987), Anglesea (Christophel et al., 1987), and Nelly Creek (Christophel et al., 1991)—are late Middle Eocene and represent the return to warmth shown on Figure 10.4. They include classical tropical rain forest taxa such as *Elaeocarpus/Sloanea*, Lauraceae, and *Gymnostoma*. These four floras show a physiognomic signature consistent with that of the litter from Webb's Complex Notophyll Vine Forest or his Complex Mesophyll Vine Forest (Christophel and Greenwood, 1988). A little-documented site at Dean's Marsh (Douglas and Ferguson, 1988) (Figure 10.5), approximately 80 km east of the Anglesea locality, displays a physiognomic signature almost identical to that of the Anglesea locality. Whereas the Anglesea fossils occur high in the Eastern View Formation and are considered late Middle Eocene, the Dean's Marsh locality is basal Eastern View Formation and is late Early Eocene (Douglas and Ferguson, 1988). An examination of the taxa collected from the two localities has shown no taxa held in common. Although the deposits of the late Middle Eocene scattered across southern Australia show similar mixes of families and in many cases genera, the Dean's Marsh locality shows no such correlation. One explanation that fits nicely with the known data is that although they occur in very similar environmental situations, and hence have similar physiognomic signatures, the two deposits occur on either side of the 8-m.y. cooling period discussed by McGowran (1989); hence, the evolution and natural selection that occurred during that time selected totally different taxa for the same niches.

The major cooling event at the end of the Eocene is much easier to document. Palynological studies have highlighted it for many years (e.g., Kemp, 1978)—the most obvious signal being the shift from the tropical *Nothofagus brassii* pollen type to the Fusca and Menzesii types that

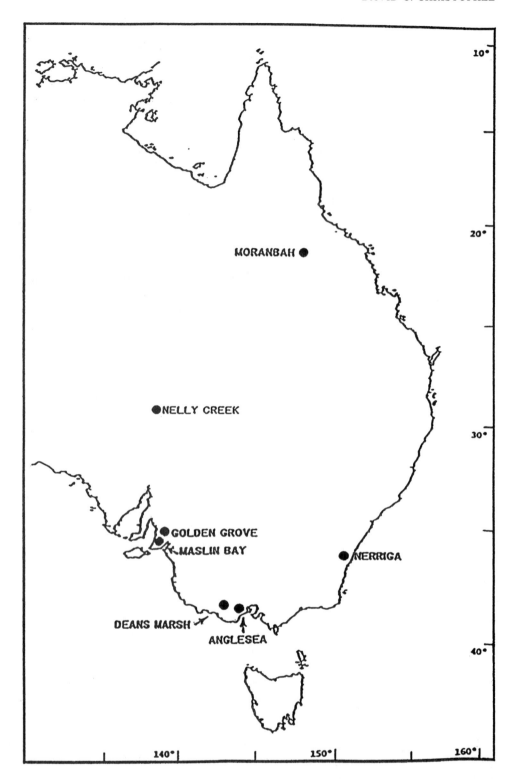

FIGURE 10.5 Map of eastern Australia showing Eocene megafossil localities cited in the text.

represent the small toothed-leaf Antarctic beeches known from the modern cool temperate floras in New Zealand, southern South America, and Tasmania. The only megafossil record of leaves and cupules of the *N. brassii* type comes from the Eocene of Tasmania (Hill, 1987), whereas the small-leafed types are prevalent in sediments from then onward. Hill and Carpenter (1991) have also provided data suggesting that some of the smaller-leafed conifers may also show a leaf size reduction across this crucial boundary.

In general the late Middle and Upper Eocene floras mentioned above all show physiognomic signatures indicating warm, wet climates, whereas those known from the Oligocene-Miocene show much reduced, sclerophyllous signatures. Localities displaying these features include Kiandra in New South Wales as well as Berwick and Bacchus Marsh in Victoria (Figure 10.6).

Not only does the physiognomic signature change, but so does the taxonomic composition of the flora. As can be seen in Table 10.1, the domination by the Gondwanic Proteaceae, Lauraceae, and *Gymnostoma* evident in the Eocene floras has been lost by Oligocene time and is replaced by elements of the flora now found in different modern communities—namely, *Eucalyptus*, *Acacia*, and Epacridaceae. The Proteaceae and Casuarinaceae are still prevalent, but the Proteaceae is now dominated by *Banksia* and sclerophyllous forms, and *Allocasuarina* and *Casuarina* (sensu Johnson, 1982) have now replaced the more mesic *Gymnostoma*. All of the above changes are well documented in the pollen record, but frustratingly, very few occurrences of *Eucalyptus* and *Acacia* are known from the megafossil record. Thus, two distinct significant global climatic events (Figure 10.4) can be seen to be reflected in the Australian megafossil record and, when considered floristically, appear to have had a major effect on the vegetation development of the continent.

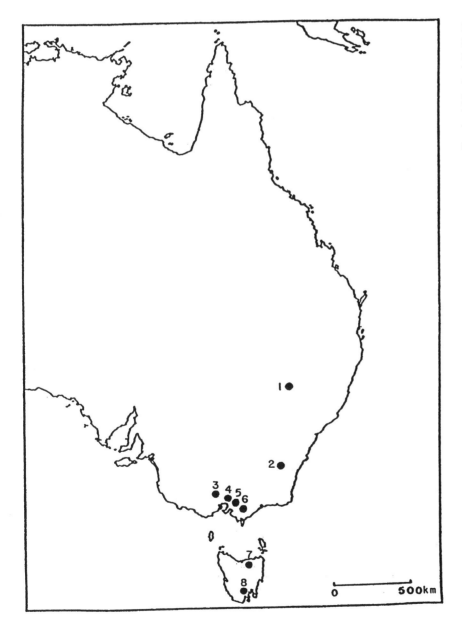

FIGURE 10.6 Map of eastern Australia showing Oligocene-Miocene plant megafossil localities cited in the text: (1) Warrumbungle Mountains, New South Wales; (2) Kinadra, New South Wales; (3) Bacchus Marsh, Victoria; (4) Berwick, Victoria; (5) Morewell, Victoria; (6) Yallourn, Victoria; (7) Pioneer, Tasmania; and (8) New Norfolk, Tasmania.

TABLE 10.1 Main Plant Groups Represented in Australian Megafossil Deposits

PALAEOGENE	NEOGENE
Proteaceae	*Banksia/Hakea*
Myrtaceae (non-*Eucalyptus*)	*Eucalyptus*
Gymnostoma	*Casuarina*
Podocarpaceae	Cupressaceae
Araucariaeae	Epacridaceae
Nothofagus	Chenopodiaceae
Elaeocarpaceae	Asteraceae
Restionaceae	Poaceae
Lauraceae	*Acacia/Cassia*

NOTE: Taxa underlined in the right-hand column may be considered as likely direct replacements (either taxonomically or vegetationally) for those opposite them in the left-hand column.

DISCUSSION

If the four major types currently found in Australia are examined once more, they can be viewed in the light of major Tertiary climatic changes to the continent. The closed forests of the north and east of the continent clearly have their affinities with the greenhouse and near-greenhouse phases of Eocene Australia. Gondwanic families dominate, and in many cases the relationships are reflected at the generic level. Although there is no Eocene evidence for *Eucalyptus*, *Acacia*, or other taxa listed as Neogene in Table 10.1, Beadles's hypothesis suggests that they nonetheless could have evolved under depauperate soil conditions, but in quantities too small to be observed in the fragmentary fossil record. Then too, they may well have been initiated or survived in the mid-Eocene cooling represented by McGowran's (in Frakes *et al.*, 1987) 8-m.y. "hole" in Australia's fossil record.

Whatever their Eocene status, the sclerophyllous plant elements that dominate the open forest and the heath scrub today either evolved or spread during the Oligocene-Miocene refrigeration. This of course was aided by the inability of the Gondwanic closed forest components to survive over large areas during this climatic deterioration. The mid-Miocene warming suggested by Frakes *et al.* (1987) may well have guaranteed the survival of some of those Gondwanic elements that struggled through the refrigeration, and may also be reflected in the mixture of floral provinces in some vegetation types such as the forests at Wilson's Promontory near the southern tip of Victoria.

As might be expected, the macrofossil record for arid floras is poor, although Chenopodiaceae and Mimosaceae pollen is well documented from the late Miocene and Pliocene of several localities (Martin, 1981). Although for Charles Darwin the flowering plants represented the "abominable mystery," for Australian researchers it is perhaps *Acacia*. Although the genus is one of the few to occur in all major Australian habitats, and contains more than 650 species in Australia (Morley and Tolkein, 1983), there is only one confirmed report of fossil leaves from the late Miocene (Christophel, 1989), and pollen is not common. Thus, the explanation for the origin and spread of a genus whose distribution suggests it to be Gondwanic, and hence ancient, remains shrouded but is almost certain to be related, when once unraveled, to the changing Tertiary climates.

ACKNOWLEDGMENTS

Much of the research for this project was supported by grants from the Australian Research Council, Alcoa of Australia, and the Adelaide University/CSIRO Granting Scheme. The figures for this chapter were prepared by Linda Allen and Leonie Jane Scriven.

REFERENCES

Bailey, I. W., and E. W. Sinnott (1916). The climatic distribution of certain kinds of angiosperm leaves, *American Journal of Botany 3*, 24-39.

Beadle, N. C. W. (1966). Soil phosphate and its role in molding segments of the Australian flora and vegetation, with special reference to xeromorphy and sclerophylly, *Ecology 47*, 992-1007.

Christophel, D. C. (1981). Tertiary megafossil floras as indicators of floristic associations and paleoclimate, in *Ecological Biogeography of Australia*, A. Keast, ed., W. Junk Publishers, The Hague, pp. 379-390.

Christophel, D. C. (1989). Evolution of the Australian flora through the Tertiary, *P. Syst. Evol. 162*, 63-78.

Christophel, D. C., and D. T. Blackburn (1978). Tertiary megafossil flora of Maslin Bay, South Australia: A preliminary report, *Alcheringa 2*, 311-319.

Christophel, D. C., and D. R. Greenwood (1987). A megafossil flora from the Eocene of Golden Grove, South Australia, *Transactions of the Royal Society of South Australia 111*, 155-162.

Christophel, D. C., and D. R. Greenwood (1988). A comparison of Australian tropical rainforest and Tertiary fossil leafbeds, in *The Ecology of Australia's Wet Tropics*, R. Kitching, ed., Proceedings of the Ecological Society of Australia 15, Surrey Beatty & Sons Pty. Ltd., Chipping Norton, New South Wales, pp. 139-148.

Christophel, D. C., and D. R. Greenwood (1989). Changes in climate and vegetation in Australia during the Tertiary, *Review of Palaeobotany and Palynology 58*, 95-109.

Christophel, D. C., W. K. Harris, and A. K. Syber (1987). The Eocene flora of the Anglesea locality, Victoria, *Alcheringa 11*, 303-323.

Christophel, D. C., L. J. Scriven, and D. R. Greenwood (1991). The Eocene Nelly Creek megafossil flora, *Transactions of the Royal Society of South Australia*.

Douglas, J. G., and J. A. Ferguson (1988). *Geology of Victoria*, Geological Society of Australia, 664 pp.

Frakes, L., B. McGowran, and J. M. Bowler (1987). Evolution of the Australian environments, in *Fauna of Australia, General Articles*, Australian Government Publishing Service, Canberra.

Galloway, R. W., and E. M. Kemp (1981). Late Cainozoic environments of Australia, in *Ecological Biogeography of Australia*, A. Keast, ed., W. Junk Publishers, The Hague, pp. 51-80.

Hill, R. S. (1987). Discovery of Nothofagus fruits corresponding to an important Tertiary pollen type, *Nature 327*, 56-57.

Hill, R. S., and R. Carpenter (1991). *Acomopyle* in the Tertiary record of Australia, *Alcheringa*.

Johnson, L. A. S. (1982). Notes on Casuarinaceae 2, *Journal, Adelaide Botanical Garden 6*, 73-87.

Kemp, E. M. (1978). Tertiary climatic evolution and vegetation history in the SE Indian Ocean region, *Palaeogeography, Palaeoclimatology, Palaeoecology 24*, 169-208.

Kemp, E. M. (1981). Tertiary paleogeography and the evolution of Australian climate, in *Ecological Biogeography of Australia*, A. Keast, ed., W. Junk Publishers, The Hague, pp. 31-50.

Kershaw, A. P. (1976). A Late Pleistocene and Holocene pollen diagram from Lynch's Crater, northeastern Queensland, Australia, *New Phytologist 77*, 469-498.

Kershaw, A. P. (1981). Quaternary vegetation and environments, in *Ecological Biogeography of Australia*, A. Keast, ed., W. Junk Publishers, The Hague, pp. 81-102.

Kershaw, A. P., and I. R. Sluiter (1982). Late Cenozoic pollen spectra from the Atherton Tableland, northeastern Queensland, Australia, *Australian Journal of Botany 30*, 279-295.

Martin, H. A. (1981). The Tertiary flora, in *Ecological Biogeography of Australia*, A. Keast, ed., W. Junk Publishers, The Hague, pp. 391-406.

McGowran, B. (1986). Cainozoic oceanic events: The Indo-Pacific biostratigraphic record, *Palaeogeography, Palaeoclimatology, Palaeoecology 55*, 247-265.

McGowran, B. (1989). The later Eocene transgressions in southern Australia, *Alcheringa 13*, 45-68.

Morley, B. D., and H. R. Toelken (1983). *Flowering Plants in Australia*, Rigby Press, 416 pp.

Nix, H. (1982). Environmental determinants of biogeography and evolution in Terra Australis, in *Evolution of the Flora and Fauna of Arid Australia*, W. R. Barker and P. J. M. Greenslade, eds., Peacock Publications, South Australia, chapter 5.

Ollier, C. D. (1986). The origin of alpine land forms in Australia, in *Flora and Fauna of Alpine Australasia: Ages and Origins*, B. Barlow, ed., CSIRO Press, Melborne, pp. 3-25.

Specht, R. L. (1981a). Major vegetation types in Australia, in *Ecological Biogeography of Australia*, A. Keast, ed., W. Junk Publishers, The Hague, pp. 163-298.

Specht, R. L. (1981b). Evolution of the Australian flora: Some generalizations, in *Ecological Biogeography of Australia*, A. Keast, ed., W. Junk Publishers, The Hague, pp. 783-806.

Webb, L. J. (1959). A physiognomic classification of Australian rainforests, *Journal of Ecology 47*, 551-570.

Wolfe, J. A. (1990). Palaeobotanical evidence for a marked temperature increase following the Cretaceous/Tertiary boundary, *Nature 343*, 153-156.

11

Global Climatic Influence on Cenozoic Land Mammal Faunas

S. DAVID WEBB and NEIL D. OPDYKE
University of Florida

ABSTRACT

The Cenozoic succession of continental mammalian faunas reflects the climatic shift from a "greenhouse" to an "icehouse" world. In midcontinental North America, where the most nearly complete record of terrestrial biotic succession has been assembled, floral and faunal evidence chronicle a broad progression of biome change from tropical forest to savanna to steppe. Thanks to improved land mammal chronologies, the tempo of faunal turnover is now seen to be strikingly syncopated. Stately chronofaunas, made up of stable sets of slowly evolving taxa, persist on the order of 10^7 yr. Chronofaunas are terminated by rapid turnover episodes in which some native taxa experience rapid evolutionary rates and other taxa appear as intercontinental immigrants.

We focus here on pulses of immigration and their correlation with major climatic shifts. The Tertiary record of North American land mammals registers a total of 140 immigrant genera. The chronology of most immigrant arrivals can be placed within the nearest million years (m.y.) or less. The pattern of immigration is decidedly nonrandom, with the largest two episodes involving 15 and 14 genera each within about 1 m.y. We categorize immigrant arrivals into first-, second-, and third-order episodes, with the first consisting of at least nine genera within less than 1 m.y. The seven first-order episodes occur at 58.5, 57, 40, 20, 18.5, 5, and 2.5 m.y. ago (Ma). The first-order episode at ~40 Ma may represent an artifact due to tallying at the end of a long interval of poorly fossiliferous deposits (Uintan). A striking generalization is that the six other first-order episodes occur as three pairs separated by only about 2 m.y.

Review of land mammal immigration and rapid turnover episodes on other continents tends to confirm the "universality" of major immigration episodes at ~20, ~5, and ~2.5Ma.

We consider two hypotheses that might causally relate high intercontinental immigration rates to global climate change. The first proposes a correlation between high immigration rates and the availability of land bridges during low sea-level intervals. In this view, the marine record of positive oxygen isotopes excursions can be read as a proxy for eustatic sea-level lowering. The second hypothesis relates first-order immigration episodes to high faunal turnover intervals and considers them to represent fundamental reorganizations of the continental ecosystem. In this view, the largest positive oxygen isotope excursions record major cooling episodes, which also radically alter terrestrial ecosystems. We consider both hypotheses as we compare the immigration data from the North American mammal record with oxygen isotope data from the marine realm.

The predicted correlation between first-order immigration episodes and rapid climatic cooling or broader land bridges, as represented by the oxygen isotope record, is corroborated in some cases but miscorrelated in others. Two clearly contradictory episodes are the Oligocene (35 to 30 Ma) and the Middle Miocene (16 to 6 Ma) when, despite major global cooling events, immigration rates were exceedingly low in North America. Possibly no land bridges were accessible in the Oligocene, but this explanation is less likely during the Middle Miocene when Beringian routes were probably accessible. It is more likely that these intervals carried few immigrants because they represent the two most stable chronofaunal phases in North American mammalian history. The Clarendonian chronofauna of the Middle Miocene, despite several episodes of global cooling and severe sea-level excursions, remained nearly closed to intercontinetal immigrants. During such robust chronofaunal intervals the continental mammal fauna stood near its ecological capacity.

This analysis of the land mammal record indicates that climatic shifts represented by first-order changes in the isotopic record are necessary but not sufficient causes of first-order immigration episodes into the North American continent. Clearly the question of why the continental ecosystem was open to immigration during certain times of isotopic excursions and resilient to others has fundamental significance in understanding the stability of present and future continental ecosystems.

INTRODUCTION

The rich record of Cenozoic land mammals provides not only a panoramic perspective on evolutionary processes, but also a valuable history of environmental change on the continents. As presently known, Cenozoic mammals number about 4500 genera, including 1051 living genera (M. McKenna, American Museum of Natural History, personal communication, 1992). Excluding bats, the number of extinct genera exceeds that of living genera by a factor of about four. Taxonomically and stratigraphically, land mammals are especially well documented in North America and Europe; on most other continents they are moderately well documented but with significant chronological gaps. Even on Antarctica, fossil mammals are now represented, albeit by only a few Paleogene genera from Seymour Island. Improved methods for sampling smaller forms (e.g., screenwashing and flotation) and continued success in discovering fossils in previously unsampled regions have greatly strengthened the fabric of the global mammalian succession.

Similarly, the chronostratigraphic framework for recording the Cenozoic history of land mammals has vastly improved in the past two decades. The succession of fossils, including evolutionary events, immigrations, and extinctions, provides the fundamental data set, but these biostratigraphic data are integrated via the stratigraphic context with other methodologies, most importantly with radiometric dates (principally K/Ar and more recently Ar/Ar) and paleomagnetic chronology (Lindsay et al., 1984). In North America, the Cenozoic Era is divided into 19 land mammal ages, including 52 (mainly informal but widely recognized) subdivisions (Woodburne, 1987). Thus 66 m.y. of Cenozoic mammalian history in North America can be divided into units that average less than 1.3 m.y. In the late Cenozoic the precision of mammal faunal dates approaches 0.5 m.y., and in favorable sections, wherein paleomagnetic and radiometric data are integrated with biostratigraphic data, particular strata may be resolved chronologically to less than 0.1 m.y. Such sections are not uncommon in western North America (Woodburne, 1987).

As the history of mammalian evolution has become more precisely calibrated, it has become increasingly evident that its tempo is strongly syncopated. Nowhere is that more evident than in the record of land mammals that immigrated to North America. We show that about 70% of immigrant genera were concentrated during about 10% of the Tertiary Period. Of the seven first-order immigration episodes that we define, one may be artifactual, and the six "real" ones occur in closely yoked pairs. In this chapter we investigate the global and regional environmental significance of such major land mammal immigration episodes.

Present societal concerns about global climatic change focus primarily on possible drastic alterations of continental ecosystems. Paradoxically, the most useful perspective on climatic history comes from the marine record of the Cenozoic Era, particularly from stable isotopic analyses of deep-sea cores. Such marine records do not directly yield the greater extremes of temperature and other environmental signals that continents experience. Accordingly, a critical area of current research concerns avenues for linkage between marine climatic records and the behavior of terrestrial ecosystems. That is why the record of land mammal faunas, by providing sufficient geochronological precision to be correlated with the marine record, holds special relevance to current considerations about global change.

Land mammal faunal succession may reflect climatic change in two somewhat distinct manners. First, the adaptive features of a given assemblage may be viewed in relation to the presumed conditions for which it was adapted; also, by analogy with living communities of similar structure, its composition may indicate the kind of landscape in which it lived. Secondly, in a largely independent and nonbiological manner, quantitative analysis of faunal turnover helps pinpoint times of ecosystem stability or, alternatively, of rapid reorganization. Rapid faunal turnover episodes may be correlated with key events in floristics, sedimentary regimes, isotopic signals, or other independent evidence of climate change episodes.

We pursue both of these approaches in the context of Cenozoic land mammals of North America. First, we review the land mammal record of North America and, together with other botanical and sedimentological evidence, indicate the general succession of ecosystems and landscapes. Second, we quantify the record of intercontinental immigration to North America, placing special emphasis on first-order episodes. Finally, we correlate this continental record with the marine record of global climate change. We show that the continental ecosystem did not respond uniformly to climatic episodes indicated in the marine record. We conclude that such shifts were a necessary but not a sufficient condition for faunal turnover and terrestrial ecosystem reorganization.

LAND MAMMAL FAUNAS AND THEIR RESPONSE TO CLIMATIC CHANGE

The richest continental record of Cenozoic mammals is that of North America. We concentrate on that record and methods of recognizing its response to climatic change. Several recent reviews of the mammalian succession may be noted. Woodburne (1987) offers a modern geochronological revision of the North American mammalian succession, and Lindsay *et al.* (1990) refine and update the biostratigraphy of Neogene mammal faunas in Europe. Savage and Russell (1983) provide comprehensive coverage of all continental and marine mammal faunas. Stucky (1990) reviews paleoecology and diversity change in North American Paleogene faunas, and Webb (1977, 1983a, 1989) treats Neogene faunal dynamics in North America. Figure 11.1 presents the geochronological and biostratigraphic framework of North American land mammals during the Tertiary.

Chronofaunas and Turnover Pulses

Vertebrate paleontologists refer to persistent faunas of essentially uniform taxonomic composition and stable diversity as *chronofaunas* following Olson (1952). In ecological terms, chronofaunas are thought to represent stable coadapted sets of species, and their detailed histories (under various perturbations) shed light on the theory of how communities or community complexes are structured (Olson, 1983; Webb, 1987). Five chronofaunas account for most of the stable intervals of the North American Cenozoic. They are the Eocene chronofauna (Krause and Maas, 1990); the White River chronofauna (Emry, 1981; Krishtalka *et al.*, 1987); and in the Miocene, the Runningwater, the Sheep Creek, and the Clarendonian chronofaunas (Webb, 1983a; Tedford *et al.*, 1987). Only in the extreme phases of the icehouse world, during the Pliocene and the Pleistocene, did the stability and persistence of mammalian chronofaunas breakdown.

These stable chronofaunas are followed by relatively brief episodes of rapid faunal turnover. Such rapid turnover episodes (RTE) have been recognized and studied during the past decade, following the establishment of relatively precise geochronometry. Several students of mammalian faunas have documented that even during an RTE the correlation between numbers of first appearances and last appearances is maintained, indicating an equilibrium process in which faunal interactions maintain a balance between gains and losses (Gingerich, 1984; Webb, 1989; Stucky, 1990). Gingerich (1984) proposed that the high peak of land mammal extinctions in the late Pleistocene of North America was an equilibrium response compensating for the high number of immigrations in the Late Pliocene and early Pleistocene; the 2-m.y. time lag tended to conceal the correlation. Others have explained RTE patterns in terms of climatic forcing. Webb (1983b) distinguished two types of rapid turnover episodes, the criterion being whether extinction waves or immigration waves peaked first, and suggested that the former (E-type RTE) were probably caused by climatic changes. He also showed that during the Neogene, six land mammal extinction episodes (including the two largest in North America) were correlated with positive oxygen isotope excursions in

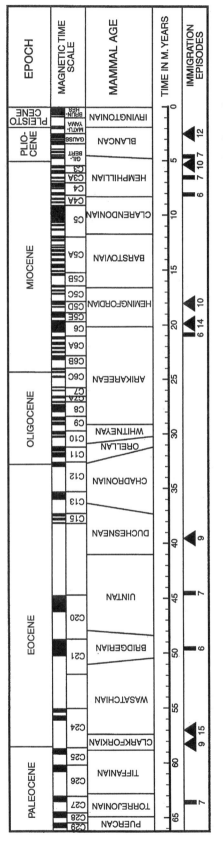

FIGURE 11.1 Land mammal immigration episodes in the Cenozoic of North America. Mammal ages, as correlated with Cenozoic Epochs, magnetic time scale, and time in millions of years, follow Woodburne (1987). First-order immigration episodes (with eight or more genera) are represented by triangles; second-order episodes (with five to seven genera) are indicated by rectangles.

the marine record and thus with global climatic episodes (Webb, 1984). When chronofaunal stability prevails, member taxa generally experience stabilizing selection. Only when climatic change perturbs the terrestrial ecosystem sufficiently does it produce an RTE.

Wing and Tiffney (1987) described similar patterns in floristic stability and turnover. They found that "the chief mechanism for drastic reorganization in community structure" was "severe climatic change." A similar view was formally proposed as the *turnover-pulse hypothesis* by Vrba (1985a, p. 232) in the following terms:

> Speciation does not occur unless forced (initiated) by changes in the physical environment. Similarly, forcing by the physical environment is required to produce extinctions and most migration events. Thus, most lineage turnover in the history of life has occurred in pulses, nearly (geologically) synchronous across diverse phylogenies, and in synchrony with changes in the physical environment.

This view underlies our analysis of immigration episodes in land mammal faunas. We attempt to test such terrestrial turnover patterns by comparing them with the isotopic record of Cenozoic climate change, which is derived independently from marine sediments.

Importance of Immigrants

In the following analysis we take the number of immigrant genera in each land mammal age as a key to faunal turnover episodes. Immigrants (first appearances) are recognized on the basis of two criteria: (1) the absence of that genus or closely related genera in prior North American records and (2) the presence of related forms (sister group) earlier in another continental fauna. Mammalian biostratigraphers have increasingly focused on immigrant taxa as the best diagnostic basis for defining land mammal ages, as well as informal subages or zones (Woodburne, 1987). Thus, for both its environmental and its stratigraphic significance, the record of immigrant mammals in North America has been well documented.

Figure 11.1 identifies major episodes of mammalian immigrations into North America during the Tertiary; we designate as first-order immigration episodes those that number nine or more genera. Second-order episodes are those with five to seven genera. These data are more fully explained in the following sections on results. Biostratigraphy and chronostratigraphy follow Woodburne (1987). These data will continue to be refined, by addition of immigrant genera, by extension of their known ranges into earlier intervals, and by refinements in the geochronological framework.

Three land bridges controlled the access of Cenozoic land mammals to North America. In the latest Paleocene and early Eocene the North Atlantic Thulean route between Europe and North America provided the basis for a strong faunal resemblance between these two continents. The next major immigration episode into North America took place in the Late Eocene. The mammals of this episode may have come across the Bering Strait, but more probably they again tracked the Thulean land bridge, which did not founder to great depth until the Oligocene (Rowley and Lottes, 1988). Subsequent episodes of faunal dispersal from the Old World must have crossed via the Bering land bridge. The very high latitude of this route (at the North Pole in the Cretaceous) placed increasingly severe ecological constraints on the immigrants that traversed it. Even as Beringian access shifted to somewhat lower latitudes during the Neogene, climatic regimes became more severe, so that Beringia continued to function as a narrow ecological filter. During the Pliocene (Blancan land mammal age) the Bering land bridge provided passage for temperate woodland species, such as an extinct deer (*Bretzia*) and an extinct panda (*Parailurus*), to the Pacific Northwest (Tedford and Gustafson, 1974). By the Pleistocene, however, the Bering filtered out all but steppe and steppe-tundra species, with the exception of humans who by then had developed highly specialized cultural adaptations to life in Beringian steppe-tundra.

A third route for intercontinental access to North America came from the south via the Panama (isthmian) land bridge. By Late Pliocene time it facilitated immigrations from South America into North America, mainly representing species adapted to subtropical savanna (Stehli and Webb, 1985).

In the following section we review the environmental history of mammalian faunas in North America, with an indication of the predominant biome during each chronofauna. We also briefly discuss the immigration (and faunal turnover) episodes that punctuate the chronofaunal succession. A major unresolved problem stems from the different durations of the land mammal ages in current use. The Duchesnean, for example, may span as much as 4 m.y. and the preceding Uintan as long as 6 m.y. Although it might be appropriate to treat immigrations as a rate by making the assumption that immigration events were uniformly distributed through a mammal age, in the present context we make no a priori assumptions about how immigrations may have been distributed within a mammal age subdivision, placing them simply at the stratigraphic level advocated by the best available empirical studies. We follow with the caveat that long intervals are more likely to produce artificially high immigration numbers, and note the need for continued refinement of fossiliferous local sections, especially in the mid-Cenozoic. Before we examine the possible correlations between immigration episodes and global climatic shifts, we briefly review the history of changing biomes in North America.

Recognition of Land Mammal Biomes

Several decades before the evolution of horses was documented in western North America, Kowalevsky (1873), working in Europe, recognized that hypsodonty in hipparionine horses reflected the spread of grasslands in the Miocene. Such reconstructions of environmental history from fossil mammal assemblages rest on two fundamental premises. The first is that the fossil record adequately samples the kinds and numbers of land mammal genera that lived during successive stages of the Cenozoic. The second premise is that samples of once-living mammals, through their morphologies, distributions, and numbers, and also through their sedimentary and taphonomic contexts, bear evidence of their environments. Such relationships are well established between living mammals and present-day environments; for example, the diversity and structure of African ungulate faunas reflect the vegetative cover and rainfall in which they range (Coe *et al.*, 1976; Sinclair and Norton-Griffiths, 1979; Vrba, 1985b; McNaughton and Georgiadis, 1986). Webb (1984) and Janis (1984) independently analyzed the distribution of body size and hypsodonty in fossil ungulate faunas from North America to develop convincing analogues with modern ungulate faunas in Africa. More generally, Andrews *et al.* (1979) empirically determined the patterns of body size and other gross features that characterize modern mammalian faunas from various terrestrial biomes regardless of the continent on which they occur.

One particular method of faunal inference that has been applied with considerable success in recent years is called the cenogram. A cenogram displays the whole spectrum of estimated body sizes for all species from a relatively complete faunal sample, for comparison with other such cenograms representing known biomes. Legendre (1986) convincingly used this method to show that the land mammal cenogram for the Late Eocene Phosphorites of Quercy in France had the same pattern as a modern equatorial rain forest fauna, notably a predominance of small-bodied (arboreal) forms, and Gingerich (1989) applied the same method with excellent results to still earlier Eocene (Wasatchian) faunas in Wyoming. Other examples are reviewed in Behrensmeyer *et al.* (1992).

Actualistic studies of taphonomic processes in the deposition of mammal remains provide the critical link between modern and fossil ecosystems (Behrensmeyer and Hill, 1980; Damuth, 1982; Behrensmeyer *et al.*, 1992). Thus, a legitimate train of logic and experiment connects the comparison of modern ecosystems with depositional settings in the recent past. It should be stressed that such relationships must be interpreted cautiously, and that data from ancient mammalian faunas apply most reasonably at the biome level in well-studied parts of the fossil record.

Another largely independent approach involves the study of ecomorphology. For example, the teeth of mammals are among their most important adaptations, and many features of their energetic and often highly specialized life styles depend to a great extent on their masticatory capabilities. The relationship between tooth morphology and ecology is particularly clear among large herbivores that must adaptively reconcile their long lives and high metabolic activities with relatively poor-quality protein resources (often further complicated by high fiber and high toxin contents). Fortunately, the fossil record of herbivore teeth is excellent and provides a clear history in diverse mammalian taxa of progressive change in their capacity to process increasingly coarse fodder. The dental adaptations of various mammalian herbivores are cited frequently in the following discussion as evidence of the prevailing conditions in North American environments. Such studies were well summarized by Scott (1937).

The most obvious such dental adaptation that many herbivore groups develop is hypsodonty (i.e., high-crowned teeth). A hypsodont animal may be defined as one in which several of its teeth (typically the molars but not uncommonly a whole battery of cheek teeth) have the crown height greater than one dimension of the crown wear surface (length, width, or the average of the two). If one imagines a cubic tooth in an unworn condition it would have a hypsodonty index of one and would be on the threshold of attaining hypsodonty. In modern ungulates and rodents it is not uncommon to find teeth with crown heights many times greater than their crown surface dimension, and some groups develop *hypselodonty* (i.e., ever-growing teeth).

The adaptations of large mammalian herbivores also contribute strongly to shaping their environments. For example, large mixed herds of grazers facilitate the food resources of one another and also help maintain optimum savanna settings, sometimes by grazing and sometimes by fostering fires (Sinclair and Norton-Griffiths, 1979; McNaughton *et al.*, 1988; Owen-Smith, 1988).

With these interpretive tools in hand we scan the history of land mammal faunas and associated floras to envision the major succession of biomes in North America.

MAMMALIAN FAUNAL SUCCESSION IN NORTH AMERICA

Paleocene Chronofauna: Tropical Forest

The earliest mammal faunas of the North American Cenozoic consisted of diverse small to medium-sized arboreal and scansorial forms. Most were frugivorous, omnivorous, or insectivorous; specialized carnivores and herbivores of larger body sizes came somewhat later. Four orders, Multituberculata, Insectivora, Primates, and Condylarthra, held over from the Cretaceous, comprise almost the entire Paleocene fauna. Analysis of virtually complete skeletal remains of the multituberculate genus *Ptilodus* demonstrated that its arboreal adaptations converged closely on those of a tree squirrel (Jenkins and Krause, 1983). The Condylarthra doubled and redoubled their generic numbers every few million years, and soon accounted for several distinct families including some moderately large herbivores (Van Valen, 1978). The distant forerunners of Carnivora (sometimes placed in their own extinct order, Creodonta) appeared by mid-Paleocene. By Late Paleocene, three relatively rare orders of larger mammals immigrated from Asia: the Pantodonta were partly amphibious molluscivores; the Taeniodonta were precociously hypsodont herbivores; and the Dinocerata were the first horned, browsing, herd animals of the Cenozoic. The Torrejonian mammal age included seven immigrant genera if one combines Torrejonian stages 2 and 3, as registered by Stucky (1990).

Predominant environments of the Paleocene land mammals were multistratal evergreen forests and cypress swamps extending north to at least 70°N latitude (Wolfe, 1985; Wing and Tiffney, 1987). During the Tiffanian (ca. 63 Ma), cooler climates—evidenced by increased percentage of deciduous trees and decreased diversity—produced "dramatically lower species richness and evenness" (Rose, 1981, p. 386), presumably because there were less reliable vegetational resources on a year-round basis. Krause and Maas (1990) introduced an enlarged data set on this interval that may weaken this conclusion.

The cooler climate and lower mammal diversity persisted into the Clarkforkian. Small mammals that had previously predominated became relatively rare, while larger forms such as phenacodont condylarths and carnivorous mammals increased in abundance. More importantly, the beginning of the Clarkforkian registers a first-order immigration episode, including Asiatic origins of the extinct order Tillodontia, the modern order Rodentia, and the pantodont family Coryphodontidae (Rose, 1981; Krause and Maas, 1990). Nine immigrant genera reached North America, if one combines Stucky's (1990) records for units 2 and 3. Although the Clarkforkian records the earliest first-order immigration episode in the history of North American land mammals, it is soon superseded by an even larger episode.

Eocene Chronofauna: Subtropical Forest

The wave of immigrations that began in the Clarkforkian intensified in the earliest Eocene or Wasatchian land mammal age. Rose (1981, p. 379) commented as follows:

The most striking aspect of the Wasatchian assemblages is their domination by several taxa that were unknown or exceedingly rare before the Wasatchian. Most of these taxa appeared for the first time in North America at or near the beginning of the Wasatchian and rapidly became the most common members of the fauna. . . . They include the first known members of the orders Perissodactyla (*Hyracotherium*) and Artiodactyla (*Diacodexis*), the primate families Adapidae (*Pelycodus*) and Omomyidae, and the creodont family Hyaenodontidae.

Land mammal immigrants of the early Wasatchian (see Figure 11.3 and Table 11.2) constitute the largest immigration cohort in the North American record. In Wasatchian 1 alone Stucky (1990) records 10 immigrant genera. Krause and Maas (1990, p. 90) more cautiously acknowledge that "many of the early Wasatchian first appearances appear to be immigrants from other continents." Mammals evidently crossed the North Atlantic freely in both directions via the Thulean bridge. According to Savage and Russell (1983), at least 50% of mammalian genera in the Sparnacian of Europe were shared with the Rocky Mountain region, by far the closest degree of transatlantic resemblance that occurred at any time during the Cenozoic.

The early stages of the Wasatchian are well constrained by stratigraphic studies in the Bighorn Basin of Wyoming. There the base of the earliest Wastachian lies within the reversed magnetic interval below stratigraphic anomaly 24. For this reason, its age lies between 56 and 57 Ma (Butler *et al.*, 1991).

Climatic conditions inferred from faunas and floras of this time suggest that warm equable conditions had returned during the Early Eocene (Rose, 1981; Wolfe, 1985). Primates and other arboreal mammals reach peak richness and abundance by the Middle Eocene, declining thereafter (Stucky, 1990). Rodents and Perissodactyla diversify profoundly, each accounting for about 20% of known mid-Eocene species (Savage and Russell, 1983). Even within the Arctic Circle on Ellesmere Island, diverse arboreal, frugivorous mammals, such as prosimian primates and dermopterans (frugivores distantly related to Asiatic flying foxes), persist along with a subtropical flora (West and Dawson, 1978).

White River Chronofauna: Woodland Savanna

In the continental record for North America a major faunal and environmental shift took place within the Late Eocene (late Duchesnean). Chronometric control on this episode needs further refinement, but it lies between 37 and 42 Ma, and so on an interim basis we call it 40 Ma. For a full discussion of the problem, see Krishtalka *et al.* (1987).

In the Middle Eocene the first indications of seasonal aridity appeared in the Rocky Mountain region. The great lakes of the Green River region, such as Lake Gosiute, formed extensive evaporites and were encroached on by deeply oxidized redbed deposits. By the Late Eocene, woodland savanna had become the predominant biome in the midcontinent (Webb, 1977; Wing and Tiffney, 1987). The dramatic shift from subtropical forest to predominantly woodland savanna in the Rocky Mountain region soon led to major faunal changes. It is unfortunate that this important turnover interval is not well documented by continuously fossiliferous sedimentary sequences.

The Duchesnean land mammal age is marked by a major faunal turnover episode, including the last appearances of such archaic groups as Condylarthra, Tillodontia, and Dinocerata. More important are the arrivals of the first eubrontothere (*Duchesneodus*) and several more modern taxa evidently dispersed from Asia (Emry, 1981; Krishtalka et al., 1987). Stucky (1990) recognizes a total of nine genera in Duchesnean 1 and 2. After the Duchesnean the number of browsing herbivore genera rose from 8 to about 40 (Stucky, 1990), and the species numbers of all herbivores cited in Savage and Russell (1983) rose from less than 40 to about 90 during the Eocene-Oligocene transition and remained at this level throughout the Oligocene (Webb, 1989). Emry (1981) labeled this persistent mammalian fauna of the Late Eocene and Oligocene the "White River chronofauna."

The Duchesnean immigration was preceded by a number of other immigration events apparently spread through the Uintan. Unfortunately this interval is about 6 m.y. long, and particular first appearance data are not tightly correlated. We arbitrarily place a second-order episode near the mid-Uintan. The following modern mammal families appear in North America during the Late Eocene: soricid insectivores; sciurid, castorid, cricetid, and heteromyid rodents; leporid lagomorphs; canid and mustelid carnivores; and ungulates—camelids, tayassuids, and rhinocerotids. Several of the newly appearing groups can be shown to have entered North America from Asia, among them the rabbits (*Mytonolagus*), the amynodont rhinocerotids, and most of the selenodont artiodactyls including camelids, hypertragulids, leptomerycids, and oreodonts (Webb, 1977; Webb and Taylor, 1980; Emry, 1981).

Most of the Late Eocene immigrant herbivores are characterized by adaptations for masticating coarse fodder. At least some members of the following groups developed hypsodont dentitions during the Late Eocene: taeniodonts, leporids, castorids, eomyids, rhinocerotids, hypertragulids, oromerycids, and "oreodonts" (Webb, 1977). For example, Wood (1980, p. 38) characterized the cheek teeth of the eomyid genus *Paradjidaumo* as "... more hypsodont and progressively more lophodont than in *Adjidaumo*." Also, although most oromerycids are brachydont, Prothero (1986, p. 461) observed that in the new genus *Montanatylopus* "the molars are much more hypsodont than in any other oromerycid." Thus it is fair to recognize the Late Eocene and Oligocene White River chronofauna as the first in North America to sustain a substantial diversity of hypsodont herbivorous mammals.

The larger mammalian herbivores may be divided into two broadly distinct habitat groups: one group lived primarily along watercourses; the other lived mainly on the interfluves. The flat-skulled leptaucheniine "oreodonts" and the trunk-bearing amynodont rhinos were short-legged, semiamphibious forms that cropped lush vegetation in or near stream courses (Scott, 1937; Wall, 1982). On the other hand, most selenodont artiodactyls, as well as the horse *Mesohippus*, had relatively long slender limbs and ranged widely in open habitats. Several studies of faunal facies in the White River deposits, reviewed in Webb (1977), provided direct statistical evidence that selenodont artiodactyls and rabbits occurred predominantly in open-country or upland habitats.

Several of the White River ungulates, notably *Leptomeryx*, ranged together in herds (Clark et al., 1967). The adaptive relationships among social behavior, body size, and feeding mode, developed by Estes (1974), Jarman (1974), and others on the basis of the modern African ungulate fauna, suggest that the appropriate comparison for leptomerycids and other selenodonts is with moderate-sized herding forms such as the gazelles (Jarman's category C). Such forms are mixed feeders, relying on grasses only in their most nutritious new-growth stages and shifting to browsing in the dry season (Janis, 1982).

A third adaptive zone, that of the small burrowing herbivore (rhizovore) and insectivore, is extensively occupied during the White River chronofauna. This implies extensive development of well-drained soils supporting shrubby vegetation. At the same time the diversity of arboreal mammal genera declines (Webb, 1977; Stucky, 1990).

With such mammalian evidence in mind, one may look for other indications that savannas were opening the landscape of the Late Eocene and Oligocene. Hutchison (1982) recognized the severe impact of increasing aridity and seasonality on the aquatic reptile fauna during that interval in the Rocky Mountain region, and in Early Oligocene floras of North America, notably the Florissant in Colorado, the dramatic decrease in the percentage of entire-margined leaves indicates approximately a 10°C drop to about 12.5°C mean annual temperature (MacGinitie, 1962; Wolfe, 1985). Retallack's (1983) pedological studies of White River sediments provide a fascinating look at local paleosols underlying various habitats, and hint at an overall trend toward increasingly grassy and shrubby environments following the Early Oligocene climatic deterioration.

In attempting to understand ecosystems of the Eocene-Oligocene, one may ask which plants the abundant browsing and mixed-feeding herbivores might have eaten. Since grass is still rare in the Eocene-Oligocene, it may be necessary to postulate an emphasis on woody shrubs such as *Sarcobatus*, *Ephedra*, and *Equisetum*. Huber (1982) postulates that just such herb-dominated savannas precede an abundance of true grass savannas in Tertiary floristic history. On the other hand, grass was present and may be underrepresented due to taphonomic conditions in paleofloral sites. Grasses were established in the mid-Eocene of Australia (Truswell and Harris, 1982, cited in Wolfe, 1985) and Europe (Litke, 1968). On present evidence, Wolfe (1985, p. 364) recognized ". . . a distinctive Eocene savanna (i.e., shrubs and widely spaced trees but no grass on the interfluves)." The extent of grasses within this chaparral and woodland savanna setting remains in doubt.

The quantum jump in mixed-feeding herbivores noted for the White River chronofauna probably helped expand the extent of woodland savanna. The new fauna and flora probably formed a positive feedback loop: large herbivores preferred feeding in more open woodlands; in turn, expansion of open formations facilitated evolution of mixed-feeding herbivores. Among the subtle changes within the history of that chronofauna ("chronofaunal creep") was a decline in the rich array of browsers (e.g., loss of titanotheres after the Chadronian), and an increase in the body sizes and numbers of shrub eaters (quasi-grazers). Such changes indicate an increasing percentage of open-country patches, and tend to be corroborated by the limited botanical and pedological data.

Runningwater Chronofauna: Transitional Savanna

The next major immigration episode in the record of North American mammals takes place in the Early Miocene. Miocene immigration patterns in North American mammals were thoroughly revised by Tedford *et al.* (1987). Their analysis recognizes six immigrant genera in the latest Arikareean and 14 genera in the earliest Hemingfordian (20 Ma). Native members characteristic of the Runningwater chronofauna of the late Arikareean and early Hemingfordian include flat-incisored beavers (notably *Palaeocastor* with its deep corkscrew-shaped burrows), hypsodont oreodonts such as *Merychyus* and *Merycochoerus*, oxydactyline and protolabine camels (especially *Michenia*), and the mixed-feeding horse, *Parahippus*. Several immigrant groups also constitute a vital and characteristic part of the Runningwater chronofauna, notably the primitive bear, *Phoberocyon*; such genera of amphicyonids (an extinct family also known as "bear-dogs") as *Cynelos* and *Amphicyon;* raccoons including *Edaphocyon* and *Amphictis*; and several mustelid carnivores including *Leptarctus*. The first representatives of three ruminant families arrived from Asia at this time: *Blastomeryx* represents the Moschidae (family of the living Chinese musk deer); *Paracosoryx* is the scion of the autochthonous North American family Antilocapridae; and *Barbouromeryx* establishes the presence of the giraffe-like family Palaeomerycidae. Each of these immigrant ruminant groups evolves and diversifies in the course of the Miocene.

The paleoecology of Arikaree assemblages in the midcontinent is thoroughly analyzed by Hunt (1990). Earlier in the Arikareean, widespread eolian distribution of tuffaceous sediments in a seasonally arid climate played an important role in preserving rich samples of mammals in burrows and other nontransported settings. Both the mammalian taxa and the sedimentological evidence indicate ". . . well vegetated stream border communities, spatially separated by open interchannel reaches occupied by grassland, or possibly low shrub savanna or open savanna-parkland environments" (Hunt, 1990, p. 107).

Sheep Creek Chronofauna: Park Savanna

The trend toward more open country habitats continued from the Early into the Middle Miocene at least in the midcontinent. Another substantial wave of immigrants from Asia altered the makeup of the North American mammal fauna and spurred the course of its evolution. It is awkward from a nomenclatural view that this turnover episode falls within the Hemingfordian land mammal age and is not very far removed in time from the previous turnover episode. Tedford *et al.* (1987) comment as follows:

> It is important to note that an important faunal discontinuity exists within the Hemingfordian, signaling an abrupt shift in the Great Plains from the chronofauna characteristic of the late Arikareean and early Hemingfordian to one typifying the late Hemingfordian and early Barstovian.... The differences between the faunas of the Box Butte through Sheep Creek interval and the chronofaunally related younger faunas here referred to the Barstovian (Observation Quarry through Lower Snake Creek) can be ascribed mainly to anagenetic change in persistent lineages, making it necessary to distinguish these ages at the specific rather than the generic level.

Major advances within the autochthonous fauna of Hemingfordian age indicate continued evolution toward more open-country adaptations. The best example is the transition from *Parahippus*, a browsing horse, to *Merychippus*, an early grazer (Hulbert and MacFadden, 1991). Similar trends are suggested by the increased hypsodonty and quantum increase in diversity of pronghorn antelopes.

The mid-Hemingfordian wave of immigrants from Asia, like the preceding episode, represents a broad ecological spectrum. Among the diverse rodents are flying squirrels of Old World (petauristine) type, eomyid rodents, and the very significant cricetid rodent, *Copemys*. New carnivores include weasel-like and otter-like mustelids, while the first true cats in the New World are represented by the genus *Pseudaelurus* (see Table 11.2). Immigration of four very large mammals had a major impact on the Middle Miocene ecosystem: these were two rhinocerotids (short-legged, aquatic *Teleoceras* and long-legged, terrestrial *Aphelops*) and also two proboscideans (the browsing mammutid *Miomastodon* and the probable mixed-feeding gomphotheriid, *Gomphotherium*). It is now generally supposed that such "megaherbivores" played a formative role in modifying the landscape, as do elephants in modern savanna ecosystems (Owen-Smith, 1988). Detailed evidence on this point, however, has not been developed in the Miocene of North America. The mid-Hemingfordian immigration episode took place about 18 to 17 Ma and was the last major episode until the latest Miocene, a span of about 12 m.y. (Tedford *et al.*, 1989).

Clarendonian Chronofauna: Grassland Savanna

Land mammal diversity in North America reached its zenith during the Barstovian mammal age (Webb, 1989). Savage and Russell (1983) recognized 16 families with 60 genera and 141 nominal species of land mammals in the Barstovian. The next highest numbers occurred in the Clarendonian (the next mammal age), with 55 genera and 117 species. These mammal ages are thought to indicate a savanna optimum in North America, with a rich mosaic of trees, shrubs, and grasses supporting an extraordinary variety of large and small, grazing and browsing, ungulates. It is not uncommon during this savanna acme to collect in a single site 20 genera of ungulates of which half are Equidae (Webb, 1983a; Voorhies, 1990). Savage and Russell (1983, p. 300) noted that during this interval "the mammalian fauna . . . appears singularly homogeneous throughout its geographic range." The Clarendonian chronofauna spanned three mammal ages, an interval of more than 10 m.y. (Tedford *et al.*, 1987).

The array of ungulates in the Late Miocene of North America is comparable in many respects to that living in Africa today. The resemblance is purely convergent: the two faunas have no genera in common, and members of the few shared families play substantially different roles. For example, the body form and paleoecological context of *Teleoceras*, the most distinctive rhinocerotid perossodactyl in the Clarendonian chronofauna, indicate a niche analogous to that of *Hippopotamus* (an artiodactyl) in Africa. Voorhies and Thomasson (1979) confirmed by direct evidence from the hyoid (throat) bones that *Teleoceras* consumed grasses and, from the biocenosis in pond deposits at Poison Ivy Quarry, that herds of young and old frequented water. Likewise in North America the role of the African giraffe is played by the native giraffe-camel, *Aepycamelus* (Webb, 1983a). The most plausible explanation for these convergences would seem to be that the North American fauna evolved in a broadly comparable manner because of environmental conditions similar to those in Africa, namely, grassland savanna.

Webb (1983a) and Janis (1984) independently attempted comparisons between North American Miocene and African Recent ungulate faunas, with emphases on body size and height (or volume) of cheek teeth. They found similar numbers of ungulate species distributed as browsers, roughage feeders, and mixed feeders. A rough estimate of biomass distribution among these categories, based on quarry censuses from major North American Miocene sites, also gave results that were remarkably similar to actual census data from African game reserves: browsers accounted for less than 10% of all ungulate biomass, whereas roughage feeders (with high hypsodonty indices) made up some 60 to 80% of the total. In a related study, Hulbert (1982) showed that population structure in a Clarendonian species of the grazing horse *Neohipparion* indicates that it underwent seasonal migrations, an essential feature of the strategy of African grazing ungulates.

African ecological studies show that savanna biotas live where annual rainfall ranges between about 400 and 1000 mm and that, within this range, biomass is positively correlated with rainfall (Coe *et al.*, 1976). Furthermore the number of ungulate species and the relative abundance of larger-bodied species interact with and are roughly correlated with primary productivity (McNaughton *et al.*, 1988). These general observations in Africa shed light on mechanisms governing species richness and size-frequency distribution of ungulates in Miocene local faunas of midcontinental North America before and after the savanna optimum in the Barstovian. In midcontinental North America the Late Miocene ungulate diversity decline presumably tracks a decrease in mean annual rainfall.

Floristic evidence supports the faunal evidence of widespread grassland savanna in the Middle Miocene of North America and increasingly arid climate in the Late Miocene. The Tehachapi flora of Hemingfordian age in California, the Stewart Valley flora of Barstovian age in Nevada, and the Kilgore flora of Barstovian age in Nebraska each has distinctive regional features, yet together they justify the conclusion (Wolfe, 1985, p. 371) that "during the Early to Middle Miocene, savanna developed in low-latitude areas that are presently dry." Studies of as many as 18 species of fossil grasses directly associated with mid-Hemphillian ungulates in the Minium Quarry in Kansas also strongly indicate a grassland savanna biome (Thomasson, 1986; Thomasson *et al.*, 1990). In the same floras are leaves with stomata and other detailed features indicative of the subfamily Chloridoideae and of C_4 photosynthesis characteristic of tropical floras growing at an optimum temperature of 30 to 32°C (Thomasson et al., 1986). Such floristic studies suggest that the Late Miocene climatic regime in midcontinental North America resembled that of seasonally arid subtropical savannas in South and Central America today.

Early Pliocene: Spread of Steppe

About 5 Ma, at the end of the Miocene, the long-term trend toward increasing aridity in midlatitudes led to the final breakup of the extensive North American savanna biome and its replacement in midcontinental North America by extensive grasslands or true steppe. Ungulates experienced a series of decimations that spanned the Hemphillian; and the first major decline (at about 12 Ma) affected browsers more than grazers (Webb, 1983a, 1984; see also Figure 11.4). By the end of the Hemphillian the number of ungulate genera was cut to less than half of its former peak. No new chronofauna was established, but a major immigration episode coincided with the final late Hemphillian extinctions.

Floristic changes at the end of the Miocene clearly record the expansion of grasslands in midcontinental and western North America. For example, Wolfe (1985, p. 371) notes that in the Hemphillian of Kansas "the fructifications include abundant grass, and the leaf-assemblages are the low-diversity type that typically occurs along streams in unforested grassland regions." In western Wyoming, Barnosky (1984) recognizes xeric shrub floras during the latest Miocene. The Mt. Eden flora in southern California suggests the beginning of steppe conditions (Wolfe, 1985). Axelrod's (1992) floristic review shows that in the Great Basin and the midcontinent, the most arid interval of the Tertiary occurred at about 6 Ma. Renewed Cordilleran tectonics during the Miocene-Pliocene greatly increased rain shadow effects and provincialism in many regions of temperate North America.

The vast herds of ungulates themselves probably accelerated the decline of savanna and the expansion of steppe in the Pliocene. In Africa during extended droughts, elephants and/or equids devastatingly transform savanna to steppe (Sinclair, 1983; Owen-Smith, 1988). In North America, many late Hemphillian and Blancan fossil quarries (e.g., the Hagerman Horse Quarry in Idaho) yield abundant samples of only one or two grazing equids. These suggest (but do not demonstrate) scenes of massive overgrazing. A positive feedback loop between faunal and floral evolution may have accelerated the process that established steppe conditions in the Early Pliocene.

It should be noted that the Early Pliocene expansion of steppe environments did not encompass all of North America. Provincialism is far more evident than in the Miocene. In the Pacific Northwest, faunal and floral evidence clearly indicates continuation of a mesic forest biome although it interdigitated with semiarid savannas in the Great Basin (Tedford and Gustafson, 1974). The biota around the Gulf of Mexico also indicates the continuation there of savanna conditions (Webb, 1989). The remarkable resemblances between such Hemphillian faunas as the Yepomera local fauna in Chihuahua, Mexico, and the Bone Valley fauna in Florida give evidence of a broad subtropical savanna south of latitude 30°N, which was perpetuated by the persistence of summer monsoons (Webb, 1977). Thus, the Pliocene shift to steppe biomes in the midcontinent produced greater provincialism among diverse regions of North America.

Onto such a scene of regional desiccation and increased provincialism burst a new wave of Asiatic immigrants. As indicated below (Table 11.2), these immigrations appear to be concentrated in the latest Hemphillian, about 5 Ma. Key groups of land mammals that crossed the Bering at this time were vole-like rodents, both terrestrial and aquatic (Repenning, 1987); at least six carnivores, including exotic groups such as hyaenids (*Chasmaporthetes*) and pandas (*Parailurus*), as well as more familiar groups such as *Lynx* and *Ursus*; and also at least two genera of deer, which gave rise to a substantial radiation in North and South America.

Late Pliocene and Pleistocene: Further Continentality and Provincialism

The secular trend of the Cenozoic toward colder, drier, and in other ways more extreme conditions led to the environments of the later Pliocene and Pleistocene. As the first Laurentide glaciation settled over North America about 2.5 Ma, forest tundra biomes became widespread in the Arctic Circle (Funder et al. 1985; Repenning, 1985). Mountain building in the Cordillera extended tundra corridors along high-elevation routes into temperate latitudes. Rain shadows helped extend cold-steppe habitats into many basins of western North America. In the Snake River Plain (as represented by the Hagerman fauna and flora), there were still broad-leaved deciduous forests. To the north and at higher elevations, however, conifers predominated (Lundelius et al., 1987). The southern plains (as represented by the Mt. Blanco fauna and the Rita Blanca flora) were characterized by a seasonally arid climate that produced grassy scrubland and an impoverished ungulate fauna dominated by grazers. Widespread caliche deposits in Late Pliocene deposits of the Great Plains have long been recognized as indicators of seasonal aridity (Hibbard and Taylor, 1960).

In Florida and presumably the rest of the southeastern United States, vertebrate faunas of late Blancan age carry a number of "holdover taxa"—such as the peccary, *Mylohyus*; the protoceratid, *Kyptoceras*; and the three-toed horse, *Cormohipparion*—which were already extinct in the midcontinent. The ecological balance of these holdovers suggests persistence of seasonal savannas and extensive mesophytic forests in warmer, more productive settings than in the High Plains (Robertson, 1976; Webb, 1978; Hulbert, 1987). In general, the Late Pliocene array of biomes in North America approximated the degree of provincialism seen at present, although the extremes of freezing winters and arid deserts were absent.

In the Late Pliocene (Blancan 2) the first large wave of immigrants from South America appeared, including glyptodonts, chlamytheres, armadillos, ground sloths, capybaras, and porcupines. At the same time, another large wave of immigrants came from Asia; among them were the spectacled bear (*Tremarctos*) and several rodents (*Synaptomys*, *Pliopotamys*, and *Mictomys*) (Lundelius et al., 1987). This final immigration episode of the Tertiary coincided with expansion of the Bering bridge. It clearly correlates with the onset of Northern Hemisphere glaciation, sea-level lowering, and evidence of pre-Nebraskan till in western Iowa (Lundelius *et al.*, 1987; Repenning, 1987).

The late Hemphillian and early Blancan immigration episodes just considered adumbrate an increasingly active series of Quaternary immigration episodes. High-latitude Quaternary studies on both sides of the Beringia land bridge record trends among microtine rodents, as well as many large mammals, such as mammoths and musk oxen, toward steppe-tundra adaptations (Sher, 1974; Repenning, 1985; Herman, 1989). The complex pattern of faunal dynamics among Quaternary mammals is well summarized by Tedford *et al.* (1987, p. 192): "The accelerating pace of immigration from Hemphillian into Pleistocene time reflects the availability of dispersal routes to North America and increasing environmental instability, particularly at high latitudes, that provided the goad for the movement of mammals."

RESULTS

We analyzed the record of immigrations in successive North American land mammal faunas, as a possible signal of major environmental changes in the history of that continent. We selected this continent because it has the most complete Cenozoic record of terrestrial mammalian history. We tallied the number of land mammal genera that apparently immigrated in each of 52 informal subdivisions of 19 land mammal ages following the biostratigraphic scheme presented in Woodburne (1987). In this review, we have not dealt with other diversity changes due to extinctions and endemic radiations. We found that the two largest episodes (Wasatchian 1 and Hemingfordian 1) involved 15 and 14 genera, respectively; five other episodes also involved nine or more genera (Table 11.1). We arbitrarily designate these seven largest as first-order immigration episodes. We categorize another cluster of seven episodes, consisting of five or seven genera each, as second-order episodes. All smaller counts were considered third-order or "background" immigration patterns. The numbers of genera in first-order and second-order episodes are documented in Table 11.1. The only third-order episodes we were able to tally accurately are the five from the Miocene. The genera counted in the seven first-order immigration episodes are listed in taxonomic order in Table 11.2.

An unexpected result of this analysis is the paired pattern among six of the seven first-order immigration episodes. Except for the Duchesnean, each first-order episode occurs within about two million years of another first-order episode. The three pairs are Clarkforkian with Wasatchian, Hemingfordian 1 and 2, and Hemphillian 3 and Blancan 2. The first-order status of the Duchesnean immigration episode may be artifactual. This interval, and the preceding Uintan, span about 10 m.y., much of it represented by sparsely fossiliferous sediments. The nine immigrant genera recognized here are thought to be clustered in the late Duchesnean (Emry, 1981; Krishtalka *et al.*, 1987), but this may well be an artifact of a relatively

TABLE 11.1 Land Mammal Immigration Episodes

Mammal Age	Genera	Age (Ma)
BLANCAN 2	13	2.5
Blancan 1	7	4.5
HEMPHILLIAN 3	12	5.0
Hemphillian 2	7	7.0
Hemphillian 1	7	8.0
Clarendonian	3	10.0
Barstovian 3	2	12.0
Barstovian 2	3	14.5
Barstovian 1	2	16.0
HEMINGFORDIAN 2	10	18.0
HEMINGFORDIAN 1	14	20.0
Arikareean 2	6	21.0
Arikareean 1	4	23.0
DUCHESNEAN	9	39.5
Uintan	7	44.5
Bridgerian	6	49.5
WASATCHIAN	15	57.0
CLARKFORKIAN	9	58.5
Torrejonian	7	63.5

NOTE: First-order episodes in capital letters; second- and third-order in lowercase.

TABLE 11.2 First-Order Immigration Episodes of North American Land Mammal Genera

Mammal Age	Order	Genera
Clarkforkian (ca. 59 Ma)	Cimolesta	*Apatemys*
	Tillodontia	*Esthonyx*
	Rodentia	*Paramys*
		Microparamys
	Creodonta	*Palaeonictis*
	Carnivora	*Uintacyon*
	Pantodonta	*Coryphodon*
	Dinocerata	*Prodinoceras*
	Condylarthra	*Hyopsodus*
Wasatchian (ca. 57 Ma)	Cimolestida	*Didelphodus*
	Soricomorpha	*Macrocranion*
		Scenopagus
	Primates	*Pelycodus*
		Cantius
		Teilhardina
	Creodonta	*Prototomus*
		Arfia
	Carnivora	*Miacis*
		Vulpavus
	Mesonychia	*Pachyaena*
	Perissodactyla	*Homogalax*
		Hyracotherium
	Artiodactyla	*Diacodexis*
		Bunophorous
Duchesnean (ca. 39 Ma)	Rodentia	*Ardinomys*
	Creodonta	*Hyaeodnon*
	Carnivora	*Nimravus*
		Eusmilus
		Procynodictis
	Perissodactyla	*Amynodon*
		Menodus
	Artiodactyla	*Bothriodon*
		Elomeryx
Hemingfordian 1 (ca. 20 Ma)	Insectivora	*Antesorex*
		Plesiosorex
	Lagomorpha	*Oreolagus*
	Carnivora	*Amphicyon*
		Amphictis
		Cephalogale
		Edaphocyon
		Potamotherium
		Phoberocyon
		Cynelos
	Perissodactyla	*Brachypotherium*
	Artiodactyla	*Blastomeryx*
		Paracosoryx
		Barbouromeryx

TABLE 11.2 *Continued*

Mammal Age	Order	Genera
Hemingfordian 2 (ca. 18 Ma)	Rodentia	*Blackia*
		Copemys
		Eomys
		Petauristodon
	Carnivora	*Miomustela*
		Mionictis
		Plithocyon
		Pseudaelurus
		Sthenictis
	Perissodactyla	*Aphelops*
Hemphillian 3 (ca. 5 Ma)	Rodentia	*Mimomys*
		Nebraskomys
		Promimomys
	Lagomorpha	*Ochotona*
	Carnivora	*Agriotherium*
		Chasmaporthetes
		Lynx
		Megantereon
		Parailurus
		Ursus
	Artiodactyla	*Bretzia*
		Odocoileini
Blancan 2 (ca. 2.5 Ma)	Edentata	*Dasypus*
		Holmesina
		Glyptotherium
		Glossotherium
		Eremotherium
	Rodentia	*Erethizon*
		Neochoerus
		Mictomys
		Pliopotomys
		Synaptomys
	Carnivora	*Canis*
		Tremarctos
	Artiodactyla	*Bovinae*

impoverished record. If one examines the distribution of first- and second-order episodes throughout the Tertiary, one notes a tendency for clustering of several episodes, notably in the Early Miocene, when the Arikareean associates with the two Hemingfordian episodes, and in the Late Miocene where a string of three second-order episodes is associated with the final pair of first-order episodes.

DISCUSSION

During the past decade, mammalian paleontologists have discovered, somewhat to their collective surprise, that the pulse of the Cenozoic succession in North America is strongly syncopated. In particular, publication of a detailed mammalian biochronology for North America (Woodburne, 1987) underlined the unevenness of faunal turnover. The rhythm of long-stable chronofaunal intervals punctuated by rapid turnover episodes has emerged as a clear pattern (Vrba, 1985a; Webb, 1989). At the level of a continental ecosystem this pattern may be referred to as "syncopated equilibrium." The need now is for closer analysis of rapid turnover episodes to gather new insights into the mechanisms and modalities that translate environmental change into radical reorganization of terrestrial ecosystems.

In the following discussion we address the question of what "trigger mechanism" might bring about the first-order immigration episodes that we have tallied above. An unexpected result of this analysis is the close association of the six first-order immigration episodes into three pairs. This is not conceivably a random pattern. These three pairs of immigration episodes pack a majority of North American immigrants into a time frame equal to only about one-tenth of the Cenozoic. Perhaps the second episode in each pair represents delayed discovery, due to poor recovery of rare fossils during early phases of colonization. Thus, the second immigration wave could be an "echo" of the first. After careful review of the stratigraphic and geographic details of each instance, we find this hypothesis unlikely, although it is worthy of more rigorous analysis. We are left with the straightforward view that the physical and ecological conditions that triggered the first of each pair (or cluster) of major immigration episodes were reiterated within one or two million years, thus bringing about paired and clustered sets of immigration episodes. If correct, this pattern provides an important clue as to the sustained (or recurrent) force of whatever conditions "triggered" the immigration episodes into North America.

In exploring possible causal mechanisms for immigration and rapid turnover episodes on the North American continent, we turned to two other large bodies of correlative evidence. First, we scanned the record of land mammal faunal turnover on other continents in search of corroborative evidence. Secondly, we looked at the marine record of global climatic change (as represented by oxygen isotope signals) for correlative patterns during the Cenozoic.

Mammal Turnover on Other Continents

As the land mammal successions on other continents become more continuously documented and their chronologies become more fully correlated with the absolute time scale, it is important to determine whether they respond synchronously with the immigration and rapid turnover episodes observed in the North American record. We summarize a few highlights from other continental mammal faunas in Europe, China, the Indian subcontinent, and Africa. South America and Australia were almost completely isolated during most of the Tertiary; therefore, their land mammal faunas need not be involved in this immigration analysis.

European Land Mammal Record

The European record of Cenozoic land mammals closely approaches that of North America both in chronological continuity and in taxonomic diversity. Unfortunately, no complete revision of European mammal invasions is available. Mein (1989) lists many generic first appearances for Neogene mammal ages, but the lists do not distinguish between allochthonous and autochthonous first appearances. Such distinctions are more challenging in western Europe than in North America because it is simply a portion of Eurasia and the proportion of immigrant genera in successive faunas is probably much greater in Europe than in North America. Thus we are unable to derive an authoritative analysis of immigrants in the European record to compare with that for North America.

Opdyke (1990) recognized a strong correlation between Neogene faunal turnover episodes on the two continents, using a somewhat indirect approach. Since the land mammal ages were established independently in Europe and North America, on the basis of empirical biostratigraphy, he compared the biochronological boundaries. Synchronous boundaries indicate synchronous pulses of faunal turnover, which must be governed by global climatic events. If, on the other hand, the mammalian turnover (and immigration) episodes were independent and governed only by local effects, there should be little or no correlation between the two systems. As Opdyke (1990) suggests, there is a remarkable degree of correspondence between the two independently derived continental land mammal chronologies during the Neogene.

The status of land mammal ages in Europe was recently reviewed by Lindsay *et al.* (1990). Although the methods of mammal age definition differ considerably from those used in North America (Woodburne, 1987), the succession of mammal ages on the two continents appears nearly isochronous from Early Miocene through Late Pliocene. We highlight some of the more favorable comparisons below.

Legendre (1986) indicates three major immigration episodes among Paleogene land mammals of Europe. First is the strong continuity with North America in the earliest Eocene; this has long been recognized as the time of arrival of modern orders such as Primates, Perissodactyls, and Artiodactyls. The second is a small immigration wave in the Lutetian about 45 Ma. The third and most important is the rapid turnover episode long known as the "Grande Coupure," at approximately 32 to 34 Ma. What began mainly as an extinction event, led soon to a modernized fauna, characterized for example by new families of perissodactyls. The landscape in western Europe shifted from predominantly rain forest to savanna during this Early Oligocene transition (Legendre, 1986).

In the European Neogene system, the beginning of the Orleanian (MN 4) is marked by the immigration of the browsing horse, *Anchitherium*; two genera of proboscidea from Africa; and at least five genera of ruminants, possibly from Africa (Mein, 1989; Ginsburg, 1989; Tassy, 1989).

Although *Anchitherium* came from North America, it reached Europe by way of Asia. Early Orleanian correlates with lower Hemingfordian and with the first-order immigration episode at about 20 Ma in North America.

The Vallesian-Astaracian boundary has been the subject of controversy, and that controversy is entwined with evidence regarding the first appearance of *Hipparion* and other hipparionine horses from North America via Asia. The genus *Hipparion* appears on the margins of the Mediterranean at 11.5 Ma in Chron C5R, which would make the base of the Vallesian at least this old. This date correlates precisely with the base of the Clarendonian assigned by Tedford *et al.* (1987) on the basis of North American biostratigraphy and many radiometric dates. Few immigrants, however, appear in North America at that time.

The base of the Turolian is marked by many immigrant appearances such as the rabbit, *Alilepus*, and the proboscidean, *Anancus*, and is well dated at 9 Ma (Berggren *et al.*, 1985; Mein, 1989). This date coincides with the beginning of the Hemphillian in the North American record and with the second-order immigration episode here recognized as Hemphillian 1 (Table 11.2).

The beginning of the European Ruscinian mammal age correlates with the latest Hemphillian in North America (Tedford *et al.*, 1987). The Ruscinian mammal age, approximating the Miocene-Pliocene boundary, is now well dated in the Mediterranean area, where that sea refilled. Ruscinian is defined by the many new appearances that occurred as a consequence of the Messinian sea-level cycle and the severe climatic deterioration that accompanied it. In North America, we recognize Hemphillian 3 as a first-order immigration episode, correlated with the Messinian sea-level cycle (Webb, 1983b).

The early Villafranchian is marked in Europe by a major turnover episode including the arrival of *Equus* there (as well as in Asia), accompanied by the appearance of *Elephas*, various large deer, and also as Azzaroli (1989, p. 341) notes, "the demise of a forest assemblage characterized by *Tapirus*, *Sus*, and *Ursus*." This episode may be correlated with the first-order immigration episode known as Blancan 2 in North America. Presumably these equivalent events were driven by the same climatic mechanisms, namely, the onset of Northern Hemisphere glaciation.

By the Pleistocene, North America and Europe lack any meaningful correspondence between their respective mammal ages. Possibly the frequency of faunal migrations and climatic shifts after the Gauss Chron will require a much more refined system than the present broadly defined mammal ages. Repenning (1987) and others have found much finer-tuned rodent zonations to be quite useful in relating mammal faunal turnover to global climatic changes in the Quaternary.

Chinese Land Mammal Record

Correlations of the rich land mammal succession in China are rapidly improving and already can test, in some instances, the hypothesis of global synchroneity of rapid faunal change episodes among land mammals. Current studies of Oligocene faunas in China suggest that Asia may have been the source of many immigrants in the European Grande Coupure after the marine seaway retreated from the Turgai Straits. In his review of Neogene mammalian biochronology in China, Qiu (1989) recognizes four major immigration episodes that alter the character of the Asian fauna. First, the nearly simultaneous immigration of *Anchitherium* from North America and *Gomphotherium* from Africa marks a major change in the Early Miocene fauna. Qiu (1989) also notes the brachyerycine insectivores as another trans-Beringian immigrant group that appears in the Early Miocene of China.

The second major Miocene immigration episode from outside Eurasia was the appearance of *Hipparion* from North America. The importance of the *Hipparion* datum is widely recognized throughout Eurasia and may occur at about 12 Ma.

Thirdly, the Gaozhuang fauna of Miocene-Pliocene age in the Yushe Basin is under intensive study and represents one of the best understood and best-dated immigration episodes in Asia. Among key immigrants are an arvicoline rodent *Germanomys*; the rabbit *Hypolagus*; the canid *Nyctereutes*; the pig *Sus*; and the camelid *Paracamelus* (Qiu, 1989; Tedford *et al.*, 1989). There is a fairly close correlation between the Gaozhuang fauna in China and the Ruscinian in Europe.

Finally, the Late Pliocene immigration episode in China closely corresponds to that in Europe, key immigrations being those of *Elephas* from Africa and *Equus* from North America. As in North America, this episode occurs at about 2.5 Ma near the base of the Matuyama Chron (Qiu, 1989). Ding *et al.* (1992) show that *Hipparion*-bearing red clays were transitional to the loess sequence in which these immigrants appear and that these events were coupled with major uplift of the Himalayan Plateau.

Indian Land Mammal Record

One of the best-studied mammal-bearing sequences in the world is that of the Siwalik Hills of Pakistan and India. Long, nearly continuous sections, extensively exposed and richly fossiliferous, yield a virtually complete magnetostratigraphic column for the Miocene. It is quite interesting, therefore, that Barry and Flynn (1989, p. 567) attribute the pulse of faunal turnover in that region primarily to immigration episodes triggered by land bridges; they summarize that relationship as follows:

In the Siwaliks in situ evolution appears to occur in only a few lineages and is therefore thought to be unimportant in most groups. Immigration and extinction events tend to be correlated and together were the principal cause of faunal change. As immigration events often precede extinctions, and in some cases can be inferred to have caused them, immigration and the resulting ecological disruption may have been the primary cause of community change. Many of the immigrant species probably originated in Africa, Europe, or other parts of Asia. Faunal turnovers are thus also intervals of faunal exchange and indicate times when land connections were established. Nearly all of these episodes show approximate correlations to global climatic, oceanographic, and tectonic events, and these through their effects on sea-level, intercontinental connections, and vegetation, may have controlled movement of mammals into the Siwalik province.

The earliest major immigration episode had already occurred before the first Neogene mammalian fauna is recorded in the Siwalik Hills, and led to the establishment of bovids and other ruminants (the dominant large herbivores and muroid rodents). Barry *et al.* (1985) estimate that this first episode occurred between 18 and 22 Ma, and Barry and Flynn (1989) correlate it approximately with the TB 2.1 sea-level fall of Haq *et al.* (1988). This episode coincides with similar (first-order) episodes of mammal immigration between Eurasia and North America, and between Europe and Africa.

A second major immigration episode (between 15 and 13 Ma) involving muroid rodents and ruminant artiodactyls was accompanied by an abrupt rise in species numbers (Barry *et al.*, 1991). This coincides approximately with the high diversity "savanna optimum" in North America, but does not correlate with any major immigration episode in North America.

The next major faunal turnover episode observed in the Siwalik mammal succession occurs at 9.5 Ma when hipparionine horses appear and murids become dominant over cricetid rodents. Barry *et al.* (1985) suggested that this probably represents one of the major Middle Miocene isotopic and sea-level events discussed by Barron (1985).

A major climatic shift toward increasing aridity and seasonality took place between 7.5 and 7.0 Ma, and caused a major turnover in Siwalik faunas, notably the replacement of large hominoids by cercopithecoid monkeys. The pattern of increased aridity, however, was time transgressive, so that contemporaneous faunas 300 miles to the southeast retained a more tropical woodland aspect (Barry and Flynn, 1989). Unfortunately, the Siwalik record does not adequately record the major faunal turnover episodes at the end of the Miocene.

African Land Mammal Record

In Africa, much of the Cenozoic record of mammalian immigrations and faunal turnover episodes is interrupted by too many hiatuses to represent a coherent record of immigration episodes. Nevertheless, during selected intervals, Africa provides an exceedingly valuable sample of a continent notable for being the evolutionary cradle of such important mammalian groups as proboscideans, ruminants, and hominoids. At the beginning of the Neogene, for example, Savage (1989, p. 592) recognizes in the Early Miocene at Gebel Zelten in Libya " . . . a tropical regime with open shrubland faunas . . . communicating across Tethys with Eurasia." Among the mammal groups with shared genera of about 20 Ma (Orleanian age of the European land-mammal scheme) are proboscidea, creodonts, carnviores, rhinoceroses, suids, and bovids (Vrba, 1985b).

Particularly significant immigration episodes occurred in Africa at about 5.0 and 2.5 Ma. At the former time, apparent immigrants are the antelope groups (bovid subfamilies) Ovibovini, Bovini, and probably Reduncini; at the latter time, caprines (goats) and the genus *Oryx* constitute key immigrant groups (Vrba, 1985b).

Coincidence of first-order immigration and rapid turnover episodes across several continents adds great force to the significance of North American episodes. At present, however, such generalizations can be developed securely only for three intervals in the Neogene. In absolute terms, these times are about 20, 5, and 2.5 Ma. These three well-corroborated rapid turnover episodes correspond to the following first-order immigration episodes in North America: Hemingfordian 1, Hemphillian 3, and Blancan 2.

Oxygen Isotopes and Mammal Immigrations

In an influential paper, Fischer (1983) introduced the concept of two Phanerozoic supercycles in which the Earth's climate has alternated between a *greenhouse state* and an *icehouse state*. In the Cenozoic the Earth has experienced its most recent change from greenhouse to icehouse. The broad outlines of this change are well known and have been revealed by shifts in the oxygen isotope ratios extracted from tests of foraminifera in marine sediment cores. The Cenozoic record for benthic foraminifera clearly displays a secular trend to more positive isotopic values, reflecting major cooling of bottom waters in the world ocean. There is general agreement that in the Early Eocene, $\delta^{18}O$ values were zero or negative; that these values became more positive throughout the Eocene; and that they declined sharply in latest Eocene, and again in Middle Miocene and latest Miocene.

Another independent approach to the history of Cenozoic climates derives from seismic stratigraphic studies of passive continental margins. Haq *et al.* (1988) provided a recent summary of the apparent eustatic curve compiled from many well-dated seismic stratigraphic studies. The Cenozoic is represented by parts A and B of the Tejas megasequence and includes seven second-order supercycles. Each of these supercycles spans about 10 Ma and terminates at a sequence boundary of major magnitude. Shackleton *et al.* (1988), Williams (1988), Christie-Blick *et al.* (1990), and Miller *et al.* (1991) have carefully considered the relationships between isotopic and seismic studies as they bear on the history of Cenozoic climate. Potential pitfalls for the seismic method are confounding the effects of local sedimentary load or local tectonism with more global eustatic effects. For these reasons, we focus on the isotopic record as the more reliable signal of global climatic change during the Cenozoic. It should be noted, however, that for many intervals, there is substantial correlation between the seismic and isotopic approaches.

Shackleton and Opdyke (1977) showed that during the late Pleistocene, oxygen isotope ratios covaried between tropical planktic and benthonic foraminifera. They interpreted the simultaneous changes in surface and bottom-dwelling foraminifera as evidence of growth and decay of ice sheets on the continents. Therefore, the $\delta^{18}O$ record represents change in both ocean temperature and ice volume. The problem has been to partition these two effects. Miller *et al.* (1991) recently extended the covarying isotopic studies to sediments of mid-Tertiary age, recognizing key intervals of ice buildup by covariant increases in heavy oxygen ratios in Miocene and even Oligocene cores.

Prentice and Matthews (1988) have attempted to monitor Cenozoic sea-level change by analyzing oxygen isotope ratios in planktic foraminifera from equatorial regions. The efficacy of this approach depends on the assumption that equatorial sea-surface temperatures (away from upwellings) have not changed during the Cenozoic. Thus, they reason that observed isotopic changes wholly reflect waxing and waning of glacial ice. Figure 11.2 juxtaposes North American land mammal immigration episodes with the trace of Cenozoic oxygen isotope ratios based on planktic forams from equatorial regions. We discuss each of the continental first-order immigration episodes in relation to the marine record.

The earliest pair of first-order immigration episodes are the Clarkforkian and early Wasatchian, straddling the Paleocene-Eocene boundary. Although they fall generally within the warmest climatic interval of the Cenozoic, they correlate with small cooling events in the isotope curve at about 59 and 56 Ma. The Clarkforkian immigration episode correlates particularly well with the abrupt cooling episode at the end of the Paleocene (59 Ma), demonstrated by Kennett and Stott (Chapter 5, this volume) in high-resolution data based on planktonic forams from the southern ocean. Miller *et al.* (1987) postulated a sea-level drop at this time, and Haq *et al.* (1988) recognized a second-order drop in sea level (TA 222-24) in the Thanetian.

The very large Wasatchian land mammal immigration episode reflects plate tectonic effects in the North Atlantic that established a very broad, low-latitude corridor across the Thulean route and thus produced extremely close faunal resemblance between North America and Europe as discussed above.

The next first-order episode occurs in the early Duchesnean (Late Eocene) at about 40 Ma (Emry, 1981; Krishtalka *et al.*, 1987). This episode corresponds well to a number of global Late Eocene events. Miller *et al.* (1987) show a major oxygen isotope increase at about 40 Ma based on benthic forams, and this event is also seen in planktonic forams from equatorial cores presented by Prentice and Matthews (1988). These isotopic events correlate with the beginning of the major sea-level drop TA4 (Priabonian) of Haq *et al.* (1988). According to Hallam (1984) and others cited therein, this latest Eocene sea-level drop is greater than that of the Late Oligocene. The profound global climatic shift of the Late Eocene correlates with a strong increase in the Earth's thermal gradient due to cooling in the southern ocean (Kennett and Barker, 1990).

During the Neogene, land mammal immigration episodes increased markedly in North America. Because of their importance and frequency, these Miocene and Pliocene immigrations were subjected to detailed analysis by Tedford *et al.* (1987). In Figure 11.3, we juxtapose the full array of Neogene land mammal immigration episodes (including third-order episodes) with the $\delta^{18}O$ excursions numbered by Miller *et al.* (1991). As suggested by Opdyke (1990) the episodes correlate remarkably closely with oxygen isotope events in the marine record. Only one isotope event (namely Miocene 3) fails to correlate with an immigration episode in North America.

On the other hand, several immigration episodes in the Miocene of North America fail to correlate with any of the positive isotope excursion numbered by Miller *et al.* (1991). One first-order immigration episode (Arikareean 2 at 20 Ma) fails this correspondence test; as noted below it does correlate with a small (unnumbered) positive isotope excursion. Blancan 1 (at 4.8 Ma) also appears to be unrequited, but in fact it may correspond with Pliocene 1 of Miller *et al.* (1991). Other lesser immigration episodes at 14.5, 7.0, and 6.0 Ma do not correspond to numbered isotope excursions.

The Early Miocene records the largest set of generic immigrations in the history of the North American land mammal fauna. Three land mammal dispersal episodes fall near the boundary between Arikareean and Hemingfordian: they begin with a second-order episode at 21 Ma

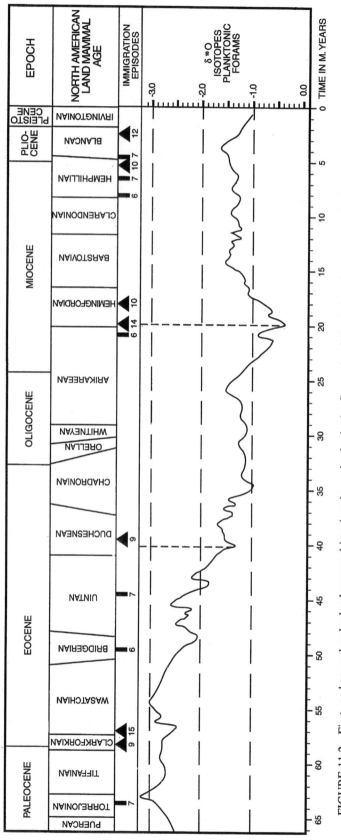

FIGURE 11.2 First- and second-order land mammal immigration episodes in the Cenozoic of North America juxtaposed with the marine record of $\delta^{18}O$ from equatorial planktic foraminifera after Prentice and Matthews (1988). If sea-surface temperatures have not changed in equatorial regions during the Cenozoic, this curve is thought to reflect ice buildup and downward shifts in global sea level. Note especially the significant correlations at about 40 and 20 Ma.

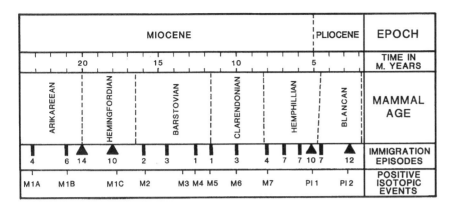

FIGURE 11.3 Land mammal immigrations in the Miocene of North America, including first-order (triangles) as well as second- and third-order (rectangles) episodes, juxtaposed with $\delta^{18}O$ positive excursions as numbered by Miller et al. (1991) based on detailed study of the benthic foram record.

followed by first-order episodes at 20 and 18 Ma. The episodes at 21 and 18 Ma coincide with isotope events Mi1b and Mi1c of Miller et al. (1991). The immigration episode at 20 Ma corresponds to a positive isotope excursion in both benthic and planktonic forams as identified by Prentice and Matthews (1988). According to the latter, sea level at 20 Ma reached depths not attained again until the Pleistocene. Miller et al. (1987) show a possible erosional interval at 20-21 Ma, and Haq et al. (1988) record a first-order drop in the seismic curve between 20 and 21 Ma (TB2). On the other hand, the episode at 18 Ma does not correspond to any major sea-level drop noted in the seismic curve.

Early Miocene land mammal immigrants, including three groups of ruminants, participated in the transition to savanna biomes in midcontinental North America, accompanied by a shift to seasonally arid climates and loess deposition, as discussed above. Presumably this Early Miocene savanna transformation was driven by a fundamental change of state in the global climatic pattern, most notably intensified deep water flow with opening of the Drake Passage and concomitant development of the Antarctic Circumpolar Current (Kennett and Barker, 1990).

Further major shifts are evident in the Middle Miocene, but in North America, these do not include significant numbers of immigrants. Instead, the acme of land mammal diversity, dominated by horses and other savanna herbivores, is attained in the Barstovian. This savanna optimum on the continent correlates with the largest Middle Miocene shift in the Neogene $\delta^{18}O$ record. The strong string of positive isotope shifts in the marine record is widely believed to represent the buildup of permanent glaciers on East Antarctica (Kennett and Barker, 1990). Haq et al. (1988) record a major sea-level drop at 16 Ma with additional drops at 14 and 12 Ma.

In Figure 11.4, we juxtapose the Neogene portion of the benthic foram isotope record after Miller et al. (1987) with the detailed record of large herbivore diversity and feeding modes in North America. The latter record, more fully discussed by Webb (1983a), shows the acme of herbivore diversity in the Barstovian, followed by a steep stepwise decline during the Clarendonian and early Hemphillian. The losses affected primarily browsing ungulates and, secondarily, grazing ungulates. Environmentally, these Middle and Late Miocene faunal changes represent the shift from savanna to steppe biomes, presumably a result of declining rainfall. A remarkably similar transition occurs concur-

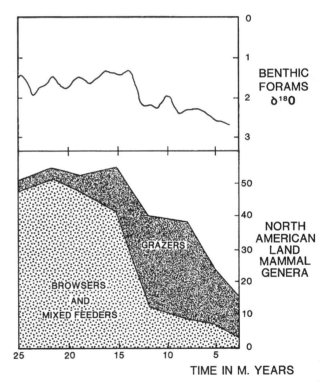

FIGURE 11.4 Decline of ungulate genera in Miocene of North America (after Webb, 1983a) compared with Miocene record of $\delta^{18}O$ based on benthic foraminifera (after Miller et al., 1987). Note that the major Middle Miocene cooling episode coincides with the major extinction of browsers and mixed feeders about 12 Ma.

rently in the land mammals of the Indian subcontinent: from the herbivore acme in woodland savanna environments, a stepwise decline takes place that also represents a shift toward more arid environments between 16 and 12 Ma (Barry et al., 1991). The Middle Miocene patterns on land and sea strongly support an interpretation of climatic forcing (driven by attainment of permanent glaciation in East Antarctica) of land mammal faunal evolution in temperate latitudes.

In the Late Miocene of North America, beginning with the Clarendonian/Hemphillian boundary at about 8.5 Ma, three second-order immigration episodes occur. The episode at 8 Ma coincides with isotope event Mi7 of Miller et al. (1991) and roughly approximates the drop in sea level recognized as TB3.2 of Haq et al. (1988). Of the two further migratory episodes at roughly 7 and 6 Ma, only the latter corresponds to a reported marine event, namely that shown by Miller et al. (1991) at 5.8 Ma. Kennett and Barker (1990) suggest that the later Miocene was the interval during which West Antarctica became permanently glaciated.

The Miocene-Pliocene boundary is associated with a first-order immigration episode in land mammal faunas of both North America and Europe. This episode correlates well with a positive shift in $\delta^{18}O$ (Hodell et al., 1986), but does not seem to correlate with any episode of sea-level lowering in the Haq et al. (1988) curve. Possibly sea level is beginning to fluctuate at high frequency in the Late Miocene and Pliocene, with the consequence that it becomes difficult to define any particular pulse with seismic stratigraphy. On the other hand, the Messinian is well known in the Mediterranean region and in the isotopic record as a major glacioeustatic cycle (Adams et al., 1977; Kennett, 1985; Hodell et al., 1986). As emphasized by Webb (1983b), the Messinian can be tightly correlated with the mid-Hemphillian extinction and faunal turnover episode in temperate North America.

The last second-order episode of the Tertiary sweeps through land mammal faunas on most continents at 2.5 Ma in the Late Pliocene. Such major immigrations are clearly related to the onset of Northern Hemisphere glaciation recognized in the $\delta^{18}O$ record and also indicated by glacial sediments (Ruddiman and Raymo, 1988; Shackleton et al., 1988; Raymo and Ruddiman, 1992). All of these features correlate with marked cooling and sea-level lowering indicated by the oxygen isotope record (Shackleton and Opdyke, 1977).

Generally speaking, land mammal immigration episodes in North America and other well-studied continental areas correlate with positive excursion of the oxygen isotope record from tropical marine forams. Other continental regions clearly corroborate the significant turnovers recorded at about 20, 5, and 2.5 Ma.

In some cases, however, the terrestrial turnover pattern gives no evident response to major cooling events. The two most glaring examples are the Early Oligocene (White River chronofauna) and the Middle Miocene (Clarendonian chronofauna) in North America when, despite major global changes, the mammal faunas show remarkable stability and continuity. Possibly no land bridges were available during the Early Oligocene. In the Middle Miocene, this excuse cannot serve, because a few Asiatic immigrants did trickle in and Beringia was emergent. Evidently, during these two stable chronofaunal stages the North American mammalian fauna was resilient and not open to immigrants.

CONCLUSIONS

The rich record of Cenozoic land mammals in North America documents a large-scale succession of terrestrial ecosystems. In the midcontinent the predominant biomes changed from subtropical forest in the Eocene through woodland savanna in the Oligocene, to park savanna and then grassland savanna in the Miocene, to steppe in the Pliocene, when increased provincialism became evident. This broad march of faunal change has been inferred not only from dental adaptations and body-size distributions in the mammalian faunas, but also from independent lines of terrestrial evidence including floral succession, sedimentology, and stable isotopes. Clearly the continental biota in North America broadly tracks the global climatic trends of the Cenozoic.

More specific faunal analyses of the well-calibrated North American record reveal a remarkably irregular progression of faunal change. Immigration of land mammals from Europe, Asia, and South America, totaling some 140 genera in the course of the Cenozoic, are concentrated mainly in seven first-order episodes. Pairs of first-order episodes bracketing 57, 19, and 4 Ma suggest that a major interval favoring immigration somehow produced an "echo" effect. The largest of all immigration sets occurred in the Early Miocene when two first-order episodes and one second-order episode arrived between 21 and 18 Ma.

Land mammal records from other continents tend to replicate strong mammal immigration episodes at 20, 5, and 2.5 Ma. Other resemblances include a Middle Miocene acme in the Siwaliks, possibly correlated with the Barstovian acme in North America, but this interval has a notably low rate of intercontinental immigrations.

Causal considerations lead to two possible hypotheses linking mammal immigrations to patterns of global climatic change. The first hypothesis postulates greater availability of land bridges due to low sea levels, given the correlation between positive oxygen-isotope excursions and eustatic drops in the marine record. The second, more purely terrestrial hypothesis links each first-order immigration episode to wholesale reorganization of the continental ecosystem triggered by climatic shifts.

Two major contradictions to the first hypothesis appear in the Oligocene (35 to 30 Ma) and the Middle Miocene (16 to 6 Ma). Despite major global cooling events, the North American mammal fauna admitted very few immigrants. Possibly no land bridges were available at those times, although the appearance of a few immigrants in the Middle Miocene suggests that there was some physical access. More probably, there were *ecological* barriers to immigrants at these times. During these two intervals the land mammal faunas experienced high diversity and long stable community development (chronofaunal evolution). This lends credence to the view that the ecosystem was near capacity during the Barstovian acme and perhaps during other chronofaunal intervals. Other more disturbed intervals, on the other hand, were open to major immigration episodes that coincided with positive isotopic excursions in the marine realm.

Evidently first-order changes in the oxygen isotope record are necessary but not sufficient causes of first-order immigration episodes into the North American land mammal fauna. The record of continental mammal faunas itself offers a strong signal of major global climatic change. The question of why the continental ecosystem was open to immigrations during certain positive isotopic excursions and not others has fundamental significance to our understanding of the stability of present continental ecosystems. It therefore demands further investigation.

ACKNOWLEDGMENTS

We have benefited from discussions with Catherine Badgely, John Barry, David Hodell, Everett Lindsay, Bruce MacFadden, Malcolm McKenna, Ken Miller, Nick Shackleton, Richard Tedford, Elizabeth Vrba, and Michael Woodburne. This research was partly supported by grant number BSR 89 18065 to S.D.W. from the National Science Foundation. This is contribution number 431 in Paleobiology from the Florida Museum of Natural History.

REFERENCES

Adams, C. G., R. H. Benson, R. B. Kidd, W. B. F. Ryan, and R. C. Wright (1977). The Messinian salinity crisis and evidence of Late Miocene eustatic changes in the world ocean, *Nature* 269, 383-386.

Andrews, P., J. M. Lord, and E. M. Newsbit Evans (1979). Patterns of ecological diversity in fossil and modern mammalian faunas, *Biol. Jour. Linnean Soc. 11*, 177-205.

Axelrod, D. I. (1992). Climatic pulses, A major factor in legume evolution, in *Advances in Legume Systematic: The Fossil Record* (part 4), P. S. Herendeen and D. L. Dilcher, eds., Royal Botanical Gardens, Kew, London, pp. 259-279.

Azzaroli, A. (1989). The genus *Equus* in Europe, in *European Neogene Mammal Chronology*, Plenum Press, New York, pp. 339-356.

Barnosky, C. W. (1984). Late Miocene vegetational and climatic variations inferred from a pollen record in northwest Wyoming, *Science 223*, 49-51.

Barron, E. J. (1985). Explanations of the Tertiary global cooling trend, *Palaeogeography, Palaeoclimatology, Palaeoecology 50*, 45-61.

Barry, J. C., and L. J. Flynn (1989). Key biostratigraphic events in the Siwalik sequence, in *European Neogene Mammal Chronology*, E. H. Lindsay, V. Fahlbusch, and P. Mein, eds., Plenum Press, New York, pp. 557-572.

Barry, J. C., N. M. Johnson, S. M. Raza, and L. L. Jacobs (1985). Neogene mammalian faunal change in southern Asia: Correlations with climatic, tectonic, and eustatic events, *Geology 13*, 637-640.

Barry, J. C., L. J. Flynn, and D. R. Pilbeam (1991). Faunal diversity and turnover in a Miocene terrestrial sequence, in *Causes of Evolution: A Paleontological Perspective*, R. Ross and W. Allmon, eds., University of Chicago Press, Chicago.

Behrensmeyer, A. K., and A. P. Hill (1980). *Fossils in the Making: Vertebrate Taphonomy and Paleoecology*, University of Chicago Press, Chicago.

Behrensmeyer, A. K., J. D. Damuth, W. A. DiMichele, R. Potts, H.-D. Sues, and S. L. Wing (1992). *Terrestrial Ecosystems Through Time*, University of Chicago Press, Chicago, 568 pp.

Berggren, W. A., D. V. Kent, J. J. Flynn, and J. A. Van Couvering (1985). Cenozoic Geochronology, *Geological Society of America Bulletin 96*, 1407-1418.

Butler, R. F., P. D. Gingerich, and E. H. Lindsay (1991). Magnetic polarity, stratigraphy and biostratigraphy of Paleocene and lower Eocene continental deposits, Clark's Ford Basin, Wyoming, *Journal of Geology 89*, 299-316.

Christie-Blick, N., G. S. Mountain, and K. G. Miller (1990). Seismic stratigraphic record of sea-level change, in *Sea-Level Change*, Studies in Geophysics, National Research Council, National Academy Press, Washington, D.C., pp. 116-140.

Clark, J., J. R. Beerbower, and K. K. Kietzke (1967). Oligocene sedimentation, stratigraphy, paleoecology and paleoclimatology in the Big Badlands of South Dakota, *Fieldiana*, Geology Memoirs 5, 1-158.

Coe, M. J., D. H. Cumming, and J. Phillipson (1976). Biomass and production of large African herbivores in relation to rainfall and primary production, *Oecologia 22*, 341-354.

Damuth, J. (1982). Analysis of the preservation of community structure in assemblages of fossil mammals, *Paleobiology 8*, 434-446.

Ding, Z., N. Rutter, H. Jingtai, and L. Tungsheng (1992). A coupled environmental system formed at about 2.5 Ma in East Asia, *Palaeogeography, Palaeoclimatology, Palaeoecology 94*, 223-242.

Emry, R. J. (1981). Additions to the mammalian fauna of the type Duchesnean, with comments on the status of the Duchesnean "age," *Journal of Paleontology 55*, 563-570.

Estes, R. D. (1974). Social organization of the African Bovidae, in *Behavior of Ungulates in Relation to Management*, V. Geist and F. Walther, eds., I.U.C.N., Morges, Switzerland.

Fischer, A. G. (1983). The two Phanerozoic supercycles, in *Catastrophes and Earth History*, W. A. Berggren and J. A. Van Couvering, eds., Princeton University Press, Princeton, N.J., pp. 129-150.

Funder, S., N. Abrahamsen, P. Bennike, and R. W. Feyling-Hannsen (1985). Forested Arctic: Evidence from North Greenland, *Geology 13*, 542-546.

Gingerich, P. D. (1984). Pleistocene extinctions in the context of origination-extinction equilibria in Cenozoic mammals, in *Quaternary Extinctions*, P. S. Martin and R. G. Klein, eds., University of Arizona Press, Tucson, pp. 211-222.

Gingerich, P. D. (1989). New earliest Wasatchian mammalian fauna from the Eocene of northwestern Wyoming: Composition and diversity in a rarely sampled high-floodplain assemblage, *University of Michigan Papers in Paleontology 28*, 1-97.

Ginsburg, L. (1989). The faunas and stratigraphical subdivisions of the Orleanian in the Loire Basin (France), in *European Neogene Mammal Chronology*, E. H. Lindsay, V. Fahlbusch, and P. Mein, eds., Plenum Press, New York, pp. 157-176.

Hallam, A. (1984). Pre-Quaternary sea-level changes, *Annual Reviews of Earth and Planetary Sciences 12*, 205-243.

Haq, B. U., J. Hardenbol, and P. R. Vail (1988). Mesozoic and Cenozoic chronostratigraphy and eustatic cycles, in *Sea Level Changes: An Integrated Approach*, Society of Economic Paleontologists and Mineralogists Special Publication 42, pp. 71-108.

Herman, Y., ed. (1989). *The Arctic Seas: Climatology, Oceanography, Geology and Biology*, Van Nostrand Reinhold, New York, 888 pp.

Hibbard, C. W., and D. W. Taylor (1960). Two late Pleistocene faunas from southwestern Kansas, *University of Michigan Museum of Paleontology Contribution 16*, 1-223.

Hodell, D. A., K. M. Elmstron, and J. P. Kennett (1986). Latest Miocene benthic $\delta^{18}O$ changes, global ice volume, sea level and the "Messinian salinity crisis," *Nature 320*, 411-414.

Huber, O. (1982). Significance of savanna vegetation in the Amazon Territory of Venezuela, in *Biological Diversification in the Tropics*, G. T. Prance, ed., Columbia University Press, New York, pp. 221-224.

Hulbert, R. C., Jr. (1982). Population ecology of *Neohipparion* (Mammalia: Equidae) from the Late Miocene Love Bone Bed of Florida, *Paleobiology 8*, 159-167.

Hulbert, R. C., Jr. (1987). A new *Cormohipparion* (Mammalia, Equidae) from the Pliocene (latest Hemphillian and Blancan) of Florida, *Journal of Vertebrate Paleontology 7*, 451-468.

Hulbert, R. C., Jr., and B. J. MacFadden (1991). Morphological transformation and cladogenesis at the base of the adaptive radiation of Miocene hypsodont horses, *Amer. Mus. Novitates*.

Hunt, R. M., Jr. (1990). Taphonomy and sedimentology of Arikaree (lower Miocene) fluvial, eolian, and lacustrine paleoenvironments, Nebraska and Wyoming: A paleobiota entombed in fine-grained volcaniclastic rocks, in *Volcanism and Fossil Biotas*, M. G. Lockley and A. Rice, eds., Geological Society of America Special Paper 244, pp. 69-111.

Hutchison, J. H. (1982). Turtle, crocodilian, and champsosaur diversity changes in the Cenozoic of the north-central region of western United States, *Palaeogeography, Palaeoclimatology, Palaeoecology 37*, 149-164.

Janis, C. (1982). Evolution of horns in ungulates: Ecology and paleoecology, *Biol. Rev. 57*, 261-318.

Janis, C. (1984). The use of fossil ungulate communities as indicators of climate and environment, in *Fossils and Climate*, P. Brenchley, ed., John Wiley & Sons, New York, pp. 85-104.

Jarman, P. J. (1974). The social organisation of antelope in relation to their ecology, *Behavior 48*, 215-267.

Jenkins, F. A., Jr., and D. W. Krause (1983). Adaptations for climbing in North American Multituberculates (Mammalia), *Science 220*, 712-715.

Kennett, J. P., ed. (1985). *The Miocene Ocean: Paleoceanography and Biogeography*, Geological Society of America Memoir 163.

Kennett, J. P., and P. F. Barker (1990). Latest Cretaceous to Cenozoic climate and oceanographic developments in the Weddell Sea, Antarctica: An ocean-drilling perspective, in *Proceedings of the Ocean Drilling Program 113*, P. F. Barker, J. P. Kennett et al., eds., College Station, Texas, pp. 937-960.

Kowalevsky, W. (1873). Sur *l'Anchitherium aurelianense* Cuvier et sur l'Histoire paleontologique des Chevaux, *Mem. de l'Acad. Imperiale des Sciences St. Petersburg, 7e ser. 20*(5),1-73.

Krause, D. W., and M. C. Maas (1990). The biogeographic origins of Late Paleocene-Early Eocene mammalian immigrants to the Western Interior of North America, in *Dawn of the Age of Mammals in the Northern Part of the Rocky Mountain Interior*, T. M. Brown and K. D. Rose, eds., Geological Society of America Special Paper 243, pp. 71-105.

Krishtalka, L., R. M. West, C. C. Black, M. R. Dawson, J. J. Flynn, W. D. Turnbull, R. K. Studky, M. C. McKenna, T. M. Bown, D. J. Golz, and J. A. Lillegraven (1987). Eocene (Wasatchian through Duchesnean) biochronology of North America, in *Cenozoic Mammals of North America: Geochronology and Biostratigraphy*, M. O. Woodburne, ed., University of California Press, Berkeley and Los Angeles.

Legendre, S. (1986). Analysis of mammalian communities from the Late Eocene and Oligocene of southern France, *Palaeovertebrata 16*, 191-212.

Lindsay, E. H., N. D. Opdyke, and N. M. Johnson. (1984). Blancan-Hemphillian land mammal ages and late Cenozoic mammal dispersal events, *Annual Review Earth and Planetary Science 12*, 445-488.

Lindsay, E. H., V. Fahlbusch, and P. Mein, eds. (1990). *European Neogene Mammal Chronology*, Plenum Press, New York, 658 pp.

Litke, R. (1968). Uber den nachweis tertiaer Gramineen, *Monatsber. Deut. Akad. Wiss. Berlin 19*, 462-471.

Lundelius, E. L., Jr., T. Downs, E. H. Lindsay, H. A. Semken, R. J. Zakrzewski, C. S. Churcher, C. R. Harington, G. E. Schultz, and S, D. Webb (1987). The North American Quaternary sequence, in *Cenozoic Mammals of North America: Geochronology and Biostratigraphy*, M. O. Woodburne, ed., University of California Press, Berkeley and Los Angeles.

MacGinitie, H. D. (1962). The Kilgore flora: A Late Miocene flora from northern Nebraska, *University of California Publications in Geological Sciences 35*, 67-158.

McNaughton, S. J., and N. J. Georgiadis. (1986). Ecology of African grazing and browsing mammals, *Annual Review of Ecological Systems 17*, 39-65.

McNaughton, S. J., R. W. Ruess, and S. W. Seagle (1988). Large mammals and process dynamics in African ecosystems, *Bioscience 38*, 794-800.

Mein, P. (1989). Updating of MN zones, in *European Neogene Mammal Chronology*, E. H. Lindsay, V. Fahlbusch, and P. Mein, eds., Plenum Press, New York, pp. 73-90.

Miller, K. G., R. G. Fairbanks, and G. S. Mountain. (1987). Tertiary oxygen isotope synthesis, sea level history, and continental margin erosion, *Paleoceanography 2*, 1-19.

Miller, K. G., J. D. Wright, and R. G. Fairbanks (1991). Unlocking the ice house: Oligocene-Miocene oxygen isotopes, eustasy, and margin erosion, *Journal of Geophysical Research 96*(B4), 6829-6848.

Olson, E. C. (1952). The evolution of a Permian vertebrate chronofauna, *Evolution 6*, 181-196.

Olson, E. C. (1983). Coevolution or coadaptation? Permo-Carboniferous vertebrate chronofauna, in *Coevolution*, M. H. Nitecki, ed., University of Chicago Press, Chicago, pp. 307-338.

Opdyke, N. D. (1990). Magnetic stratigraphy of Cenozoic terrestrial sediments and mammalian dispersal, *Journal of Geology 98*, 621-637.

Owen-Smith, R. N. (1988). *Megaherbivores: The Influence of Very Large Body Size on Ecology*, Cambridge University Press, Cambridge, 369 pp.

Prentice, M. L., and R. K. Matthews. (1988). Cenozoic ice-volume history: Development of a composite oxygen isotope record, *Geology 16*, 963-66.

Prothero, D. R. (1986). A new Oromerycid (Mammalia, Artiodactyla) from the Early Oligocene of Montana, *Journal of Paleontology 60*, 458-465.

Qiu, Z. (1989). The Chinese Neogene mammalian biochronology: Its correlation with the European Neogene mammalian zonation, in *European Neogene Mammal Chronology*, E. H. Lindsay, V. Fahlbusch, and P. Mein, eds., Plenum Press, New York, pp. 527-556.

Raymo, M. E., and W. F. Ruddiman (1992). Tectonic forcing of late Cenozoic climate, *Nature 359*, 117-122.

Repenning, C. A. (1985). Pleistocene mammalian faunas: Climate and evolution, *Acta Zool. Fennica 170*, 173-176.

Repenning, C. A. (1987). Biochronology of the microtine rodents of the United States, in *Cenozoic Mammals of North America: Geochronology and Biostratigraphy*, M. O. Woodburne, ed., University of California Press, Berkeley and Los Angeles.

Retallack, G. J. (1983). A paleopedological approach to the interpretation of terrestrial sedimentary rocks: The mid-Tertiary fossil soils of Badlands National Park, South Dakota, *Geological Society of America Bulletin 94*, 823-840.

Robertson, J. R. (1976). Latest Pliocene mammals from Haile XV A, Alachua Co., Florida, *Bulletin of the Florida State Museum, Biological Sciences 20*, 111-186.

Rose, K. D. (1981). The Clarkforkian land mammal age and mammalian faunal composition across the Paleocene-Eocene boundary, *University of Michigan Papers in Paleontology 26*, 1-196.

Rowley, D. B., and A. L. Lottes (1988). Plate-kinematic reconstructions of the North Atlantic and Arctic Late Jurassic to Present, *Tectonophysics 155*, 73-120.

Ruddiman, W. F., and M. E. Raymo (1988). Northern Hemisphere climate regimes during the last three million years: Possible tectonic connections, *Philosophical Transactions of the Royal Society of London B318*, 411-430.

Savage, D. E., and D. E. Russell. (1983). *Mammalian Paleofaunas of the World*, Addison-Wesley Publishing Co, Reading, Mass.

Savage, R. J. G. (1989). The African dimension in European Early Miocene faunas, in *European Neogene Mammal Chronology*, E. H. Lindsay, V. Fahlbusch, and P. Mein, eds., Plenum Press, New York, pp. 587-599.

Scott, W. B. (1937). *A History of Land Mammals in the Western Hemisphere*, Macmillan, New York.

Shackleton, N. J., and N. D. Opdyke. (1977). Oxygen isotope and paleomagnetic stratigraphy of equatorial Pacific core V28-239; Oxygen isotope temperatures and ice volumes on a 10^5-10^6 year scale, *Quaternary Research 3*, 39-55.

Shackleton, N. J., R. G. West, and D. Q. Bowen (1988). The past three million years: Evolution of climatic variability in the North Atlantic region, *Philosophical Transactions of the Royal Society of London B318*, 409-688.

Sher, A. V. (1974). Pleistocene mammals of the far northeast USSR and North America, *International Geology Review 16*, 1-89.

Sinclair, A. R. E. (1983). The adaptation of African ungulates and their effects on community function, in *Tropical Savannas*, F. Bourliere, ed., Elsevier, Amsterdam.

Sinclair, A. R. E., and M. Norton-Griffiths (1979). *Serengeti: Dynamics of an Ecosystem*, University of Chicago Press, Chicago and London.

Stehli, F. G., and S. W. Webb, eds. (1985). *The Great American Biotic Interchange*, Topics in Geobiology, Plenum Press, New York.

Stucky, R. (1990). Evolution of land mammal diversity in North America during the Cenozoic, *Current Mammalogy 2*, 375-432.

Tassy, P. (1989). The "Proboscidean Datum Event": How many proboscideans and how many events, in *European Neogene Mammal Chronology*, E. H. Lindsay, V. Fahlbusch, and P. Mein, eds., Plenum Press, New York, pp. 237-252.

Tedford, R. H., and E. Gustafson (1974). First North American record of the extinct Pand, *Parailurus, Nature 265*, 621-623.

Tedford, R. H., M. F. Morris, R. W. Fields, J. M. Rensberger, D. P. Whistler, T. Galusha, B. E. Taylor, J.R. Macdonald, and S. D. Webb (1987). Faunal succession and biochronology of the Arikareean through Hemphillian interval (Late Oligocene through earliest Pliocene Epochs) in North America, in *Cenozoic Mammals of North America: Geochronology and Biostratigraphy*, M. O. Woodburne, ed., University of California Press, Berkeley and Los Angeles.

Tedford, R. H., L. J. Flynn, and Z. Qiu (1989). Neogene faunal succession, Yushe Basin, Shanxi Province, PRC, *Journal of Vertebrate Paleontology 9*, 41A.

Thomasson, J. R. (1986). Tertiary fossil plants found in Nebraska, *National Geographic Society Research Report 19*, 553-564.

Thomasson, J. R., M. E. Nelson, and R. J. Zakrzewski (1986). A fossil grass (Gramineae: Chloridoideae) from the Miocene with Kranz anatomy, *Science 233*, 876-878.

Thomasson, J. R., R. J. Zakrsewski, H. E. Lagarry, and D. E. Mergen (1990). A Late Miocene (late early Hemphillian) biota from northwestern Kansas, *National Geographic Society Research 6*, 231-244.

Truswell, E. M., and W. K. Harris (1982). The Cainozoic palaeobotanical record in arid Australia: Fossil evidence for the origins of an arid-adapted flora, in *Evolution of the Flora and Fauna of Arid Australia*, W. R. Barker and P. J. M. Greenslade, eds., Peacock Publications, Frewville, Australia, pp. 57-76.

Van Valen, L. M. (1978). The beginning of the age of mammals, *Evolution Theory 4*, 45-80.

Voorhies, M. R. (1990). *Vertebrate Paleontology of the Proposed Norden Reservoir Area, Brown, Cherry and KeyaPaha Counties, Nebraska*, Bureau of Reclamation, Denver, Colo., 944 pp.

Voorhies, M. R., and J. R. Thomasson (1979). Fossil grass anthoecia within Miocene rhinoceros skeletons: Diet in an extinct species, *Science 206*, 331-333.

Vrba, E. S. (1985a). Environment and evolution: Alternative causes of the temporal distribution of evolutionary events, *South African Journal of Science 81*, 229-236.

Vrba, E. S. (1985b). African bovidae: Evolutionary events since the Miocene, *Suid-Arikaanse Tydskrif vir Wetenskap 81*, 263-266.

Wall, W. P. (1982). Evolution and biogeography of the Amynodontidae (Perissodactyla, Rhinocerotoidea), *Third North American Paleontological Convention, Proceedings 2*, 563-567.

Webb, S. D. (1977). A history of savanna vertebrates in the New World. Part I: North America, *Annual Reviews of Ecological Systems 8*, 355-380.

Webb, S. D. (1983a). The rise and fall of the Late Miocene ungulate fauna in North America, in *Coevolution*, M. H. Nitecki, ed., University of Chicago Press, Chicago, pp. 267-306.

Webb, S. D. (1983b). On two kinds of rapid faunal turnover, in *Catastrophes and Earth History: The New Uniformitarianism*, W. A. Berggren and J. A. Van Couvering, eds., Princeton University Press, Princeton, N.J., pp. 417-436.

Webb, S. D. (1984). Ten million years of mammal extinctions in North America, in *Quaternary Extinctions: A Prehistoric Revolution*, P. S. Martin and R. G. Klein, eds., University of Arizona Press, Tucson, pp. 189-210.

Webb, S. D. (1989). The fourth dimension in North American terrestrial mammal communities, in *Patterns in the Structure of Mammalian Communities*, D. W. Morris *et al.*, eds., Special Publications, Museum of Texas Technical University, pp. 181-203.

Webb, S. D., and B. E. Taylor (1980). The phylogeny of hornless ruminants and a description of the cranium of *Archaeomeryx*, *Bulletin of the American Museum of Natural History 167*, 117-158.

West, R. M., and M. R. Dawson (1978). Vertebrate paelontology and the Cenozoic history of the North Atlantic region, *Polarforschung 48*, 103-119.

Williams, D. F. (1988). Evidence for and against sea-level changes from the stable isotopic record of the Cenozoic, in *Sea-Level Changes: An Integrated Approach*, Society of Economic Paleontologists and Mineralogists Special Publication 42, pp. 31-38.

Wing, S. L., and B. H. Tiffney (1987). The reciprocal interaction of angiosperm evolution and tetrapod herbivory, *Reviews of Palaeobotany and Palynology 50*, 179-210.

Wolfe, J. A. (1985). Distribution of major vegetational types during the Tertiary, in *The Carbon Cycle and Atmospheric CO_2 Natural Variations Archean to Present*, E. T. Sundquist and W. S. Broecker, eds., Geophysical Monograph 32, American Geophysical Union, Washington, D.C., pp. 357-374.

Wood, A. E. (1980). The Oligocene rodents of North America, *Transactions of the American Philosophical Society 70*, 1-68.

Woodburne, M. O., ed. (1987). *Cenozoic Mammals of North America: Geochronology and Biostratigraphy*, University of California Press, Berkeley and Los Angeles, 336 pp.

12

Biotic Responses to Temperature and Salinity Changes During Last Deglaciation, Gulf of Mexico

BENJAMIN P. FLOWER and JAMES P. KENNETT
University of California, Santa Barbara

ABSTRACT

Understanding the biotic response to past global change provides insights into past and present Earth systems. Marine sedimentary records of the last deglaciation centered at 11,000 ^{14}C years ago (ka) are particularly promising in contributing to understanding and predicting the biotic effects of anthropogenic environmental changes. Fossil assemblages of plankton preserved in marine sediments represent an often underutilized source of paleoenvironmental information. Relative abundances of different species and their morphology vary dynamically in response to environmental change. Numerous studies have exploited the sensitivity of marine plankton assemblages to environmental changes in the last glacial ocean by transforming downcore relative abundances directly into quantitative estimates of surface temperature, salinity, and other parameters.

In some areas with unique oceanographic characteristics, transform functions do not provide reliable surface temperature and salinity estimates during times of great environmental change in the geologic past, so-called "no analog situations." For instance, planktonic foraminiferal assemblages during the last deglaciation in the Gulf of Mexico were influenced considerably by surface water salinities much lower than at present. Detailed oxygen isotopic and faunal analyses of radiometrically dated cores from the Orca Basin, Gulf of Mexico, illustrate the response of planktonic foraminifera to temperature and salinity changes in a marginal basin that amplified deglacial environmental change. Warm-water planktonic foraminifera began to replace cold-water forms in response to early deglacial warming at 14 ka, and the euryhaline form *Globigerinoides ruber* dominated during an episode of low-salinity meltwater influx into the Gulf of Mexico from 14 to 11.4 ka. Warm-water forms increased in abundance at 13 ka. A brief reappearance of cool species assemblages associated with the last glacial episode documents the presence of Younger Dryas cooling between 11.4 and 10.2 ka in the Gulf of Mexico. Late deglacial warming fostered the appearance of Holocene assemblages at 10.2 ka. Further warming occurred at 6 ka.

Faunal assemblage analysis offers an independent approach to understanding paleoenvironmental change. Information contained in census data bears on entire communities throughout the year and is not limited to certain times and/or depths. By comparison, oxygen isotopic analysis of the pink form of *Gs. ruber*, an inferred summer dweller in the Gulf of Mexico, does not show evidence for Younger Dryas cooling. Faunal changes further demonstrate that the onset of Younger Dryas cooling was very rapid, occurring in less than 200 yr. The integration of high-resolution faunal and geochemical records of the last deglaciation helps constrain climate models of the geographic extent and rapidity of climatic changes. Faunal analysis also has the potential for resolving smaller-scale events and perturbations that cannot be confirmed by geochemical methods but are essential to an understanding of climate systems.

INTRODUCTION

The chapters presented in this volume explore the close linkages between global environmental and biotic changes in the geologic past. The sensitivity of the Earth's biota to Phanerozoic environmental events is illustrated by a range of responses, including migrations, extinctions, evolutionary turnover, and morphological variation. Little is known, however, about the potential responses of the modern biota to recent and future anthropogenically forced environmental changes. Understanding the faunal response to past global change provides insights into past and present Earth systems, and can assist with the prediction of the effects of global warming, consequent sea-level rise, changes in oceanic and atmospheric circulation patterns, and climatic variability.

Prediction of the biotic response to global change requires an understanding of the faunal response to similar changes in the recent geological past. Excellent records exist of the changes associated with the last major climatic transition, from the last glacial episode (oxygen isotope stage 2) to the Holocene, centered at ~11 ka. Marine sedimentary sequences of this age are accessible, have been extensively cored, are well-preserved, and provide continuous records. As a result, much is known about global environmental changes during this time.

Marine microfossils in deep-sea sedimentary sequences are often abundant enough for statistically reliable faunal analysis and geochemical work. Much information about past environments has been inferred from microfossil assemblages composed of extant species whose ecology is well known. These assemblages not only respond to major environmental changes inferred by using independent approaches, but also reflect smaller-scale changes that are not resolved by geochemical and other methods. Morphological responses of marine plankton to environmental change are large and have also been underutilized. First exploited primarily as a stratigraphic tool and termed "the new paleontology" by Emiliani (1969), the analysis of ecophenotypic variation has important applications to paleoenvironmental problems (for a summary, see Kennett, 1976).

Considerable work on Quaternary marine sediments over the past two decades has highlighted the sensitivity of marine plankton communities to major oceanographic changes in the North Atlantic associated with the transition from the last glacial episode to the Holocene. A selection of papers representing work on different plankton groups and using different approaches is discussed here. Foraminifera (Imbrie and Kipp, 1971; Duplessey et al., 1981; Ruddiman and McIntyre, 1981; Kellogg, 1984), radiolaria (Imbrie and Kipp, 1971; Morley and Hays, 1979), and coccolithophorids (McIntyre et al., 1976) followed deglacial shifts in water mass locations and geochemical parameters in the North Atlantic. Ruddiman and McIntyre (1981) traced the latitudinal migrations of planktonic foraminifera associated with the North Atlantic polar front from 15 to 9 ka. Boltovskoy (1990) showed how planktonic foraminiferal assemblages followed temperature changes in the western equatorial Pacific. Deglacial oceanographic changes in the North Pacific and the Bering Sea produced changes in community structure in diatoms (Sancetta, 1979) and radiolaria (Morley and Hays, 1983; Heusser and Morley, 1985). The CLIMAP project fully exploited the sensitivity of marine plankton to environmental change and utilized empirical transfer functions to translate downcore plankton assemblages into quantitative estimates of temperature, salinity, and other parameters for the glacial ocean at 18 ka (CLIMAP, 1976; Cline and Hays, 1976; McIntyre et al., 1976; Moore et al., 1980). Recent work has also focused on the biotic response to other water mass property changes in addition to temperature. Molfino and McIntyre (1990) interpreted abundance changes of the deep-dwelling *Florisphaera profunda* relative to other coccolithophorids over the past 20,000 yr in terms of nutricline depth changes in the equatorial Atlantic associated with variations in equatorial divergence. Other studies have examined the response of both planktonic and benthic biota. Planktonic and benthic foraminiferal assemblages were affected by salinity and circulation changes in the Red Sea and the Gulf of Aden (Locke and Thunell, 1988) and in the Mediterranean (Thunell *et al.*, 1977; Muerdter *et al.*, 1984). Pedersen *et al.* (1988) documented the response of benthic foraminiferal populations to deglacially induced changes in surface productivity in the Panama Basin.

Radiometrically dated, high-resolution work in some areas has produced highly detailed faunal and isotopic records of deglaciation. Oxygen isotopic and planktonic foraminiferal assemblage records from the Sulu Sea (Linsley and Thunell, 1990; Kudrass *et al.*, 1991) and the Gulf of Mexico (Kennett *et al.*, 1985; Flower and Kennett, 1990) illustrate the sensitivity of planktonic foraminifera to rapid oceanic environmental change and document the expression of the Younger Dryas cooling event. Labracherie *et al.* (1989) showed coherence between faunal and oxygen isotopic changes in the Indian sector of the Southern Ocean and documented a brief return to near-glacial temperatures from ~12 to 11 ka. Studies such as these integrate high-resolution faunal and isotopic evidence to define in detail the character of rapid deglacial environmental and biotic changes.

The integration of Quaternary faunal and geochemical records from deep-marine sequences has demonstrated the extensive influence of North Atlantic oceanography in disparate areas of the world's oceans and has led to the concept of a "conveyor belt" mode of circulation (Broecker *et al.*, 1985). In this model, the Quaternary ocean circulation system is paced by the formation of North Atlantic deep water (NADW), which flows as a deep current through the South Atlantic Ocean, is entrained into the circumpolar current, and flows through the Indian Ocean to the Pacific where it upwells, eventually returning to the North Atlantic via surface currents. A circuit is completed, on average, every 1000 yr. Turning the conveyor belt on or off may have controlled oceanographic and climatic changes in the late Quaternary, which influenced the Earth's biota.

Ecology of Modern Planktonic Foraminifera

The need to interpret fossil planktonic foraminiferal assemblages has fueled interest in the biology and ecology of modern forms. Modern planktonic foraminifera are found throughout the oceans, with diversity increasing toward the tropics. Effectively only one species is found in polar oceans, whereas about 30 species inhabit the tropics. The geographic distribution of distinctive assemblages, which follows major water mass boundaries, is shown for the North Atlantic Ocean in Figure 12.1. Plankton tow work (for example, Bé and Tolderlund, 1971; Tolderlund and Bé, 1971; Fairbanks *et al.*, 1982) has shown that most forms live in the mixed layer (upper 100 m),

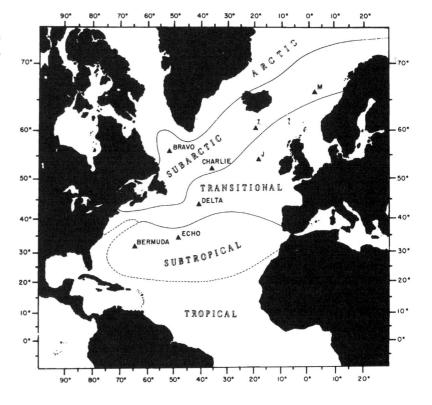

FIGURE 12.1 Distribution of planktonic foraminiferal assemblage provinces and ocean stations in the North Atlantic (from Tolderlund and Bé, 1971).

while some are associated with thermocline depths, and a few are found as deep as 600 m. Shallow-dwelling species tend to be spinose, whereas deeper-dwelling forms are nonspinose.

Time-series sampling through the year by plankton tows and sediment traps (Curry *et al.*, 1983; Thunell *et al.*, 1983a; Deuser and Ross, 1989) has demonstrated a seasonal succession of planktonic foraminiferal communities. Different depth habitats and seasonal preferences are reflected in the stable isotopic and trace metal chemistry of their calcium carbonate tests (Thunell *et al.*, 1983b; Deuser and Ross, 1989). Shallower-dwelling species can be distinguished from deeper forms on the basis of oxygen and carbon isotopic composition, other geochemical differences, and morphology. Summer-dwelling forms are marked by lower oxygen isotopic values. Morphology also varies between different environments. For example, test size and pore diameter of *Orbulina universa* in the Indian Ocean increase toward lower latitudes (Bé *et al.*, 1973).

Laboratory culture work has demonstrated the temperature and salinity tolerances of certain planktonic foraminiferal species (for example, Bijma *et al.*, 1990). Stable isotopic analysis of forms cultured under different growth conditions has furthered understanding of the effect of light, temperature, and nutrient availability on the isotopic composition of the shell (Spero and DeNiro, 1987; Spero and Williams, 1988, 1990). A valuable summary of recent work on the biology and ecology of planktonic foraminifera is provided by Hemleben *et al.* (1988).

The Last Deglaciation

The oceanwide trend of rising temperatures associated with the last deglaciation was accompanied by a poleward expansion of warm-water faunal provinces. Changes in surface circulation returned tropical and subtropical waters and their associated planktonic foraminiferal assemblages to the Gulf of Mexico, where they replaced transitional zone assemblages (Kennett and Huddlestun, 1972; Sidner and Poag, 1972; Brunner, 1982). Average sea-surface temperature (SST) rose from 22 to 24°C (CLIMAP, 1976).

The general trend of rising temperatures and higher sea level was punctuated by rapid increases, but included a brief cool interval, the Younger Dryas. Named after the reappearance of the dryas flower that flourished in northern Europe during the late glacial, the Younger Dryas is shown as an intense cooling episode in marine records of the North Atlantic region, and in terrestrial records including northern Europe. Faunal, isotopic, and sedimentological work has also documented its occurrence in areas across the Northern Hemisphere. An oceanic cool interval correlative with the Younger Dryas has recently been documented in the Sulu Sea (Linsley and Thunell, 1990; Kudrass *et al.*, 1991), the Gulf of California (Keigwin and Jones, 1990), the western North Pacific (Chinzei and Oba, 1986; Chinzei *et al.*, 1987; Kallel *et al.*, 1988), and the Gulf of Mexico (Kennett *et al.*, 1985). High resolution work in the Gulf of Mexico highlights the speed of faunal and isotopic changes associated with the Younger Dryas episode (Flower and Kennett, 1990). A brief cooling during deglaciation in the Indian sector of the Southern Ocean (Labracherie *et al.*, 1989) seems to precede the Younger Dryas cooling by 1000 yr. If the dates for this cooling are correct, it implies a diachronism between the Southern Ocean and the Northern Hemisphere for this brief reversal during general deglaciation.

Greenland ice core work has suggested that the Younger Dryas is bracketed by rapid climatic transitions of less than 300 yr. Dansgaard *et al.* (1989) have suggested that this cool interval ended in about 20 yr and involved a temperature increase of about 7°C, although Fairbanks (1989, 1990) has questioned the interpretation of the oxygen isotopic record and the chronology, and instead suggests that the Younger Dryas was an interval of slower sea-level rise. Two distinct meltwater-induced rises in sea level, centered at 13 to 11 ka and 10 to 7 ka, are separated by an episode of slower sea-level rise between 11 and 10 ka, correlative with the Younger Dryas.

Because the source of deglacial sea-level rise was largely meltwater derived from the Laurentide and Fennoscandian ice sheets, oceanic biota near continental outlets and in marginal basins were affected directly by the influx of low-salinity fresh water. Planktonic foraminiferal assemblages in the Gulf of Mexico (Kennett *et al.*, 1985; Flower and Kennett, 1990; this chapter) responded to this influx of meltwater from the Laurentide ice sheet. Periodic freshwater input to the Mediterranean in the late Quaternary also had an influence on foraminiferal assemblages. Sapropel layers in the eastern Mediterranean inferred to accumulate during episodes of freshwater influx are usually characterized by a distinctive "sapropel-related" assemblage (Thunell *et al.*, 1977; Muerdter *et al.*, 1984). Rapid faunal changes associated with deglacial freshwater influx should also be apparent in the Black Sea and the Caspian Sea, as a result of the wasting of the Fennoscandian ice sheet. High resolution biostratigraphic work in such marginal basins will expand our understanding of the faunal response in areas especially sensitive to global change.

RESULTS AND DISCUSSION

In this chapter, data are summarized on the response of key members of the planktonic foraminiferal community to deglacial warming and to freshwater influx to the Gulf of Mexico. As a semienclosed basin influenced by the

BIOTIC RESPONSES TO TEMPERATURE AND SALINITY CHANGES

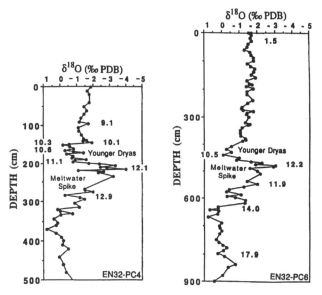

FIGURE 12.2 Oxygen isotopic records from analyses of *Globigerinoides ruber* (white variety) for EN32-PC4 (left, after Broecker *et al.*, 1989) and EN32-PC6 (right, after Leventer *et al.*, 1982), plotted against depth in core as $\delta^{18}O$ relative to the PDB standard. The cores show the late glacial to early Holocene interval, with the meltwater spike and the Younger Dryas chronozone labelled. Ages in thousands of years are accelerator radiocarbon dates, corrected for the $^{14}C/^{12}C$ difference between atmospheric CO_2 and surface water ΣCO_2, revised after Broecker *et al.* (1990).

Mississippi outlet, the Gulf of Mexico was particularly sensitive to circulation changes in the North Atlantic and to meltwater runoff from the Laurentide ice sheet (Prest *et al.*, 1968; 1970; Kennett and Shackleton, 1975). We have examined undisturbed sequences marked by high sedimentation rates from the Orca Basin with abundant planktonic microfossils, thus providing a high-resolution record. Located on the continental rise 290 km south of the Mississippi Delta (Figure 12.2), with a maximum depth of 2400 m and a sill depth of 1800 m, the Orca Basin is filled with a hypersaline brine to a depth of 2230 m. Resulting anoxic conditions in the bottom waters provide excellent preservation, eliminate dissolution, and exclude benthic organisms that mix the sediments. These conditions preserve a pristine sedimentary sequence with undisturbed laminae, allow a high-resolution stratigraphy with a sampling interval of 100 yr, and provide a clear picture of dynamic changes in faunal assemblages unaffected by dissolution or bioturbation.

^{14}C Chronology

Accelerator radiocarbon dates from two Orca Basin cores (EN32-PC4 and EN32-PC6), recently revised by Broecker *et al.* (1990a), provide excellent age control over the last deglaciation (Figure 12.3). Although calibration of the ^{14}C time scale has shown that ^{14}C dates correspond to somewhat older U-Th ages (Bard *et al.*, 1990), the ^{14}C time scale is adopted here for ease of comparison with previously published chronologies. The ^{14}C ages show no stratigraphic inversions, although recent work has shown that the radiocarbon time scale remains constant at about 10 ka for a few hundred years (Oeschger *et al.*, 1980; Andrée *et al.*, 1986; Becker and Kromer, 1986; Lowe *et al.*, 1988; Bard *et al.*, 1990). All data presented here are plotted against ^{14}C age by extrapolation between dated samples.

Stable Isotopic Records

High-resolution stable isotopic records have also been generated for these two Orca Basin samples (EN32-PC4, Broecker *et al.*, 1989; Flower and Kennett, 1990, and EN32-PC6, Leventer *et al.*, 1982; Kennett *et al.*, 1985). The two oxygen isotopic records are similar, measured on *Gs. ruber* and plotted as $\delta^{18}O$ (‰ relative to the PDB belemnite standard) against depth in Figure 12.2. The main feature is the pronounced negative spike in $\delta^{18}O$, indicating major freshwater flooding into the Gulf of Mexico. Because the isotopic composition of Laurentide glacial ice

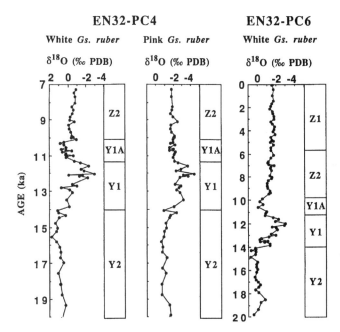

FIGURE 12.3 Plots of $\delta^{18}O$ records from analyses of *Globigerinoides ruber* plotted versus age for EN32-PC4 and EN32-PC6 from the Orca Basin. White and pink varieties of *Globigerinoides ruber* are plotted for EN32-PC4 against ^{14}C age, as discussed in text. Note the different age scale for EN32-PC6.

was about −30 to −40‰ (Shackleton, 1977), the spike centered at 12 ka has been interpreted as an influx of glacial meltwater. The deglacial meltwater spike is superimposed on the general decrease in $\delta^{18}O$ due to deglacial warming and the effect of reduced continental ice volume. The meltwater spike is followed immediately by an episode of increased $\delta^{18}O$ (Figure 12.3) correlative with the Younger Dryas cooling (Flower and Kennett, 1990).

Oxygen isotopic records for the two cores are plotted versus age in Figure 12.3. The meltwater spike began at about 14 ka, reached a peak at 12 ka, and ended abruptly at 11.4 ka. The cessation of meltwater influx to the Gulf of Mexico was followed immediately by an episode of higher $\delta^{18}O$ that lasted from 11.4 to 10.2 ka, indicating some combination of higher salinity and cooler sea-surface temperatures. Cool surface water foraminiferal assemblages in EN32-PC4 (Flower and Kennett, 1990; this chapter) and in EN32-PC6 (Kennett et al., 1985) confirm the presence of the Younger Dryas event in the Gulf of Mexico. The rapid onset of the Younger Dryas following the cessation of meltwater influx lends support to the hypothesis of a meltwater trigger in the North Atlantic shutting down the conveyor belt and causing an oceanically controlled cooling (Broecker et al., 1989; Broecker et al., 1990b; Flower and Kennett, 1990). A diversion of low-salinity meltwater away from the Mississippi toward the St. Lawrence system might have upset the density-driven production of NADW and disrupted the heat pump in the North Atlantic, plunging the region into the Younger Dryas cool episode. This hypothesis requires a reintroduction of meltwater to the Gulf of Mexico at the end of the Younger Dryas, as the ice front receded and southern outlets were reexposed. Support comes from a rapid decrease of 1.2‰ within 200 yr centered at 10.2 ka in EN32-PC4 (Broecker et al., 1989; Flower and Kennett, 1990); Spero and Williams (1990) also found evidence for seasonal low-salinity events at 9.8 ka based on isotopic analyses of single foraminifera from EN32-PC6. An oxygen isotopic shift of 1‰ occurs over 500 yr in EN32-PC6 at 10 ka (Figure 12.2). In both cores, however, oxygen isotopic values are much less negative than those of the main meltwater spike.

Further, the oxygen isotopic record for EN32-PC4 derived from the pink form of *Gs. ruber* shows no evidence for the Younger Dryas event or for a reintroduction of meltwater to the Gulf of Mexico at 10ka (Figure 12.2). Since this form is favored during summers in the North Atlantic and the Gulf of Mexico (Bé and Tolderlund, 1971; Tolderlund and Bé, 1971; Deuser and Ross, 1989; Flower and Kennett, 1990), it should have recorded an increase in meltwater flux ~10 ka, because meltwater flow from the continent almost certainly would have peaked during summer months. The constancy of the $\delta^{18}O$ values is a complication possibly explained by continued summer meltwater flux during the Younger Dryas without an increase at its conclusion (Flower and Kennett, 1990).

Faunal Response to Temperature and Salinity Changes in the Gulf of Mexico

Planktonic foraminiferal assemblages were very sensitive to the temperature and salinity changes in Gulf of Mexico surface waters associated with rapid deglacial climatic shifts. Past work in the North Atlantic has shown faunal migrations in response to changing surface water circulation (Duplessey et al., 1981; Ruddiman and McIntyre, 1981). Our high-resolution work in the Gulf of Mexico documents the deglacial faunal response not only to changing surface water temperatures, but also to other environmental stresses including the influence of low-salinity meltwater from the Laurentide ice sheet to the north.

Relative abundance changes in two Orca Basin cores (EN32-PC4 and EN32-PC6) are presented here for five temperature- and/or salinity-sensitive species of planktonic foraminifera (Figures 12.4 and 12.5); a more complete treatment is given in Flower and Kennett (1990). Species sensitive to surface water temperatures were identified by their association with late Quaternary glacial-interglacial cycles in the Gulf of Mexico (Kennett and Huddlestun, 1972; Malmgren and Kennett, 1976). The warmest surface water indicators include *Globorotalia menardii* and *Pulleniatina obliquiloculata*. Cold water forms include *Globorotalia inflata*, *Globigerina falconensis*, and *Globigerina bulloides*. The $\delta^{18}O$ stratigraphy derived from *Gs. ruber* (white variety) is shown on the same plot for comparison. Abundances of cool-water forms were generally higher, whereas warm-water forms were low during the late glacial through the early part of the meltwater spike from 20 to 13 ka, when surface water temperatures were low, nearly 2°C cooler at the glacial maximum (CLIMAP, 1976). Warm-water forms dominated from the later part of the meltwater spike beginning at 13 ka through the Holocene, except for a brief return of cold-water forms during the Younger Dryas.

Globigerinoides ruber (Figures 12.4e and 12.5e) was the dominant species in fossil assemblages throughout the late glacial and the Holocene, with abundances ranging from 20 to 70% but usually averaging 30 to 40%. In both EN32-PC4 and EN32-PC6, *Gs. ruber* averaged about 35% during the late glacial, reached maximum abundances of 70% during the early part of the meltwater spike, dropped rapidly to 30% at the cessation of meltwater influx, and increased slightly into the late Holocene.

Globorotalia inflata (Figures 12.4a and 12.5a) is an indicator species for the temperate/subarctic zone in the modern North Atlantic ocean and a clear marker of Quaternary glacial episodes in the Gulf of Mexico (Kennett

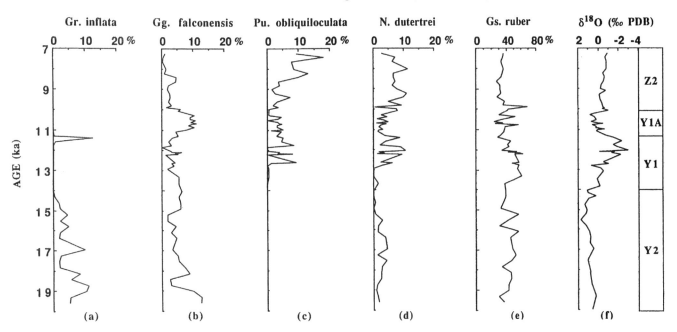

FIGURE 12.4 Percentage frequency variations of selected environmentally-sensitive species of planktonic foraminifera in the >150 μm size fraction in EN32-PC4 from Orca Basin: (a) *Globigerinoides falconensis*, (b) *Globorotalia inflata*, (c) *Pulleniatina obliquiloculata*, (d) *Neogloboquadrina dutertrei*, (e) *Globigerinoides ruber*, plotted against corrected ^{14}C age. Also shown are the δ^{18}O record (f) of *Globigerinoides ruber* (white variety) and foraminiferal subzones of Kennett and Huddlestun (1972). Subzones Y2, Y1, Y1A, and Z2 correspond to the late glacial, the meltwater spike, the Younger Dryas, and the early Holocene, respectively.

and Huddlestun, 1972; Malmgren and Kennett, 1976). This species showed high abundances during the last glacial maximum, disappeared near the beginning of the meltwater spike at 14ka, and reappeared briefly at 11.4 ka.

Globigerina falconensis (Figures 12.4b and 12.5b), a cold-water species in the Gulf of Mexico (Kennett and Huddlestun, 1972; Malmgren and Kennett, 1976), showed relatively high frequencies during the late glacial, decreased during the later part of the meltwater interval from 13 to 11.5 ka, increased between 11.0 and 10.0 ka, and decreased to its lowest frequencies in the Holocene.

Pulleniatina obliquiloculata (Figures 12.4c and 12.5c), a warm-water species in the Gulf of Mexico, was absent during the late glacial, appeared during the early part of the meltwater spike at about 13.7 ka, and showed sporadically high abundances between 12.7 and 11.7 ka, after which it decreased slowly to a minimum ~10.2 ka. It then increased in steps at 9.8 and 8.7 ka.

Globorotalia menardii (not figured), a tropical/warm subtropical species in the Gulf, was absent during the late glacial, was present sporadically during the meltwater spike, was a consistent component after 9.8ka, and underwent a further increase at 5.5 ka.

Neogloboquadrina dutertrei (Figures 12.4d and 12.5d), a marginally warm-water species in the Gulf of Mexico (Kennett and Huddlestun, 1972; Malmgren and Kennett, 1976), exhibited moderate frequencies during the late glacial. Abundances increased at 13 ka and generally remained high until 11.3 ka, decreased between 11.3 and 10.2 ka, and then increased to Holocene values.

Changing relative abundances of planktonic foraminifera with well-known environmental preferences follow closely the history of deglacial temperature and salinity changes. Late glacial assemblages until about 14 ka included *Globorotalia inflata* and *Globigerina falconensis* (Figures 12.4 and 12.5). The reappearance of a warm-water fauna at about 13 ka in the Gulf of Mexico corresponded to an increase in meltwater influx, but preceded its peak. Low salinities in the early part of the meltwater spike favored the euryhaline *Gs. ruber*. This association is supported by independent observations, which showed that *Gs. ruber* tolerates lower salinities than other planktonic species (as low as 22‰; Bijma *et al.*, 1990). Field observations in Barbados (Hemleben *et al.*, 1987) suggested that all planktonic foraminiferal species except *Gs. ruber* descend to higher-salinity waters in response to periodic appearances of low-salinity lenses derived from the Amazon River. Maximum abundances of *Gs. ruber* were reached at 13.5 ka and preceded the peak of the meltwater spike at 12 ka marking lowest salinities.

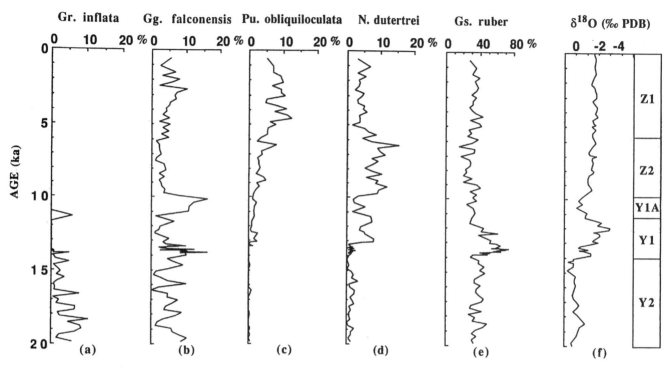

FIGURE 12.5 Percentage frequency variations of selected environmentally sensitive species of planktonic foraminifera in the >150-μm size fraction in EN32-PC6 from Orca Basin: (a)*Globigerinoides falconensis*, (b) *Globorotalia inflata*, (c) *Pulleniatina obliquiloculata*, (d)*Neogloboquadrina dutertrei*, (e) *Globigerinoides ruber*, plotted against corrected ^{14}C age. Also shown are the δ^{18}O record (f) of *Globigerinoides ruber* (white variety) and foraminiferal subzones of Kennett and Huddlestun (1972). Subzones Y2, Y1, Y1A, Z2 and Z1 correspond to the late glacial, the meltwater spike, the Younger Dryas, the early Holocene, and the late Holocene, respectively.

N. dutertrei also displays an association with low salinity, but its increase in abundance during lowest salinities from 13 to 12 ka could be due to surface water warming, because other warm-water species increased at the same time. This association was found previously in Gulf of Mexico cores (Kennett and Shackleton, 1975; Thunell, 1976) and in the North Atlantic and Mediterranean (Ruddiman, 1969; Bé and Tolderlund, 1971; Thunell et al., 1977; Thunell, 1978; Loubere, 1981). Increased abundances of *N. dutertrei* were found associated with some but not all Quaternary sapropel layers in the eastern Mediterranean, inferred to have been triggered by periodic low-salinity events (for a summary, see Muerdter et al., 1984). Surface waters in the early part of the meltwater spike are thought to have remained relatively cool until 13 ka, too cool for *N. dutertrei* to have proliferated like the more opportunistic *Gs. ruber*.

Deglacial warming at 13 ka and lowest salinities at 12 ka were marked by increased abundances of warm-water species, decreased abundances of cool-water species, and continued elevated frequencies of low-salinity tolerant species. *Neogloboquadrina dutertrei* and *Pulleniatina obliquiloculata* show simultaneous increases in both cores, while *Gg. falconensis* decreases and *Gr. inflata* disappears (Figures 12.4 and 12.5). Surface waters in the later part of the meltwater spike were then warm enough to support *N. dutertrei* in addition to *Gs. ruber*.

Our data further show that the lowest salinities during deglaciation favored the pigmented form of *Gs. ruber* over the white form. A plot of the pink:white *Gs. ruber* percentage ratio for EN32-PC4 shows a peak coincident with the meltwater spike in δ^{18}O (Figure 12.6). The reason for this is unknown and may involve different salinity tolerances, a longer summer season favoring the pink form, or other environmental conditions favorable to pigmentation. The latter may include an association with a symbiont that induces coloration.

There is no faunal evidence to indicate that surface waters were cooled directly by the meltwater influx itself, as suggested from modeling studies (Oglesby et al., 1989; Overpeck et al., 1989). Although the planktonic foraminiferal assemblages cannot be translated directly into temperatures, they do indicate warm surface waters during the interval of lowest salinity. The isotopic values of *Gs.*

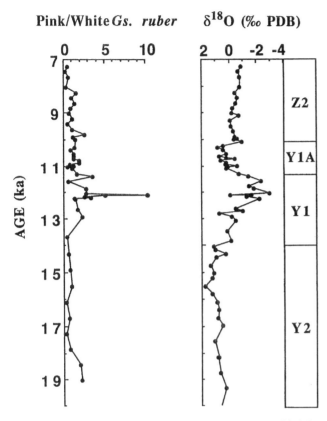

FIGURE 12.6 Percentage ratio (percent pink:percent white) for the pink and white varieties of *Globigerinoides ruber* is plotted for EN32-PC4 from the Orca Basin against ^{14}C age, as discussed in text. Also shown are the δ^{18}O record (f) of *Globigerinoides ruber* (white variety) and foraminiferal subzones of Kennett and Huddlestun (1972). Subzones Y2, Y1, Y1A, and Z2 correspond to the late glacial, the meltwater spike, the Younger Dryas, and the early Holocene, respectively.

ruber during the meltwater spike also cannot be translated into temperature because of the low-salinity overprint. However, modern field observations (Hemleben *et al.*, 1987) suggest that most planktonic foraminifera probably migrated to deeper waters below the relatively fresh surface waters that were perhaps colder.

The end of the low-salinity meltwater interval was marked by an important reappearance of cold-water forms characteristic of the last glacial episode. A brief reappearance of *Gr. inflata* occurs within a few centimeters in both cores and is dated at 11.4 ka. This species is today associated with the transition zone in the North Atlantic between the subtropical and subpolar surface water masses. Its reappearance in the Gulf of Mexico represents a major shift in the position of this boundary and is inferred to mark the onset of the Younger Dryas. The speed of the biotic response was remarkable, occurring in less than 200 yr.

The reappearance of *Gr. inflata* was followed by increased relative abundances of *Gg. falconensis* and decreased abundances of *N. dutertrei*, until about 10.2 ka. These changing abundances mark an interval between 11.4 and 10.2 ka of very cold, followed by cool, surface water conditions. The correspondence of this cool interval with the Younger Dryas centered in the North Atlantic region shows that oceanic cooling extended to the Gulf of Mexico. Its presence in the Gulf of Mexico (Kennett *et al.*, 1985; Flower and Kennett, 1990; this chapter), the Sulu Sea (Linsley and Thunell, 1990; Kudrass *et al.*, 1991), and the Gulf of California (Keigwin and Jones, 1990) suggests that expressions of the Younger Dryas occur throughout the Northern Hemisphere, if not worldwide.

The end of the Younger Dryas is marked by faunal and isotopic changes over less than 500 yr. Declining cool-water assemblages coincide with a decrease in δ^{18}O and are followed by increases in warm-water assemblages. The reappearance of consistent *Gr. menardii*, usually taken as marking the beginning of the Holocene, occurs at 9.8 ka. It is accompanied by large increases in *Pu. obliquiloculata* and *N. dutertrei*.

Further increases in the warm-water species *Gr. menardii* and *Pu. obliquiloculata* along with a decrease in the marginally warm-water species *N. dutertrei* occur at 5.5 ka and distinguish a warmer subzone in the late Holocene. This warmer late Holocene assemblage is not accompanied by any change in δ^{18}O of *Gs. ruber*, underlining the importance of independent faunal analysis in addition to geochemical methods in the investigation of paleoenvironmental change.

CONCLUSIONS

Oxygen isotopic and faunal analyses of high-resolution, radiometrically dated sediment sequences in the Gulf of Mexico demonstrate that planktonic foraminiferal communities were sensitive to deglacial environmental changes, including rapid temperature and salinity changes. Fossil assemblages reflect deglacial warming and low-salinity meltwater influx from the Laurentide ice sheet, the rapid onset and conclusion of the Younger Dryas, and stepwise warming into the Holocene. Rapid migrations characterized the dynamic response of oceanic biota and water masses in an area that amplified deglacial environmental change.

Warm-water planktonic foraminifera including *Pulleniatina obliquiloculata* and *Neogloboquadrina dutertrei* began to replace cold-water forms at 14 ka as the glacial species *Globorotalia inflata* disappeared in response to early deglacial warming. Simultaneously, the euryhaline form *Globigerinoides ruber* increased to its greatest abundances of 70% during a period of low-salinity meltwater influx. Warm-water forms increased in abundance at 13 ka as the cool-water species *Globigerina falconensis* decreased in response to warmer SSTs. A brief reappearance of the glacial species *Gr. inflata* at the expense of warm-water forms at 11.4 ka marks a rapid, temporary migration of cold surface water into the Gulf of Mexico. This event is followed immediately by an interval of increased abundances of *Gg. falconensis*, and decreased abundances of *N. dutertrei* and *Pu. obliquiloculata*, and heralds the beginning of the Younger Dryas cooling in the Gulf of Mexico. Late deglacial warming at about 10.2 ka fostered the appearance of warm-water Holocene assemblages including *Gr. menardii*. Further warming at 5.5 ka distinguishes a warmer subzone in the late Holocene.

The euryhaline species *Gs. ruber* bloomed during the early portion of the meltwater spike. After surface waters had warmed sufficiently at 13 ka, the low-salinity tolerant species *N. dutertrei* also showed higher abundances due to some combination of lower salinities and warmer temperatures. Lowest salinities at 12 ka favored the pink form of *Gs. ruber*.

There is no faunal evidence that surface waters were cooled directly by meltwater influx. In fact, warm-water assemblages are present during the interval of lowest salinity. However, field observations suggest that most planktonic foraminifera probably migrated to deeper waters below the relatively fresh surface waters that were perhaps cooler.

These results demonstrate the need for further high-resolution work on the response of oceanic fauna to rapid environmental changes associated with deglaciation, including temperature and salinity. As our understanding of past global change improves through paleontological, geochemical, and modeling efforts, the effect of particular combinations of environmental parameters becomes clearer. Insight into the controlling combinations in the past will assist in the assessment of the biotic response to present and future anthropogenically forced global change.

ACKNOWLEDGMENTS

This research was supported by National Science Foundation grants OCE88-17135 and DPP89-11554.

REFERENCES

Andrée, M., H. Oeschger, U. Siegenthaler, T. Riesen, M. Moell, B. Amman, and K. Tobolski (1986). ^{14}C dating of plant macrofossils in lake sediment, *Radiocarbon* 28, 411-416.

Bard, E., B. Hamelin, R. G. Fairbanks, and A. Zindler (1990). Calibration of the ^{14}C timescale over the past 30,000 years using mass spectrometric U-Th ages from Barbados corals, *Nature* 345, 405-410.

Bé, A. W. H., and D. S. Tolderlund (1971). Distribution and ecology of living planktonic foraminifera in surface waters of the Atlantic and Indian Oceans, in *The Micropaleontology of Oceans*, B. M. Funnell and W. R. Riedel, eds., Cambridge University Press, London, pp. 105-149.

Bé, A. W. H., S. M. Harrison, and L. Lott (1973). *Orbulina universa* d'Orbigny in the Indian Ocean, *Micropaleontology* 19, 150-192.

Becker, B., and B. Kromer (1986). Extension of the Holocene dendrochronology by the Preboreal pine series, 8800 to 10,100 BP, *Radiocarbon* 28, 961-967.

Bijma, J., W. W. Faber, Jr., and C. Hemleben (1990). Temperature and salinity limits for growth and survival of some planktonic foraminifers in laboratory cultures, *Journal of Foraminiferal Research* 20, 95-116.

Boltovskoy, E. (1990). Late Pleistocene-Holocene planktic foraminifera of the western equatorial Pacific, *Boreas* 19, 119-125.

Broecker, W. S., D. M. Peteet, and D. Rind (1985). Does the ocean-atmosphere system have more than one stable mode of operation? *Nature* 315, 21-25.

Broecker, W. S., J. P. Kennett, B. P. Flower, J. T. Teller, S. Trumbore, G. Bonani, and W. Wolfli (1989). Routing of meltwater from the Laurentide ice sheet during the Younger Dryas cold episode, *Nature* 341, 318-321.

Broecker, W. S., M. Klas, E. Clark, S. Trumbore, G. Bonani, W. Wolfli, and S. Ivy (1990a). Accelerator mass spectrometric radiocarbon measurements on foraminifera shells from deep sea cores, *Radiocarbon* 32, 119-133.

Broecker, W. S., G. Bond, M. Klas, G. Bonani, and W. Wolfli (1990b). A salt oscillator in the glacial Atlantic? The concept, *Paleoceanography* 5, 469-478.

Brunner, C. A. (1982). Paleoceanography of surface waters in the Gulf of Mexico during the late Quaternary, *Quaternary Research* 17, 105-119.

Chinzei, K., and T. Oba (1986). Oxygen isotope studies of the deep sea sediments around Japan, in *Recent Progress of Natural Sciences in Japan 11: Quaternary Research*, Science Council of Japan, Tokyo, pp. 35-43.

Chinzei, K., H. Fujioka, I. Kitazatom, T. Koizumi, M. Oba, H. Oba, T. Okada, T. Sakai, and Y. Yanimura (1987). Postglacial environmental change of the Pacific off the coast of central Japan, *Marine Micropaleontology* 11, 273-291.

CLIMAP Project Members (1976). The surface of the ice age Earth, *Science* 191, 1131-1137.

Cline, R. M., and J. D. Hays, eds. (1976). *Investigation of Late Quaternary Paleoceanography and Paleoclimatology*, Geological Society of America Memoir 145, Boulder, Colo., 464 pp.

Curry, W. B., R. C. Thunell, and S. Honjo (1983). Seasonal changes in the isotopic composition of planktonic foraminifera collected in Panama Basin sediment traps, *Earth and Planetary Science Letters* 64, 33-43.

Dansgaard, W., J. W. C. White, and S. J. Johnsen (1989). The abrupt termination of the Younger Dryas climate event, *Nature* 339, 532-534.

Deuser, W. G., and E. H. Ross (1989). Seasonally abundant planktonic foraminifera of the Sargasso Sea: Succession, deep-water fluxes, isotopic compositions, and paleoceanographic implications, *Journal of Foraminiferal Research 19*, 268-293.

Duplessey, J.-C., M. Arnold, P. Maurice, E. Bard, J. Duprat, and J. Moyes (1981). Direct dating of the oxygen-isotope record of the last deglaciation by ^{14}C accelerator mass spectrometry, *Nature 320*, 350-352.

Emiliani, C. (1969). A new paleontology, *Micropaleontology 15*, 265-300.

Fairbanks, R. G. (1989). A 17,000-year glacio-eustatic sea level record: Influence of glacial melting rates on the Younger Dryas event and deep-ocean circulation, *Nature 342*, 637-642.

Fairbanks, R. G. (1990). The age and origin of the "Younger Dryas Climate Event" in Greenland ice cores, *Paleoceanography 5*, 937-948.

Fairbanks, R. G., M. Sverdlove, R. Free, P. H. Wiebe, and A. W. H. Bé (1982). Vertical distribution and isotopic fractionation of living planktonic foraminifera: Seasonal changes in species flux in the Panama Basin, *Nature 298*, 841-844.

Flower, B. P., and J. P. Kennett (1990). The Younger Dryas cool episode in the Gulf of Mexico, *Paleoceanography 5*, 949-961.

Hemleben, C., M. Spindler, I. Breitinger, and R. Ott (1987). Morphological and physiological responses of *Globigerinoides sacculifer* (Brady) under varying laboratory conditions: *Marine Micropaleontology 12*, 305-324.

Hemleben, C., M. Spindler, and O. R. Anderson (1988). *Modern Planktonic Foramifera*, Springer Verlag, 363 pp.

Huesser, L., and J. J. Morley (1985). Pollen and radiolarian records from deep-sea core RC14-103, climatic reconstructions of northeast Japan and northwest Pacific for the last 90,000 years, *Quaternary Research 24*, 60-72.

Imbrie, J., and N. Kipp (1971). A new micropaleontological method for quantitative paleoclimatology: Application to a late Pleistocene Caribbean core, in *Late Cenozoic Glacial Ages*, K. K. Turekian, ed., Yale University Press, New Haven, Conn.

Kallel, L., D. Labeyrie, M. Arnold, H. Okaka, W.C. Dudly, and J.-C. Duplessy (1988). Evidence of cooling during the Younger Dryas in the western North Pacific, *Oceanol. Acta 11*, 369-376.

Keigwin, L. D., and G. A. Jones (1990). Deglacial climatic oscillations in the Gulf of California, *Paleoceanography 5*, 1009-1023.

Kellogg, T. B. (1984). Late-glacial-Holocene high-frequency climatic change in deep-sea cores from the Denmark Strait, in *Climatic Changes on a Yearly to Millenial Basis*, N.-A. Mörner and W. Karlén, eds., D. Reidel, Boston, pp. 123-133.

Kennett, J. P. (1976). Phenotypic variation in some Recent and late Cenozoic planktonic foraminifera, in *Foraminifera, Volume 2*, R. H. Hedley and C. G. Adams, eds., Academic Press, London.

Kennett, J. P., and P. Huddlestun (1972). Late Pleistocene paleoclimatology, foraminiferal biostratigraphy and tephrochronology, western Gulf of Mexico, *Quaternary Research 2*, 38-69.

Kennett, J. P., and N. J. Shackleton (1975). Laurentide ice sheet meltwater recorded in Gulf of Mexico deep-sea cores, *Science 188*, 147-150.

Kennett, J. P., K. Elmstrom, and N. L. Penrose (1985). The last deglaciation in Orca Basin, Gulf of Mexico: High-resolution planktonic foraminifera changes, *Palaeogeography, Palaeoclimatology, Palaeoecology 50*, 189-216.

Kudrass, H. R., H. Erlenkeuser, R. Vollbrecht, and W. Weiss (1991). Global nature of the Younger Dryas cooling event inferred from oxygen isotope data from Sulu Sea cores, *Nature 349*, 406-409.

Labracherie, M., L. D. Labeyrie, J. Duprat, E. Bard, M. Arnold, J.-J. Pichon, and J.-C. Duplessy (1989). The last deglaciation in the southern ocean, *Paleoceanography 4*, 629-638.

Leventer, A., D. F. Williams, and J. P. Kennett (1982). Dynamics of the Laurentide ice sheet during the last deglaciation: Evidence from the Gulf of Mexico, *Earth and Planetary Science Letters 59*, 11-17.

Linsley, B. K., and R. C. Thunell (1990). The record of deglaciation in the Sulu Sea: Evidence for the Younger Dryas event in the tropical western Pacific, *Paleoceanography 5*, 1025-1039.

Locke, S., and R. C. Thunell (1988). Paleoceanographic record of the last glacial/interglacial cycle in the Red Sea and Gulf of Aden, *Palaeogeography, Palaeoclimatology, Palaeoecology 64*, 163-187.

Loubere, P. (1981). Oceanographic parameters reflected in the seabed distribution of planktonic foraminifera from the North Atlantic and Mediterranean Sea, *Journal of Foraminiferal Research 11*, 137-158.

Lowe, J. J., S. Lowe, A. J. Fowler, R. E. M. Hedges, and T. J. F. Austin (1988). Comparison of accelerator and radiometric measurements obtained from late Devensian late glacial lake sements from Llyn Gwernan, North Wales, UK, *Boreas 17*, 355-369.

Malmgren, B. A., and J. P. Kennett (1976). Principal component analysis of Quaternary planktic foraminifera in the Gulf of Mexico: Paleoclimatic applications, *Marine Micropaleontology 1*, 299-306.

McIntyre, A., N. G. Kipp, A. W. H. Bé, T. Crowley, T. Kellog, J. V. Gardner, W. Prell, and W. F. Ruddiman (1976). Glacial North Atlantic 18,000 years ago: A CLIMAP reconstruction, in *Investigation of Late Quaternary Paleoceanography and Paleoclimatology*, R. M. Cline and J. D. Hays, eds., Geological Society of America Memoir 145, Boulder, Colo., pp. 43-75.

Molfino, B., and A. McIntyre (1990). Nutricline variation in the equatorial Atlantic coincident with the Younger Dryas, *Paleoceanography 5*, 997-1008.

Moore, T. C., Jr., L. H. Burckle, K. Geitzenauer, B. Luz, A. Molina-Cruz, J. H. Robertson, H. Sachs, C. Sancetta, J. Thiede, P. Thompson, and C. Wenkam (1980). The reconstruction of sea surface temperatures in the Pacific Ocean of 18,000 B.P., *Marine Micropaleontology 5*, 215-247.

Morley, J. J., and J. D. Hays (1979). *Cycladophora davisiana*; A stratigraphic tool for Pleistocene North Atlantic and inter-hemispheric correlations, *Earth and Planetary Science Letters 44*, 383-389.

Morley, J. J., and J. D. Hays (1983). Oceanographic conditions associated with high abundances of the radiolarian *C. davisiana*, *Earth and Planetary Science Letters 66*, 63-72.

Muerdter, D. R., J. P. Kennett, and R. C. Thunell (1984). Late Quaternary sapropel sediments in the eastern Mediterranean Sea: Faunal variations and chronology, *Quaternary Research 21*, 385-403.

Oeschger, H., M. Welten, U. Eicher, M. Möll, T. Riesen, U. Siegenthaler, and S. Wegmüller (1980). ¹⁴C and other parameters during the Younger Dryas cold phase, *Radiocarbon 22*, 299-310.

Oglesby, R. J., K. A. Maasch, and B. Saltzman (1989). Glacial meltwater cooling of the Gulf of Mexico: GCM implications for Holocene and present-day climates, *Climate Dynamics 3*, 115-133.

Overpeck, J. T., L. C. Peterson, N. Kipp, J. Imbrie, and D. Rind (1989). Climate change in the circum-North Atlantic region during the last deglaciation, *Nature 338*, 553-557.

Pedersen, T. F., M. Pickering, J. S. Vogel, J. N. Southon, and D. E. Nelson (1988). The response of benthic foraminifera to productivity cycles in the eastern equatorial Pacific: Faunal and geochemical constraints on glacial bottom water oxygen levels, *Paleoceanography 3*, 157-168.

Prest, V. K. (1970). Quaternary geology, in *Geology and Economic Minerals of Canada*, R. J. W. Douglas, ed., Department of Energy, Mines and Resources, Ottawa, pp. 675-764.

Prest, V. K., D. R. Grant, and V. N. Rampton (1968). *Glacial Map of Canada*, Map 1253A, Geological Survey of Canada, Ottawa.

Ruddiman, W. F. (1969). Planktonic foraminifera of the subtropical North Atlantic gyre, Ph.D. thesis, Columbia University, New York, 291 pp.

Ruddiman, W. F., and A. McIntyre (1981). The mode and mechanism of the last deglaciaton: Oceanic evidence, *Quaternary Research 16*, 125-134.

Sancetta, C. (1979). Oceanography of the North Pacific during the last 18,000 years: Evidence from fossil diatoms, *Marine Micropaleontology 4*, 103-123.

Shackleton, N. J. (1977). The oxygen isotopic stratigraphic record of the late Pleistocene, *Philosophical Transactions of the Royal Society of London 280*, 169-182.

Sidner, B. R., and C. W. Poag (1972). Late Quaternary climates indicated by foraminifers from the southwestern Gulf of Mexico, *Gulf Coast Association of Geological Societies Transactions 22*, 305-313.

Spero, H. J., and M. J. DeNiro (1987). The influence of symbiont photosynthesis on the $\delta^{18}O$ and $\delta^{13}C$ values of planktonic foraminiferal shell calcite, *Symbiosis 4*, 213-228.

Spero, H. J., and D. F. Williams (1988). Extracting environmental information from planktonic foraminiferal $\delta^{13}C$ data, *Nature 335*, 717-719.

Spero, H. J., and D. F. Williams (1990). Evidence for low salinity surface waters in the Gulf of Mexico over the last 16,000 years, *Paleoceanography 5*, 963-975.

Thunell, R. C. (1976). Calcium carbonate dissolution history in late Quaternary deep-sea sediments, Western Gulf of Mexico, *Journal of Quaternary Research 6*, 281-297.

Thunell, R. C. (1978). Distribution of Recent planktonic foraminifera in surface sediments of the Mediterranean Sea, *Marine Micropaleontology 3*, 147-173.

Thunell, R. C., D. F. Williams, and J. P. Kennett (1977). Late Quaternary paleoclimatology, stratigraphy and sapropel history in eastern Mediterranean deep-sea sediments, *Marine Micropaleontology 2*, 371-388.

Thunell, R. C., W. B. Curry, and S. Honjo (1983a). Seasonal variation in the flux of planktonic foraminifera: Time series trap results from the Panama Basin, *Earth and Planetary Science Letters 64*, 44-55.

Thunell, R. C., W. B. Curry, and S. Honjo (1983b). Seasonal changes in the isotopic composition of planktonic foraminifera collected in Panama Basin sediment traps, *Earth and Planetary Science Letters 64*, 33-43.

Tolderlund, D. S., and A. W. H. Bé (1971). Seasonal distribution of planktonic foraminifera in the western North Atlantic, *Micropaleontology 17*, 297-329.

13

Pollen Records of Late Quaternary Vegetation Change: Plant Community Rearrangements and Evolutionary Implications

THOMPSON WEBB III
Brown University

INTRODUCTION

Radiocarbon-dated records of late-Quaternary pollen data provide a unique window on biospheric dynamics. They illustrate the vegetational response to large-scale climatic forcing; provide a space-time view of community and plant population variations; and fill a gap between short-term observations of ecological patterns and dynamics and the long-term fossil records of the Phanerozoic. With temporal sequences of maps in 1000-yr intervals, the paleoecological records of the past 18,000 yr add both a temporal dimension to ecological observations and a temporally and spatially precise mapping view to the standard fossil record.

Succession, which has long dominated ecological thinking, no longer appears as the dominant mode of vegetational dynamics when viewed within the context of continental-scale maps of the late Quaternary pollen data. Rather these maps depict populations of sessile plants as mobile entities that move in response to orbitally paced glacial-interglacial climate changes (Huntley and Webb, 1989). Such movement allows the plant populations to track the climate conditions favorable for their growth and indicates that evolutionary responses among these populations are secondary to migration. Were evolution the primary response to Quaternary climate change, then plant taxa would maintain fixed populations south of the advancing or retreating ice sheets; oaks, for example, would evolve new climate tolerances rather than being replaced by pines or other trees more suited to the new climate conditions. The rates of migration are sufficiently fast that the plant taxa can match (with relatively small lags) the rates of orbitally forced climatic change; otherwise, the plant taxa would have gone extinct long ago (Webb, 1986).

Studies of modern pollen and vegetation data show that the taxonomic resolution in pollen data is good enough to allow resolution of vegetational patterns across continents, states, and counties (Figure 13.1). Maps of temporal changes in the data illustrate the independent movement of individual taxa, and this individualistic behavior leads to community breakup and rearrangement. Within the context of the periodic large changes in Quaternary climates, communities are ephemeral within cycles, though often recurrent between cycles, and represent epiphenomena that arise out of the changing co-occurrence of plant taxa (Davis, 1983; Jacobson *et al.*, 1987; Webb, 1987; Jackson and Whitehead, 1991).

Despite the large changes in late Quaternary climates and communities, most mid- to high-latitude plant genera and species changed quickly enough in abundance and location to survive. Such an observation raises interesting questions about the nature of the ecological theater within which evolution occurs (Hutchinson, 1965). Climate forces the theater to be a traveling road show with a changing group of major actors. Selection must therefore occur on a stage with continuously changing combinations of environmental variables and competitors, and the local setting (specific soils, relief, etc.), local players, and recent arrivals generally have only a short-term influence on the evolutionary play.

Within this chapter, I begin by describing pollen data and their ability to record patterns in present and past vegetation. I then summarize what recent mapping studies of late Quaternary pollen data have illustrated about past changes in vegetation and their link to climate change. Simulations of past climates by climate models have aided these studies (Barnosky et al., 1987; Webb et al., 1987). These models are being used as tools for generating hypotheses that the data can test (COHMAP, 1988; Wright et al., 1993). In the final sections, I discuss (1) how an understanding of the space and time scales for taxonomic ecological units leads to a temporal separation of ecological units from evolutionary units for many plant taxa (McDowell et al., 1990) and (2) the advantages of a 4-dimensional space-time perspective of the data.

SENSITIVITY OF POLLEN DATA TO VEGETATION PATTERNS

Pollen data are sensitive to a wide spectrum of spatial and temporal variations in the vegetation (Figure 13.1). These can vary from local succession (Janssen, 1967; Bradshaw, 1988; Edwards, 1986) up to long-term continental-scale changes in plant formations (Jacobson et al., 1987; Huntley, 1990). The data can also record how the vegetation was influenced by humans (Behre, 1988; Birks et al., 1988; McAndrews, 1988), disease (Davis, 1981; Webb, 1982; Allison et al., 1986), fire (Patterson and Backman, 1988; Clark, 1990), soils (Webb, 1974; Brubaker, 1975; Jacobson, 1979; Bernabo, 1981; Tzedakis, 1992), topography (Janssen, 1981; Gaudreau et al., 1989; Lutgerink et al., 1989; Jackson and Whitehead, 1991), and climate (Bartlein et al., 1984; Webb et al., 1987; Huntley and Prentice, 1988; Prentice et al., 1991). With such a variety of possible processes and influences, palynologists must choose appropriate methods of data collection, analysis, and display to obtain results indicative of the vegetational variations of particular interest (Webb et al., 1978, 1993; Jacobson and Bradshaw, 1981; Grimm, 1988).

A key metaphor for understanding the interpretation of pollen data is to think of them as remotely sensed vegetation data (Webb, 1981; Webb et al., 1993). Just as the current vegetation emits or reflects radiation that remote sensors on satellites intercept, so too does (and has) the current (and past) vegetation shed pollen that accumulates "remotely" (i.e., well away from the source) in lakes and bogs. Both types of "remote" sensors record data with

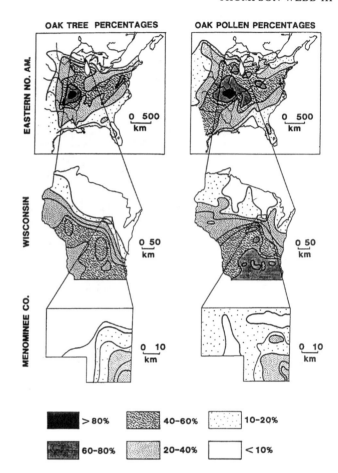

FIGURE 13.1 Relative abundance of oak trees and oak pollen at different spatial scales from that of a subcontinent, a state, and a county (from Solomon and Webb, 1985).

certain sampling characteristics (e.g., spatial and temporal resolution), and their data need calibration in terms of climate or vegetation variables. One major thrust in palynology, therefore, has been the analysis of modern pollen data to see what features of the modern vegetation are recorded (Figures 13.1 to 13.3). Palynologists have attempted to learn what the modern vegetation looks like in pollen terms (Webb, 1974), so that they can better visualize the past vegetation, which is only represented in pollen (or other fossil) terms. Palynologists have also studied modern data in order to learn how varying the sampling characteristics of the data can alter what vegetational features or processes are represented (Janssen, 1966; Andersen, 1970; Webb et al., 1978; Heide and Bradshaw, 1982; Bradshaw and Webb, 1985; Prentice et al., 1987; Prentice, 1988; Jackson, 1990, 1991).

Recent mapping studies at the subcontinental scale (Peterson, 1984; Webb, 1988; Huntley, 1990; Anderson et al., 1991) have shown how well the modern pollen data match contemporary vegetation patterns. In eastern North

FIGURE 13.2 Zoom-lens view of the vegetation from local to global showing the different levels of subdivision of the vegetation at each mapping scale (from Kutzbach and Webb, 1991).

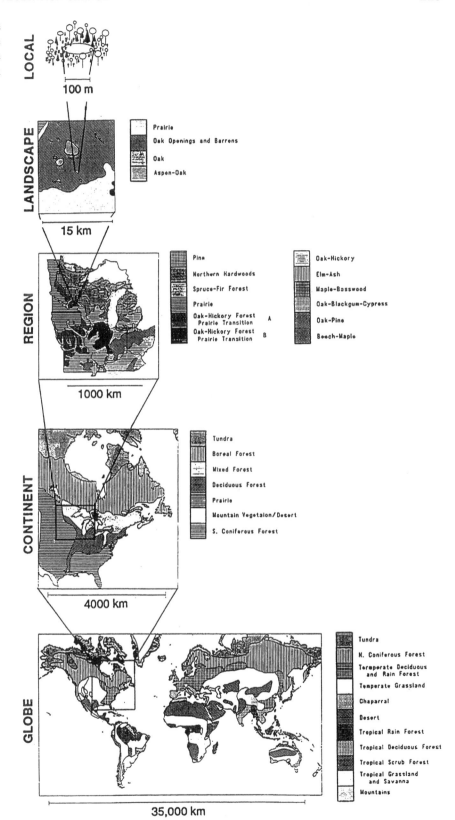

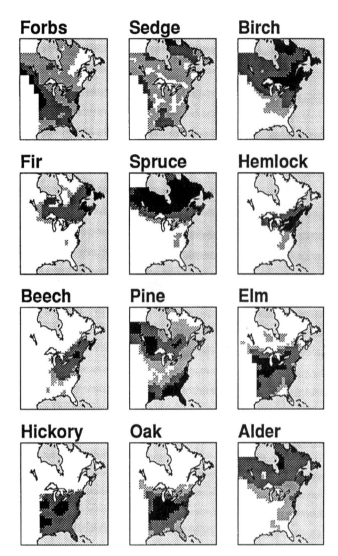

FIGURE 13.3 Maps (from Webb *et al.*, 1993) with contours of equal pollen frequency that show the current distribution for 9 pollen types. For forbs (*Artemisia*, other Compositae excluding *Ambrosia*, and Chenopodiaceae/Amaranthaceae), sedge (Cyperaceae), birch (*Betula*), spruce (*Picea*), and alder (*Alnus*), black areas are >20%, dark gray 5-20%, and light gray 1-5%. For oak (*Quercus*) and pine (*Pinus*), the values are >40%, 20-40%, and 5-20%. For fir (*Abies*), hemlock (*Tsuga*), beech (*Fagus*), and hickory (*Carya*), the values are >5%, dark gray 1-5%, and light gray 0.5-1%.

America, for example, maps for 12 major pollen categories record the patterns of the major formations from the western prairie into the eastern forests and from the northern tundra into the southeastern pine-oak forest (Figures 13.2 and 13.3). At this scale, the major vegetation gradients are aligned with temperature and moisture gradients. The pollen maps show not only the pattern of the major plant assemblages (i.e., formations) but also the steepness and position of the ecotones between them. The maps also show the changes in species composition not only among but within the major plant assemblages. Within the mixed forest, for example, a gradient exists between the dominance of birch, beech, and hemlock in the east and the dominance of pine in the west. Other mapping studies show similar ability of pollen to record vegetational patterns at state and county levels in support of the general picture shown in Figure 13.1 (Webb, 1974; Webb *et al.*, 1983; Bradshaw and Webb, 1985; Jackson, 1991). At these spatial scales, the vegetation patterns reflect differences in soils, topography, and disturbance history as well as climate gradients. Pollen data can therefore give a zoom lens view of the vegetation when spatial arrays of the samples are properly organized (Figure 13.1). By being sensitive to vegetation at various spatial scales, the pollen data are also sensitive to a variety of factors affecting the vegetation from climate at the continental scale down to disturbance history and patch dynamics at local sites (Figure 13.2; Delcourt *et al.*, 1983; McDowell *et al.*, 1990).

MAPS OF CHANGING TAXON DISTRIBUTION THROUGH TIME

Time sequence maps for the past 18,000 yr show where and when the modern patterns and vegetational regions developed. Webb (1988) and Jacobson *et al.* (1987) used time series of maps such as those in Figure 13.4 to infer that the modern boreal forest, for example, with its coincident patterns of spruce, birch, and alder, developed only after 6000 yr ago (ka). From 18 to 12 ka in the area south of the ice sheet between treeless boreal vegetation and pine-dominated mixed forests, spruce, sedge, and sagebrush pollen co-occurred, indicating widespread growth of a spruce parkland. From 11 to 7 ka, transitional forests near the ice front were dominated first by spruce and later by birch populations. Considerable compositional and structural change therefore occurred in the regions where spruce trees grew, and the modern boreal forest was one outcome of these changes. Similar types of sequences of compositional change occurred for other modern plant assemblages (Davis, 1983). Webb (1988) used time series of maps like those in Figure 13.4 to infer that the modern deciduous forests developed by 12 ka; prairie and tundra by about 10 ka; mixed forest with beech, birch, and hemlock after 8 ka; and the southeastern pine-oak forest after 8 ka. Maps of European vegetation for the past 12,000 yr illustrate similar types of broad-scale changes in the abundance, location, and association of taxa, with consequent changes in vegetational assemblages (Huntley, 1990). In the western United States, where low regional site density and topographic complexity have thus far precluded mapping studies, the data also show individualistic changes among taxa (Thompson, 1988). In each of these regions,

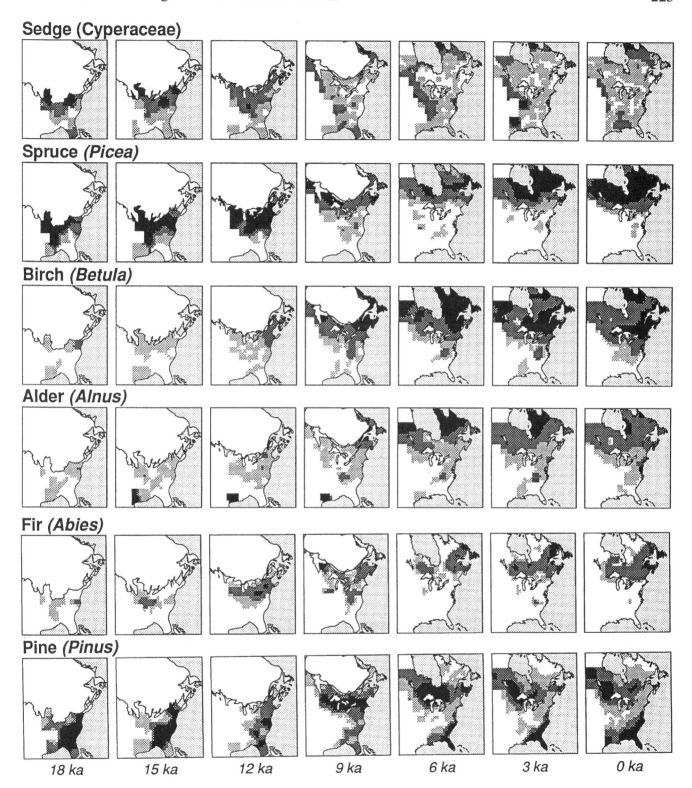

FIGURE 13.4 Maps of pollen frequency for sedge (Cyperaceae), spruce (*Picea*), birch (*Betula*), alder (*Alnus*), fir (*Abies*), and pine (*Pinus*) from 18 ka to present (0 ka). For all taxa but fir, black areas are >20%, dark gray 5-20%, and light gray 1-5%. For fir, black are >5%, dark gray 1-5%, and light gray 0.5-1%.

the climate changes were large, and simulations of past climates by climate models have reproduced many of the patterns of change in the data (COHMAP, 1988; Wright *et al.*, 1993).

Mapping studies of the fossil pollen data at regional and local scales also show variations in composition, location, and extent of vegetation. At these scales, variations in elevation and soil type become important along with climate in shaping the development of the vegetation (Davis *et al.*, 1980; Webb *et al.*, 1983; Ritchie, 1987; Gaudreau *et al.*, 1989; Woods and Davis, 1989; Jackson and Whitehead, 1991). Studies at each spatial scale show the independent behavior of individual taxa and the resultant changes in vegetational composition. Many of the past pollen assemblages have no analogues among modern pollen samples (Baker *et al.*, 1989; Overpeck *et al.*, 1992). Each of these types of vegetational change has occurred during the switch from full glacial to interglacial climates, and is likely to have occurred each time that such shifts in climate occurred in Earth history (see Figure 7.2 in Stanley and Ruddiman, Chapter 7, this volume). During the past 700,000 yr, the major changes have occurred seven times, and the estimated total change in the global mean temperature is $5° \pm 1°C$ (Webb, 1991). During times with lesser degrees of global climate change, the changes in vegetation were also less dramatic but probably still involved significant changes in community composition that produced assemblages without modern analogues. As Figure 7.2 in Stanley and Ruddiman (Chapter 7, this volume) shows, the climate has been changing continuously for millions of years; therefore, the vegetation is a continuously changing set of variables chasing a continuously changing set of other variables, namely, climate (Webb and Bartlein, 1992).

Webb (1986) and Prentice *et al.* (1991) have interpreted these compositional changes and consequent no-analogue assemblages as resulting primarily from the different climatic response of each taxon to the changing mixture of climatic variables as climates have varied temporally. They argue for the taxa being in dynamic equilibrium with climate (Prentice, 1986; Webb, 1986). Studies matching observed pollen maps with those simulated from climate model output provide support for this interpretation (Webb *et al.*, 1987; COHMAP, 1988). Other researchers argue for disequilibrium conditions between plant taxa and climate. They have given major emphasis to the role of biotic factors, such as differing dispersal rates and time lags for populations growth, when interpreting the development of no-analogue assemblages and patterns of species migration (Bennett, 1985; Birks, 1986). Recognition is now developing that a hierarchy of factors is operating, and that the importance of different factors varies with time and space scale (Davis, 1991). Biotic factors are most evident over short time and small spatial scales, and climatic impact is most evident over long time and large spatial scales.

IMPLICATIONS FOR SPECIES AND EVOLUTION

No matter which interpretation is favored (equilibrium or disequilibrium), the pollen record shows major changes in plant assemblages at all spatial scales with major plant assemblages (i.e., formations) having an average life time of ca. 10,000 yr in response to orbitally driven climate change. Consideration of the record of climate forcing for the past 2.8 million years (m.y.) (Figure 7.2 in Stanley and Ruddiman, Chapter 7, this volume) reveals that this forcing has been long-term, large, and continuous (Webb and Bartlein, 1992). The net result has been a continuously changing ecological theater for the evolutionary play (Hutchinson, 1965), and individualistic behavior has produced a continuously changing role, setting, and cast of associated characters for each taxon. Despite all this environmental and ecological change, most species have survived. Evidence from the fossil record suggests that the average longevity of species is 1 to 10 m.y. (Stanley, 1985). One reason for the longevity may be the relatively high frequency of mixing (induced by changes in species abundance, distribution, and association) that prevents long-term isolation of genetically distinct populations (Coope, 1978; Webb, 1987; Bartlein and Prentice, 1989).

Gould (1985) and Bennett (1990) discuss how "progress in life's history" may be thwarted, and selection over 10,000 years or less ("ecological time") is erased or lost by longer-term processes. In a well-argued paper, Bennett (1990) identifies orbitally forced climate change, which occurs at time scales of 20,000 to 100,000 yr, as the key longer-term process. As stated by Bartlein and Prentice (1989), "The paleoecological record of the past 20,000 years demonstrates that orbitally induced climatic changes produce changes in the distribution of organisms, leading to the quasi-cyclical alternation between allopatry and sympatry, commonness and rarity, continuous distribution and fragmentation." Furthermore, if recognition is given to the orbital control of long-term variations in monsoonal climates that depends on land-sea contrast (Kutzbach, 1981; COHMAP, 1988; Kutzbach and Webb, 1991), then the orbital pulse to climate can be seen to be possible in the absence of large ice sheets and to have a long history throughout the geological record. Crowley *et al.* (1986) and Ruddiman and Kutzbach (1989) have explored the potential implications of known tectonic changes on this mechanism for climate change, and Barnosky (1984), Olsen (1986), and others (see Berger *et al.*, 1984) have documented orbitally driven climate changes during several intervals of the Phanerozoic. The long-term occurrence of these climate oscillations indicates that the continuously changing ecological theater on orbital time scales may be a long-term feature in the geological record and hence not restricted to the Quaternary. For example, the long time series of vegetation changes from Sabana de Bogota in Colombia reveals large quasi-periodic changes dating back more than 1.5 m.y. (Hoogheimstra, 1989). Species, therefore, have evolved in the face of orbitally forced climatic changes, and orbital cycles may be as familiar to million-year-old species as the annual cycle is to 100-yr-old trees. Along with intraspecific plasticity in gene frequency, species migration and major population expansion under favorable conditions are two key mechanisms by which species respond to the orbitally induced climate changes (Dexter et al., 1987; Huntley and Webb, 1989; Bennett et al., 1991).

TIME AND SPACE SCALES OF VEGETATIONAL AND TAXONOMIC UNITS

One of the unifying figures for global change research is a scatter diagram of various earth-system processes plotted along log-scaled axes in units of time and area (e.g., Figure 2.3 in Earth System Sciences Committee, 1988). This type of plot shows the regions in space and time in which selected phenomena occur because of their characteristic time constants or spatial extent. It also provides a useful way to view the interaction of ecological and evolutionary phenomena. The diagram mixes many different phenomena. McDowell et al. (1990) recently separated the climatic, vegetational, and geomorphic phenomena and plotted them on different graphs. The graph for climate (Figure 13.5) shows the different spatial and temporal scales of

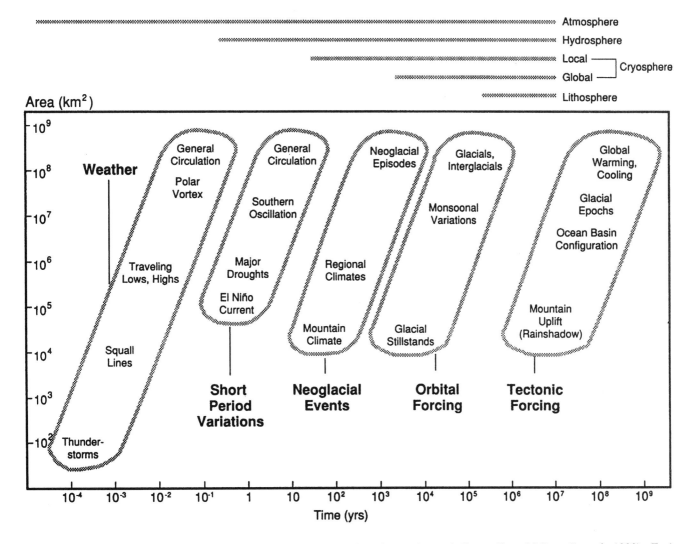

FIGURE 13.5 Characteristic spatial and temporal scales for variations in weather and climate (from McDowell et al., 1990). Each bubble encloses a group of related types of variations, in which the shorter-term, smaller-area units are part of the longer-term, larger-area units. The shortest-term units (weather) are limited primarily to the atmosphere, but the longer-term variations involve progressively more earth systems.

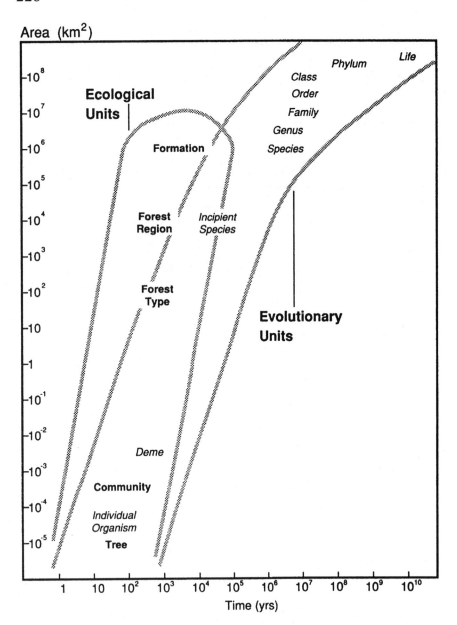

FIGURE 13.6 Characteristic temporal and spatial scales for vegetational (ecological) and taxonomic (evolutionary) units (from McDowell *et al.*, 1990). The former appear in boldface, the latter in italics. The chosen scales apply most particularly to wind-pollinated plants.

weather features up through the different spatial and temporal scales of climate phenomena on a tectonic (tens of millions of years) time scale. For each time scale, smaller-scale features have shorter time constants than large-scale features. The idea that a minimum time and area can be assigned to each feature may seem arbitrary at first because we experience large-scale features such as low pressure systems locally as well as regionally. However, we experience these large-scale features in terms of regional changes such as a frontal passage, and these regional features are included in the large-scale system. The lower limits also make sense as the smallest area or shortest time over which one is forced to recognize that a larger or longer-term feature exists.

Using the plots of climatic and oceanographic features as an example (Figure 13.5, and Haury *et al.*, 1978), Delcourt *et al.*, (1983) published the original version of the scale diagram for ecological phenomena, and McDowell *et al.* (1990) used the evidence on vegetational rearrangements (Figure 13.4, Webb, 1988) to revise the Delcourt *et al.* (1983) figure (Figure 13.6). McDowell *et al.* (1990) also added the set of evolutionary or taxonomic categories, in part inspired by Eldridge's (1985) *Unfinished Synthesis*, in which he noted that individual organisms can be organized either into communities or into taxa (i.e., into either ecological or systematic units).

The lineup of ecological phenomena begins with trees being 10 to 200 yr old and covering on average 10^{-5} km², and then shows the grouping of trees into communities whose upper size is determined by the area over which organisms interact directly with each other (Figure 13.6). Modern vegetation maps illustrate the different areas for vegetation regions (Figure 13.2), and Figure 13.4 and estimates from Webb (1988) for the time scale of recent formations fix the time scale for formations (see earlier discussion). The assumption that units with smaller areas exist for shorter times leads to the alignment of ecological units. This alignment illustrates the assumed role of orbital forcing in setting the upper age limit for the phenomena. Were the time scale of orbital forcing to change, then the age limit for formations should be changed.

An alignment of evolutionary phenomena was constructed by plotting the temporal and spatial dimensions of genetically related units from individuals up through all levels of taxonomic units. The estimates from the fossil record for the average longevity of each species is 1 to 10 m.y. This fact leads to the evolutionary units being plotted along a different slope from the ecological units. The average pollen dispersal distance sets the size of demes for wind-pollinated trees (Levin and Kerster, 1975; Bradshaw and Webb, 1985). Incipient species represent genetically distinct populations that are evident today or at any time, but disappear within 1000 to 10,000 yr and therefore never fully qualify as species. Stanley (1979) defined them as aborted species. In general, higher taxonomic units such as families will be longer lived and more widely distributed than lower units such as genera, species, populations, or demes.

A major result from constructing this figure is its illustration of how the ecological and the evolutionary axes diverge. The observations and theory underlying this divergence have been discussed in the previous section. Such figures as this should be useful in designing studies of how ecological processes influence evolution. These studies should lead to revisions in this figure as better understanding is obtained of the time constants and spatial coverage for each unit plotted.

SPACE-TIME PERSPECTIVE

When the temporal sequence of maps for spruce pollen is stacked to form a box, the contours among maps can be connected to form a three-dimensional surface in space and time (Webb, 1988; Banchoff, 1990). If contours for several abundance levels are connected, then we have a four-dimensional plot with abundance (a), varying in space (x, y), and time (t). Various cross sections can be removed from this box, such as maps and latitude-time or longitude-time plots. In terms of the space-time box, the maps representing plant distributional data for today are just an arbitrary cross section of a continuously changing four-dimensional distribution. Such a perspective should raise questions about the generality of studies of short-term selection or of plant interaction studies based on data from one such cross section. Evolution occurs in space and time, and the four-dimensional view of taxon distributions and dynamics is needed to represent the ecological theater in which speciation and evolution are part of the play. As pointed out by Bartlein and Prentice (1989) and by Bennett (1990), Holocene- and Quaternary-scale paleoecological research is key to linking short-term studies in microevolution to paleontological studies of macroevolution. Such a linkage will lead to a more complete interpretation of the fossil record and will enable the fossil record to contribute fundamental knowledge to global change research.

REFERENCES

Allison, T. B., R. E. Moeller, and M. B. Davis (1986). Pollen in laminated sediments provides evidence for a mid-Holocene forest pathogen outbreak, *Ecology 67*, 1101-1105.

Andersen, S. T. (1970). The relative pollen productivity and pollen representation of north European trees, and correction factors for tree pollen spectra, *Geological Survey of Denmark II*, Series No. 96, 1-99.

Anderson, P., P. J. Bartlein, L. B. Brubaker, K. Gajewski, and J. C. Ritchie (1991). Vegetation-pollen-climate relationships for the arcto-boreal region of North America and Greenland, *Journal of Biogeography 18*, 565-582.

Baker, R. G., J. Van Nest, and G. Woodworth (1989). Dissimilarity coefficients for fossil pollen spectra from Iowa and western Illinois during the last 30,000 years, *Palynology 13*, 63-77.

Banchoff, T. F. (1990). *Beyond the Third Dimension*, Scientific American Library, New York.

Barnosky, C. W. (1984). Late Miocene vegetational and climatic variations inferred from a pollen record in northwest Wyoming, *Science 223*, 49-51.

Barnosky, C. W., P. M. Anderson, and P. J. Bartlein (1987). The northwestern U.S. during deglaciation; Vegetational history and paleoclimatic implications, in *North America and Adjacent Oceans During the Last Deglaciation, The Geology of North America*, W. F. Ruddiman and H. E. Wright, Jr., eds., Geological Society of America, Boulder, Colo., pp. 289-323.

Bartlein, P. J., and I. C. Prentice (1989). Orbital variations, climate, and paleoecology, *Trends in Ecology and Evolution 4*, 195-199.

Bartlein, P. J., T. Webb III, and E. C. Fleri (1984). Holocene climatic change in the northern Midwest: Pollen-derived estimates, *Quaternary Research 22*, 361-374.

Behre, K-E. (1988). The role of man in European vegetation history, in *Vegetation History*, B. Huntley and T. Webb III, eds., Kluwer Academic Publishers, Dordrecht, pp. 633-672.

Bennett, K. D. (1985). The spread of *Fagus grandifolia* across eastern North America during the last 18,000 years, *Journal of Biogeography 12*, 147-164.

Bennett, K. D. (1990). Milankovitch cycles and their effects on species in ecological and evolutionary time, *Paleobiology 16*, 11-21.

Bennett, K. D., P. C. Tzedakis, and K. J. Willis (1991). Quaternary refugia of north European trees, *Journal of Biogeography* 18, 103-115.

Berger, A., J. Imbrie, J. Hays, G. Kukla, and B. Saltzman, eds. (1984). *Milankovitch and Climate*, Reidel, Dordrecht.

Bernabo, J. C. (1981). Quantitative estimates of temperature changes over the last 2700 years in Michigan based on pollen data, *Quaternary Research 15*, 143-159.

Birks, H. J. B. (1986). Late Quaternary biotic changes in terrestrial and lacustrine environments with particular reference to north-west Europe, in *Handbook of Holocene Palaeoecology and Palaeohydrology*, B. E. Berglund, ed., J.W. Wiley & Sons Ltd., London, pp. 3-65.

Birks, H. H., H. J. B. Birks, P. E. Kaland, and D. Moe (1988). *The Cultural Landscape—Past, Present, and Future*, Cambridge University Press, Cambridge.

Bradshaw, R. H. W. (1988). Spatially-precise studies of forest dynamics, in *Vegetation History*, B. Huntley and T. Webb III, eds., Kluwer Academic Publishers, Dordrecht, pp. 725-751.

Bradshaw, R. H. W., and T. Webb III (1985). Relationships between contemporary pollen and vegetation data from Wisconsin and Michigan, USA, *Ecology 66*, 721-737.

Brubaker, L. B. (1975). Postglacial forest patterns associated with till and outwash in northcentral Upper Michigan, *Quaternary Research 5*, 499-527.

Clark, J. S. (1990). Fire and climate change during the last 750 yr in northwestern Minnesota, *Ecological Monographs 60*, 135-159.

COHMAP Members (1988). The development of late-glacial and Holocene climates: Interpretation of paleoclimate observations and model simulations, *Science 241*, 1043-1052.

Coope, G. R. (1978). Constancy of species versus inconstancy of Quaternary environments, in *Diversity of Insect Faunas*, J. H. R. Gee and P. S. Giller, eds., Blackwell Scientific Publications, Oxford, pp. 421-438.

Crowley, T. J., D. A. Short, J. G. Mengel, and G. R. North (1986). Role of seasonality in the evolution of climate over the last 100 million years, *Science 231*, 579-584.

Davis, M. B. (1981). Outbreaks of forest pathogens in Quaternary history, *IV. International Palynological Conference, Lucknow (1976-77) 3*, 216-227.

Davis, M. B. (1983). Quaternary history of deciduous forests of eastern North America and Europe. *Annals of the Missouri Botanical Garden 70*, 550-563.

Davis, M. B. (1991). Research questions posed by the paleoecological record of global change, in *Global Changes of the Past*, R.S. Bradley, ed., Office for Interdisciplinary Earth Studies, Global Change Institute, Boulder, Colo., pp. 385-395.

Davis, M. B., R. W. Spear, and L. C. K. Shane (1980). Holocene climate of New England, *Quaternary Research 14*, 240-250.

Delcourt, H. R., P. A. Delcourt, and T. Webb III (1983). Dynamic plant ecology: The spectrum of vegetational change in space and time, *Quaternary Science Reviews 1*, 153-175.

Dexter, F., H. T. Banks, and T. Webb III (1987). Modeling Holocene changes in the location and abundance of beech populations in eastern North America, *Review of Palaeobotany and Palynology 50*, 273-292.

Earth System Sciences Committee, NASA Advisory Council (1988). *Earth System Science: A Closer View*, Office for Interdisciplinary Earth Sciences, Boulder, Colo.

Edwards, M. E. (1986). Disturbance histories of four snow domain woodlands and their relation to Atlantic bryophyte distributions, *Biological Conservation 37*, 301-320.

Eldridge, N. (1985). *The Unfinished Synthesis, Biological Hierarchies and Modern Evolutionary Thought*, Oxford University Press, New York.

Gaudreau, D. C., S. T. Jackson, and T. Webb III (1989). The use of pollen data to record vegetation patterns in regions of moderate to high relief, *Acta Botanica Nederl. 38*, 369-390.

Gould, S. J. (1985). The paradox of the first tier: An agenda for paleobiology, *Palebiology 11*, 2-12.

Grimm, E. C. (1988). Data analysis and display, in *Vegetation History*, B. Huntley and T. Webb III, eds., Kluwer Academic Publishers, Dordrecht, pp. 43-76.

Haury, L. R., J. A. McGowan, and P. H. Wiebe (1978). Patterns and processes in the time-space scales of plankton distributions, in *Spatial Patterns in Plankton Communities*, J. H. Steele, ed., Proc. NATO Conference on Marine Biology, Erice, Italy, Plenum, New York, pp. 277-327.

Heide, K. M., and R. H. W. Bradshaw (1982). The pollen-tree relationship within forests of Wisconsin and upper Michigan, U.S.A., *Review of Palaeobotany and Palynology 36*, 1-23.

Hoogheimstra, H. (1989). Quaternary and Upper Pliocene glaciations and forest development in the tropical Andes: Evidence from a long high-resolution pollen record from the sedimentary basin of Bogota, Colombia, *Palaeogeography, Palaeoclimatology, Palaeoecology 72*, 11-26.

Huntley, B. (1990). European post-glacial forests: Compositional changes in response to climate change, *Journal of Vegetation Science 1*, 507-518.

Huntley, B., and I. C. Prentice (1988). July temperatures in Europe from pollen data, 6000 years before present, *Science 241*, 687-690.

Huntley, B., and T. Webb III (1989). Migration: Species' response to climatic variations caused by changes in the Earth's orbit, *Journal of Biogeography 16*, 5-19.

Hutchinson, G. E. (1965). *The Ecological Theater and the Evolutionary Play*, Yale University Press, New Haven, Conn.

Jackson, S. T. (1990). Pollen source area and representation in small lakes of the northeastern United States, *Review of Palaeobotany and Palynology 63*, 53-76.

Jackson, S. T. (1991). Pollen representation of vegetational patterns along an elevational gradient, *Journal of Vegetation Science 2*, 613-624.

Jackson, S. T., and D. R. Whitehead (1991). Holocene vegetation patterns in the Adirondack Mountains, *Ecology 72*, 641-653.

Jacobson, G. L., Jr. (1979). The paleoecology of white pine (*Pinus strobus*) in Minnesota, *Journal of Ecology 67*, 697-726.

Jacobson, G. L., and R. H. W. Bradshaw (1981). The selection of sites for paleovegetation studies, *Ecology 47*, 804-825.

Jacobson, G. L., Jr., T. Webb III, and E. C. Grimm (1987). Patterns and rates of vegetation change during deglaciation of eastern North America, in *North America and Adjacent Oceans*

During the Last Deglaciation, Geology of North America, Vol. K-3, W. F. Ruddiman and H. E. Wright, Jr., eds., Geological Society of America, Boulder, Colo., pp. 277-288.

Janssen, C. R. (1966). Recent pollen spectra from the deciduous and coniferous-deciduous forests of northwestern Minnesota: A study in pollen dispersal, *Ecology 47,* 804-825.

Janssen, C. R. (1967). Stevens Pond: A postglacial pollen diagram from a small *Typha* swamp in northwestern Minnesota, interpreted from pollen indicators and surface samples, *Ecological Monographs 37,* 145-172.

Janssen, C. R. (1981). Contemporary pollen assemblages of the Vosges, *Review of Palaeobotany and Palynology 33,* 183-313.

Kutzbach, J. E. (1981). Monsoon climate of the early Holocene: Climate experiment with the Earth's orbital parameters for 9000 years ago, *Science 214,* 59-61.

Kutzbach, J. E., and T. Webb III (1991). Late quaternary climatic and vegetational change in eastern North America: Concepts, models, and dates, in *Quaternary Landscapes,* L. C. K. Shane and E. J. Cushing, eds., University of Minnesota Press, Minneapolis, pp. 175-217.

Levin, D. A., and H. W. Kerster (1974). Gene flow in seed plants, *Evolutionary Biology 7,* 139-220.

Lutgerink, R. H. P., Ch. A. Swertz, and C. R. Janssen (1989). Regional pollen assemblages versus landscape regions in Monts du Forez, Massif central, France, *Pollen et Spores 31,* 45-60.

McAndrews, J. H. (1988). Human disturbance of North American forests and grasslands: The fossil pollen record, in *Vegetation History,* B. Huntley and T. Webb III, eds., Kluwer Academic Publishers, Dordrecht, pp. 673-697.

McDowell, P. F., P. J. Bartlein, and T. Webb III (1990). Long-term environmental change, in *The Earth as Transformed by Human Action,* B. J. Turner II et al., eds., Cambridge University Press, New York, pp. 143-162.

Olsen, P. (1986). A 40-million-year lake record of early Mesozoic orbital climatic forcing, *Science 234,* 842-848.

Overpeck, J. T., R. S. Webb, and T. Webb III (1992). Mapping eastern North American vegetation change of the past 18 ka: No-analogs and the future, *Geology 20,* 1071-1074.

Patterson, W., III, and A. E. Backman (1988). Fire and disease history of forests, in *Vegetation History,* B. Huntley and T. Webb III, eds., Kluwer Academic Publishers, Dordrecht, pp. 603-632.

Peterson, G. M. (1984). Recent pollen spectra and zonal vegetation in the western USSR, *Quaternary Science Reviews 2,* 281-321.

Prentice, I. C. (1986). Vegetation responses to past climatic changes, *Vegetatio 67,* 131-141.

Prentice, I. C. (1988). Records of vegetation in time and space: The principles of pollen analysis, in *Vegetation History,* B. Huntley and T. Webb III, eds., Kluwer Academic Publishers, Dordrecht, pp. 17-24.

Prentice, I. C., B. E. Berglund, and T. Olsson (1987). Quantitative forest composition sensing characteristics of pollen samples from Swedish lakes, *Boreas 16,* 43-54.

Prentice, I. C., P. J. Bartlein, and T. Webb III (1991). Vegetation change in eastern North America since the last glacial maximum: A response to continuous climatic forcing, *Ecology 72,* 2038-2056.

Ritchie, J. C. (1987). *Postglacial Vegetation of Canada,* Cambridge University Press, Cambridge.

Ruddiman, W. F., and J. E. Kutzbach (1989). Forcing of late Cenozoic Northern Hemisphere climates by plateau uplift in southern Asia and the American West, *Journal of Geophysical Research 94,* 18,409-18,427.

Solomon, A. M., and T. Webb III (1985). Computer-aided reconstruction of late-Quaternary landscape dynamics, *Annual Review of Ecology and Systematics 16,* 63-84.

Stanley, S. M. (1979). *Macroevolution Pattern and Process,* W. H. Freeman and Co., San Francisco, Calif.

Stanley, S. M. (1985). Rates of evolution, *Paleobiology 11,* 13-26.

Thompson, R. S. (1988). Western North American, in *Vegetation History,* B. Huntley and T. Webb III, eds., Kluwer Academic Publishers, Dordrecht, pp. 415-458.

Tzedakis, P. (1992). Effects of soils on the Holocene history of forest communities, Cape Cod, Massachusetts, U.S.A., *Geographie physique et Quaternaire 46,* 113-124.

Webb, T., III (1974). Corresponding distributions of modern pollen and vegetation in lower Michigan, *Ecology 55,* 17-28.

Webb, T., III (1981). 11,000 years of vegetational change in eastern North America, *Bioscience 31,* 501-506.

Webb, T., III (1982). Temporal resolution in Holocene pollen data, *Third North American Paleontological Convention, Proceedings 2,* 569-572.

Webb, T., III (1986). Is vegetation in equilibrium with climate? How to interpret late-Quaternary pollen data, *Vegetatio 67,* 75-91.

Webb, T., III (1987). The appearance and disappearance of major vegetational assemblages: Long-term vegetational dynamics in eastern North America, *Vegetatio 69,* 177-187.

Webb, T., III (1988). Eastern North America, in *Vegetation History,* B. Huntley and T. Webb III, eds., Kluwer Academic Publishers, Dordrecht, pp. 385-414.

Webb, T., III (1991). The spectrum of temporal climatic variability: Current estimates and the need for global and regional time series, in *Records of Past Global Change,* R. Bradley, ed., Office of Interdisciplinary Earth Studies, Boulder, Colo., pp. 61-81.

Webb, T., III, and P. J. Bartlein (1992). Global changes during the last 3 million years: Climatic controls and biotic responses, *Annual Reviews of Ecology and Systematics 23,* 141-173.

Webb, T., III, R. A. Laseski, and J. C. Bernabo (1978). Sensing vegetation with pollen data: Control of the signal-to-noise ratio, *Ecology 59,* 1151-1163.

Webb, T., III, E. J. Cushing, and H. E. Wright, Jr. (1983). Holocene changes in the vegetation of the Midwest, in *Late-Quaternary Environments of the United States, Vol. 2, The Holocene,* H. E. Wright, Jr., ed., University of Minnesota Press, Minneapolis, pp. 142-165.

Webb, T., III, P. J. Bartlein, and J. E. Kutzbach (1987). Climatic change in eastern North America during the past 18,000 years: Comparison of pollen data with model results, in *North America and Adjacent Oceans During the Last Deglaciation, Geology of North America, Vol. K-3,* W. F. Ruddiman and H. E. Wright, Jr., eds., Geological Society of America, Boulder, Colo., pp. 447-462.

Webb, T., III, P. J. Bartlein, S. P. Harrison, and K. H. Anderson (1993). Vegetation, lake level, and climate change in eastern North America, in *Global Climates Since the Last Glacial Maximum*, H. E. Wright *et al.*, eds., University of Minnesota Press, Minneapolis, pp. 415-467.

Woods, K. D., and M. B. Davis (1989). Paleoecology of range limits: Beech in the upper peninsula of Michigan, *Ecology 70*, 681-696.

Wright, H. E., Jr., J. E. Kutzbach, T. Webb III, W. F. Ruddiman, F. A. Street-Perrott, and P. J. Bartlein, eds. (1993). *Global Climates Since the Last Glacial Maximum*, University of Minnesota Press, Minneapolis.

14

Climatic Forcing and the Origin of the Human Genus

STEVEN M. STANLEY
The Johns Hopkins University

ABSTRACT

With the onset of the modern ice age, climatic changes in Africa caused grasslands to expand and forests to contract. This environmental shift appears to account for the evolution of *Homo* about 2.4 million years ago (Ma). Evolution could not have established the mode of development that produces the large brain of *Homo* until human ancestors were fully terrestrial. This mode of development produces the long interval of physical helplessness that distinguishes humans from all other mammals; it yields infants that cannot cling to mothers. The more mature infants of apes can cling to their mothers, allowing the mothers to climb trees. Gracile australopithecines, from which the genus *Homo* evolved, retained an apelike pattern of development, arboreal adaptations, from their ancestors. Thus, even females with infants could have been adept tree climbers, and the need to avoid predators must have required that gracile australopithecines be habitual climbers. In contrast, the small pelvic dimensions and large brain of early *Homo* point to delayed development, helpless infants, and a totally terrestrial mode of life. The origin of *Homo*, about 2.4 Ma, appears to have resulted from the onset of the recent ice age, when climatic changes in Africa caused savannas to expand at the expense of woodlands. These changes must have had a severe impact on australopithecines, as they did on other groups of mammals. In particular, they should have compelled many australopithecine populations to abandon the arboreal activity that had maintained evolutionary stability for more than 1.5 million years (m.y.). The resulting crisis conditions presumably caused the extinction of many populations, but the evolution of a huge brain through delayed development was now possible for the first time. Selection pressure for superior intelligence fostered the development of advanced social behavior and tool manufacture, which offset the problems created by helpless infants and the loss of arboreal refugia and food resources. African antelopes experienced parallel changes, with forest-adapted species suffering heavy extinction and a variety of new species coming into existence.

INTRODUCTION

Some adaptive breakthroughs in the history of life have resulted from natural selection for improved adaptation without the influence of environmental change. Other major evolutionary transitions have resulted from selection pressures imposed by a changing habitat. Thus, not only do severe environmental changes alter the biosphere by causing migration and extinction, they also stimulate evolution. In fact, a species can, in a sense, experience both extinction and successful evolution as a result of environmental crisis. A species as it was constituted for 10^6 to 10^7 generations may die out and yet have one or more of its populations emerge as a distinctive new species adapted to the new conditions.

The onset of the recent ice age, at about 2.5 Ma, transformed habitats in many regions of the world. A variety of circumstantial evidence suggests that at this time a particular kind of environmental crisis in Africa caused the extinction of one or more species of the human family but also triggered the evolution of the modern human genus, *Homo* (Stanley, 1992). *Homo* evolved from a species of the genus *Australopithecus*. This ancestral genus, having been confined to Africa, could not have escaped the environmental changes that pervaded this continent.

The early evolution of *Homo* was of unique significance in the history of life. Most importantly, it entailed a marked increase in brain size. The early *Homo* fossil skull 1590, which belonged to a child about 6 or 7 years old, would have grown to have an adult cranial capacity well in excess of 800 cm^3, about twice that of a male belonging to the gracile australopithecine species from which early *Homo* evolved.

The term "gracile" means slender, and applies specifically to the jaws, teeth, and skull, which in both of the known gracile species *Australopithecus afarensis* and *A. africanus* were less heavily developed than in the "robust" species, which many authors now segregate into the genus *Paranthropus*. The robust forms were almost certainly evolutionary offshoots of the line of descent that led to *Homo*. Most experts now regard *A. africanus* as the gracile species most likely to have given rise to *Homo*, in part because the known range of this South African form extends to about 2.3 Ma, more than half a million years beyond that of *A. afarensis*, whose fossil record is confined to northern Africa. In any event, the two gracile species were rather similar in general morphology and presumably therefore in mode of life.

It is important to recognize that the gracile australopithecines existed for more than 1.5 m.y., from about 4 Ma to perhaps 2.3 Ma, without experiencing appreciable evolutionary change. Not only did their cranial capacity remain only slightly above the level of a chimpanzee, but their postcranial anatomy experienced approximate evolutionary stasis as well (McHenry, 1986). They were obviously successful, well-adapted creatures. It has long been understood that australopithecines walked bipedally. Their pelvic configuration is much more human than apelike in form, and their hindlimbs also in many ways resemble those of *Homo*. Furthermore, fossil bipedal footprints in Tanzania were formed at about 3.7 Ma, long before *Homo* or robust australopithecines evolved. By default, these tracks are attributed to gracile australopithecines (Leakey and Hay, 1979).

Although there remains no doubt that australopithecines moved bipedally on the ground, the past decade has seen the emergence of abundant evidence that tree climbing was also part of their normal behavioral repertoire (Stern and Susman, 1983; Susman *et al.*, 1984; inter alia). The bones of their forelimbs display numerous traits that evolved as arboreal adaptations in their ancestors. The bones of their hindlimbs were more human in form, but nonetheless retained certain traits that would have enhanced the ability to climb while at the same time restricting performance in walking and running to a level below that of modern humans. The habits of extant primates indicate that australopithecines should have been required to use their appropriate traits in order to avoid predators. The australopithecines' division of activities between terrestrial and arboreal habitats accounts for the failure of their postcranial morphology to evolve appreciably in a human direction for some 1.5 m.y.

In a less direct way, semiarboreal behavior can also account for the gracile australopithecines' failure to evolve appreciably larger brains during the same interval. In fact, it was impossible for the large brain of *Homo* to evolve until obligate arboreal activity had been abandoned. This restriction related to the developmental mechanism by which the large brain evolved: extension into the postnatal interval of the high *in utero* rate of brain growth, which in all primates maintains brain weight at about 10% of body weight. In monkeys and apes, this rate gives way to a much lower rate soon after birth. In humans, however, cranial development is delayed, so that the high fetal rate of brain growth is projected through the first year of life after birth. This produces a cranial capacity for a 1-yr-old human infant that is more than double that of an adult chimpanzee. Figure 14.1 illustrates the differing developmental patterns for apes and humans.

The reason that human ancestors could not evolve the large brain of *Homo* until they totally abandoned arboreal activity is that the developmental delay that yielded the large brain was linked to other aspects of maturation. A general retardation of development produced the large brain by projecting the high fetal rate of brain growth into the postnatal interval. In contrast to infant chimpanzees and

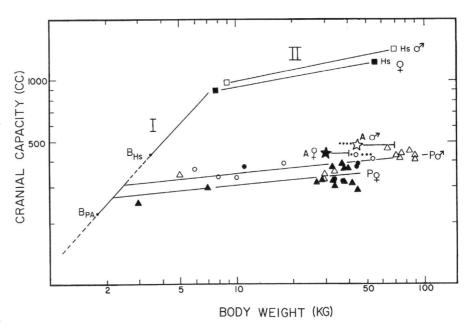

FIGURE 14.1 Differences in brain growth between apes and humans, and similarities between apes and australopithecines. Solid symbols represent males; hollow symbols, females. Circles represent chimpanzees; triangles, orangutans. The Phase I slope is shared by all taxa. Range for present estimates of gracile australopithecine cranial capacities is shown by dotted lines; the estimated male mean is placed near the upper level, and the estimated female mean near the lower level (stars); horizontal bars show ranges of body sizes estimated by less favored methods. BHs: mean birth size for the two species of apes, and also for gracile australopithecines (estimated from pelvic dimensions; Tague and Lovejoy, 1986). The postnatal segment of the Phase I slope for australopithecines was short, like that of apes, not long, like that of humans in which it is associated with a long interval of physical helplessness. (See Stanley, 1992, for details.)

orangutans, which are mature enough to cling to their semiarboreal mothers soon after birth, infants of early *Homo* had to be carried. It follows that, because a mother cannot climb while carrying an infant, a totally terrestrial life was a prerequisite for the evolution of *Homo*.

As it turns out, we can infer from fossil evidence the body and brain sizes of newborn and adult australopithecines. From these data, we can show that gracile australopithecines were apelike in their pattern of development: their infants were not physically helpless and should therefore have been able to cling to climbing mothers. Figure 14.1 illustrates how the pattern for gracile australopithecines resembles that for modern apes. On the other hand, comparable data for early *Homo* indicate a modern human pattern: infants were helpless (Stanley, 1992).

There is extensive evidence that at about 2.5 Ma, African forests shrank at the expense of grasslands, which are adapted to drier conditions. It has been suggested that these major vegetational changes that swept across Africa caused evolutionary turnover within the human family (Vrba, 1975, 1988, inter alia). In fact, the changes were precisely the kind that could be expected to have forced australopithecines to abandon habitual arboreal activities (Stanley, 1992). The result would have been extermination of many populations at the hands of large predators, but also provision of the opportunity for evolution of a large brain because physically helpless infants were now tolerable. The environmental changes should also have engendered strong selection pressures for brain expansion, given the need of hominids to cope with predation without escaping into trees and to replace tree-borne food materials.

DEVELOPMENT IN APES, HUMANS, AND AUSTRALOPITHECINES

Estimates of brain volumes, body weights, and pelvic dimensions lead to the conclusion that the infants of early *Homo* were physically helpless, like those of modern humans (Stanley, 1992). A key fact underlies calculations leading to this conclusion: all primates adhere to the same general curve when their brain weights are plotted against their body weights for the fetal interval: brain weight constitutes roughly 10% of body weight (Holt et al., 1975). Figure 14.1 shows how brain growth in humans departs from that of apes in the postnatal interval. In apes, the fetal (Phase I) slope gives way to a much lower (Phase II) slope soon after birth, so that there is only modest postnatal encephalization. In humans, the inflection is delayed to an age of about one year, so that an infant this age still has a brain that constitutes nearly one-tenth of its body weight. The remarkable fact is that the brain of a one-year-old human is more than twice as large as that of an adult chimpanzee or orangutan.

Several observations indicate that a general retardation of development produced the delay in the Phase I-Phase II transition that yielded the large brain of modern humans. Overall retardation is evident in the physical helplessness and slow maturation of human infants compared to the offspring of apes. The evolution of highly immature infants in human evolution represented a profound ecological sacrifice. Offspring that must be carried, fed, and protected for years occupy parents' time that could otherwise be spent in important activities such as acquisition of food. Such infants also complicate the avoidance of predators. For natural selection to have produced this deleterious developmental pattern, there had to be some overriding benefit. Encephalization was clearly the change that provided this benefit: not only did it have great adaptive value, but it was as profound a change as the delay in development. In degree of brain expansion immediately after birth, modern humans rank first among mammals (Count, 1947), just as they rank first in the length of their postnatal interval of physical helplessness (Krogman, 1972).

No such developmental delay characterized gracile australopithecines. Fossils reveal that these animals closely resembled apes rather than humans in pattern of brain growth. It is most meaningful here to consider the pattern for males, because their mean birth size is larger than that of females, so that its maximum value can be estimated from the size of the female pelvis. Pelvic inlet breadth is the dimension that limits cranial size and therefore body size for neonates. This dimension for the pelvis of the famous "Lucy" skeleton, which represents *Australopithecus afarensis*, indicates that male birth size for this species approximated that for a chimpanzee or orangutan (Tague and Lovejoy, 1986). Other fossils show that adult male body size averaged at least 45 kg, not far from the mean for chimpanzees (McHenry, 1991), and that adult male brain size (averaging about 480 cm^3) was only slightly above the chimpanzee mean (Figure 14.1). Apes do not give birth to neonates as large as their pelvic dimensions would allow. Thus, they do not develop brains as large as they might, even without postnatal extension of the Phase I portion of the brain-body growth curve. The probable reason is that apes' forelimbs are so heavily occupied in locomotion—arboreal climbing and terrestrial knuckle walking—that extensive tool use is impossible. Encephalization beyond the present level is therefore unwarranted, given the costs that accompany brain expansion: the demographic sacrifice that a lengthened gestation time would entail, for example, and the high energy expenditure required for growth of brain tissue. As a result, there is no reason to believe that the slight postnatal extension of the Phase I slope in gracile australopithecines resulted in helpless infants that could not cling to climbing mothers.

One might ask why natural selection should not simply have expanded the australopithecine brain without being required to retard development in general, with all the attendant problems. The primary answer is that an organ such as the brain does not develop in isolation, but is morphogenetically linked to other anatomical systems. As a result, a general delay in development was by far the simplest mechanism for brain expansion. All that was needed was prolongation into the postnatal interval of a pattern of development that was already in place. The complexity of other potential evolutionary mechanisms was so great that the probability of their occurrence was very low. A likely second problem with other mechanisms would have been their inability to expand the brain dramatically soon after birth. Delayed development offered the key advantage of producing a large brain during the first year, thus permitting the immense human learning process to proceed rapidly at a very young age.

THE LIFE OF GRACILE AUSTRALOPITHECINES

Even if clinging neonates permitted australopithecines to engage in arboreal activity, habitual climbing would have been possible only if the adults possessed appropriate adaptations. In fact, numerous of their morphological traits indicate that although these animals were adapted for bipedal locomotion on the ground, they were nonetheless much more adept climbers than modern humans. Furthermore, as I will explain below, one can make a strong case that australopithecines were compelled to put their climbing abilities to use in their everyday life.

Arboreal Traits

Modern humans can climb trees better than many members of advanced civilizations recognize, not only by shinnying but also by what amounts to walking up tree trunks. Members of certain Malaysian tribes are excellent barefoot climbers, gripping a tree trunk with a hand on each side, taking small steps upward by applying their splayed feet to the trunk, and then regripping with the hands at a higher level (Wood-Jones, 1900; Skeat and Blagden, 1906). Sometimes they climb so rapidly in this manner as to be described as "running" up trees. A variety of attributes would have made australopithecines more adept at these activities than modern humans are (Figure 14.2).

A trait of australopithecines that would have given them a substantial advantage over modern humans in "walking" up trees was their larger ratio of arm length to leg length. This is reflected in the humerofemoral index, which is the ratio of upper arm length to upper leg length (Figure 14.3). The australopithecines' relatively long arms permitted the upper torso to tilt backward, so that gravity contributed more to the strength of the hands' grip (Cartmill, 1974; Jungers and Stern, 1983). This principle is the one employed by a repairman who climbs a telephone pole by looping a strap around the pole and his waist.

FIGURE 14.2 A gracile australopithecine "walking" up a tree. Key adaptations rendering australopithecines superior to modern humans in this activity are long arms relative to leg length; powerful wrists and hands with long, curved toes; ankles capable of great dorsiflexion; and small body size.

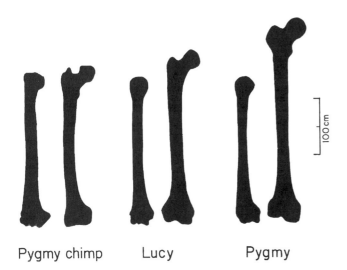

FIGURE 14.3 Relative lengths of the humerus (upper forelimb bone: left-hand member of each pair) and femur (upper hindlimb bone) for a pygmy chimpanzee, human pygmy, and Lucy (female *Australopithecus afarensis*). The length ratio for Lucy is intermediate. (See Stanley, 1992, for sources.)

The australopithecine forelimb exhibits additional traits that would have enhanced climbing ability. The glenoid cavity, the socket of the shoulder blade that receives the head of the humerus, is more upward directed than in modern humans (Vrba, 1979; Stern and Susman, 1983; Stanley, 1992). This is advantageous for suspensory activity.

Various skeletal features indicate that the australopithecines' wrists and hands were more powerful relative to body size than those of modern humans. In addition, their finger bones were long and curved, resembling those of chimpanzees (Figure 14.4). In fact, their hands were more apelike than human in form (Stern and Susman, 1983; Aiello and Dean, 1990). All of these traits would have served australopithecines well in climbing.

The hindlimbs of australopithecines also bore toes that were comparatively long and curved (Stern and Susman, 1982; Latimer and Lovejoy, 1990). Although these animals could not have grasped tree trunks or branches with

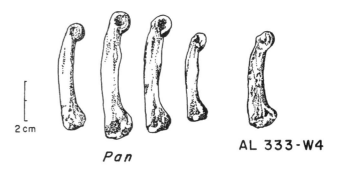

FIGURE 14.4 Strong curvature of a proximal phalanx (finger bone) from the hand of an individual of *Australopithecus afarensis* (AL 333-W4). This bone resembles the equivalent bones from the fingers of *Pan* (chimpanzees).

their hind feet in the manner of apes, which have highly prehensile big toes, they were nonetheless better equipped than modern humans to grip arboreal substrata with curled toes. In addition, their ankles allowed for much greater upward flexure of the foot than ours (Latimer *et al.*, 1987), which would have been very useful for vertical climbing (Figure 14.2). Finally, the relatively small body sizes of gracile australopithecines would have facilitated climbing by offering a higher ratio of strength to weight than characterizes modern humans. New estimates suggest that females averaged about 30 kg and males about 45 kg (McHenry, 1991).

In summary, australopithecines were certainly better climbers than modern humans. There is no question, however, that they were also less proficient at arboreal activity than modern apes. Their morphology suggests ability for upright climbing, but not extensive acrobatic activity of the sort engaged in by chimpanzees and orangutans. Once in a tree, they probably walked bipedally on limbs, gripping branches with their forelimbs for support.

The Arboreal Imperative

It is clear that the traits described in the preceding section evolved as arboreal adaptations in australopithecine ancestors, although unfortunately the fossil record of African primates for the Late Miocene and very early Pliocene is too poor to reveal the identity of these more heavily arboreal predecessors. The upright climbing posture of australopithecine ancestors was retained during the evolutionary transition to activity on the ground; there is no evidence that knuckle walking of the sort employed by modern apes had any place in human ancestry. We would not predict that a transition from a strictly arboreal mode of life to an essentially terrestrial one would occur instantaneously on a geological scale of time. Whether the change took place gradually or in steps, there should have been intermediate stages represented by taxa that divided their time between activities in trees and activities on the ground. The morphology of the australopithecines suggests that they were such taxa.

As Figure 14.1 shows, the australopithecines' pattern of development was apelike, so that the ability of infants to cling should have permitted a mother to climb. One might nonetheless hypothesize that the australopithecines had converted to a totally terrestrial life but, through some kind of evolutionary inertia, retained inherited arboreal traits. There are major difficulties with this idea, however. Traits that had evolved as arboreal adaptations persisted for a total of about 3 m.y. in the australopithecine complex of species. The robust australopithecines inherited from the graciles most of the "arboreal" traits that the latter themselves had inherited. In general, the australopithecines remained intermediate between apes and humans in a variety of locomotory features. Their forelimbs possessed many apelike traits. Their hindlimbs, having been occupied extensively with bipedal locomotion on the ground, were more human in form but seemingly remained compromised by residual arboreal activities: It is highly unlikely that evolution would have failed to improve terrestrial locomotion by eliminating these deleterious traits had they not been employed in essential arboreal activities.

Short legs were one of the traits that natural selection for improved locomotion on the ground might be expected to have eliminated if stabilizing selection were not maintaining them because of their value in climbing. The short legs of australopithecines relative to body weight must have reduced endurance in bipedal locomotion by increasing the number of strides per unit of distance traversed

(Jungers and Stern, 1983). Similarly, the relatively long toes of australopithecines, though useful for climbing, would have reduced speed and endurance in running. Toes of some minimum length, apparently approximated in modern humans, are necessary for gripping the substratum and providing balance. Longer toes, however, lengthen the moment arm about which the body's center of gravity must rotate upward and forward during the so-called toe-off stage of running (Stanley, 1992; Figure 14.5).

It is not difficult to understand why australopithecines should have retained compulsory arboreal activities. Their dental morphology has been taken to indicate that fruits and seed pods formed a large part of their diet (Kay, 1985), and climbing would have expanded the range of available food items of this type. Even more important, however, should have been the need to elude large African predators.

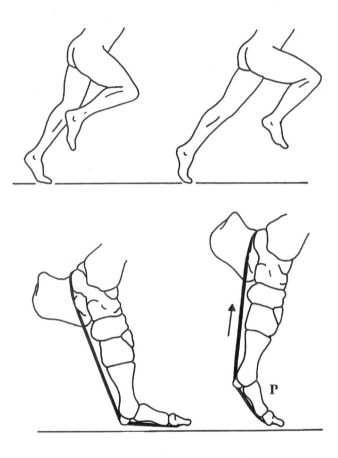

FIGURE 14.5 Illustration of the reason that lengthening of human toes would impair running. Arrow shows the force that leg muscles apply to the long tendon of the big toe in elevating the body during the toe-off stage of running. The tendon attaches to the distal bone of the toe. A longer proximal phalanx of the toe (P) would increase the lever arm of the forces opposing elevation of the body.

The fossil record reveals that the group of large mammalian predators that inhabited the African continent during Early Pliocene time was much like the one that exists today. In fact, the Early Pliocene fauna was slightly more diverse. In addition to the single living species of lion, leopard, and cheetah, and the three living species of hyenas, there were at least five species that are now extinct: two additional hyenas and three sabertooth cats. It is inconceivable that australopithecines could have withstood the predation pressure of this formidable array of carnivores had they not habitually climbed trees. Significantly, large terrestrial herbivores in modern Africa closely resemble their predators in running speed. The difficulty that modern carnivores encounter in capturing these speedy prey is indicated by two facts: first, they focus heavily on young, old, and sick animals; and second, the availability of food generally limits their population sizes (Kruuk, 1972; Schaller, 1972). Australopithecines, having been even slower than modern humans, would have been no match for predators. Lacking fire and stone weapons, they would also have had little ability to ward off attacks on the ground.

Australopithecines in trees would have faced few effective predators—perhaps only leopards and the false sabertooth, *Dinofelis*. (The true sabertooth cats, having had long, fragile canine teeth, are thought to have specialized on pachyderms; summary by Marean, 1989). Furthermore, leopards are solitary, territorial predators rather than group hunters, so that their density and hunting prowess are both relatively low. A treed australopithecine with a sharp stick might have warded off a leopard whose forelimbs were occupied in clinging to a branch, but australopithecines confined to the ground would frequently have found themselves within the ranges of several species of social predators with excellent night vision and a preference for nocturnal hunting.

Modern primates that resemble australopithecines in body size offer a test of the idea that australopithecines would have needed to climb trees in order to reduce predation pressure. Chimpanzees and baboons both habitually employ trees as arboreal refuges in two ways. They sleep in trees at night, and they flee into them during waking hours when threatened by predators. Male baboons, with their formidable canine teeth, have been seen to face down leopards, yet when lions are in the vicinity, trees are as important a limiting resource for baboons as are food and water (Devore and Washburn, 1963).

In addition, accumulations of gracile australopithecine bones in South African cave deposits appear to be the products of predation (Vrba, 1980; Brain, 1981). These primates, together with baboons, greatly outnumber ungulates and are represented primarily by cranial remains. The implication is that the australopithecines were under heavy predation pressure. In summary, for more than 1.5 m.y., australopithecines retained traits that made them much better climbers than modern humans. The fact that evolution failed to rid them of these traits despite the fact that some of the traits were deleterious to terrestrial locomotion suggests that stabilizing selection was maintaining the traits. The source of this stabilizing selection is readily found in the need to climb trees frequently to feed and especially to avoid numerous species of fast, powerful, group-hunting predators.

THE NATURE OF EARLY *HOMO*

The taxonomy of early representatives of the genus *Homo* is controversial, in part because of a patchy fossil record. "Early *Homo*" is a convenient label for fossil representatives of the genus older than *Homo erectus*, which ranges back to about 1.6 Ma. Traditionally, early *Homo* specimens have been assigned to the single species *Homo habilis*, which many workers now judge should be divided into two or more species. Details aside, it is clear that by about 2 Ma there existed some members of the genus whose brain capacities were at least twice the average for a gracile australopithecine. In addition, some members of early *Homo* had a pelvis that was not compressed from front to back, like that of a gracile australopithecine, but was instead remarkably like that of a modern human, except in having a smaller inlet, which required that babies be born at a smaller size than ours. Two known femora (thigh bones) of early *Homo* that are dated at about 1.9 Ma are also well within the length range for modern human females (Kennedy, 1983). In contrast, Lucy's femur is considerably shorter even than that of a female pygmy (Figure 14.3). All of the early *Homo* fossils mentioned above are quite similar to the equivalent skeletal parts of *Homo erectus* and modern humans, indicating a high level of adaptation to terrestrial locomotion.

The pelvic dimensions of early *Homo* indicate a small birth size; yet the remarkably large early *Homo* skull KNM-ER 1590, representing a small child, would have expanded into the *Homo erectus* range in adulthood (perhaps exceeding 900 cm^3 in cranial capacity). The enormous amount of brain growth between birth and adulthood in early *Homo* would have required a considerable extension of the Phase I interval of growth into the postnatal interval (Stanley, 1992). Thus, early *Homo* could not have been an obligate tree climber: its infants could not have clung to their mothers.

CLIMATIC FORCING

The oldest known fossils representing big-brained *Homo* are dated at 2.4 Ma (Hill *et al.*, 1992). The oldest known manufactured stone tools also date to about this time, and these are customarily attributed to early *Homo* (Harris, 1983). The tools are simple flakes that represent the so-called Oldowan culture. The youngest gracile australopithecines are not precisely dated, but the famous Taung skull of *Australopithecus africanus* is now dated at about 2.3 Ma (Delson, 1988).

Gracile australopithecines had existed for at least 1.5 m.y. without experiencing appreciable evolutionary change by the time that one of their populations turned into *Homo*. I have argued above that (1) their postcranial morphology was straitjacketed in an adaptive compromise between terrestrial and arboreal activities, and (2) obligate arboreal activity also prevented them from becoming encephalized appreciably above the level of an ape. It is reasonable to conclude that some kind of environmental change would have been required to end their nearly static evolutionary condition. In particular, what should have been required was a change that caused at least one population to abandon habitual arboreal activity.

As it turns out, the onset of the recent ice age at about 2.5 Ma produced exactly the kind of environmental change in Africa that could be expected to have shifted australopithecine behavior in the appropriate direction. Africa became markedly drier, like many other regions of the world at this time (see review by Stanley and Ruddiman, Chapter 7, this volume). As a consequence, forests shrank and grasslands expanded. Fossil pollen reveals that in the Omo Valley region of Ethiopia, climates were warmer and moister than today before about 2.6 to 2.4 Ma, but cooler and drier than today thereafter (Bonnefille, 1983). Similar changes are recorded from Algeria, Chad, and Kenya (Coque, 1962; Conrad, 1968; Bonnefille, 1976; Servant and Servant-Vildary, 1980). Carbon isotopes in soils provide more detailed evidence of this change (Figure 14.6). Samples from a large number of hominid sites reveal no canopied forests after about 2.5 Ma and also document the first occurrence of wooded grasslands at about this time (Cerling, 1992).

That the floral changes in Africa had a profound effect on mammals is well established. Close to 2.5 Ma, numerous species of antelopes that had adapted to forest conditions suffered extinction, and during the next few hundred thousand years, there appeared a variety of new savanna-dwelling species, most of which survive as elements of the modern African fauna (Vrba, 1974, 1975, 1985a; Vrba *et al.*, 1989). Micromammals underwent similar changes (Wesselman, 1985).

It has been suggested that the climatic changes may also in some way have promoted evolutionary turnover within the human family (Vrba, 1975, 1985b; Vrba *et al.*, 1989). The changes of behavior and ontogenetic development that I have attributed to the origin of *Homo* suggest

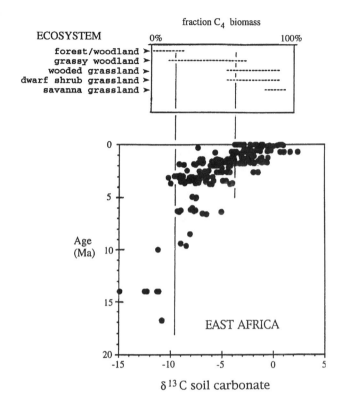

FIGURE 14.6 Stable carbon isotope composition and inferences about floral composition for paleosol carbonates from East African fossil localities. Isotopic compositions for modern biomes are shown above. Figure after Cerling (1992).

a particular mode of climatic forcing. Before the shrinkage of forests, troops of australopithecines probably occupied woodlands, which consist of groves or copses of trees separated by small areas of grassland. They could not have climbed well enough to have moved into and through the tall canopies of dense forests. Presumably, they used groves as home bases, sleeping in trees and occasionally feeding in them during the day (Rodman and McHenry, 1980). They may well have spent most of their waking hours on the ground, but only by remaining close enough to the home base to seek arboreal refuge when predators threatened. Modern baboons use trees in this way, even though they feed primarily on grass.

Saddled with the low intrinsic rate of natural increase that characterizes species of large primates because of solitary births (as opposed to litters) and lengthy generation times, australopithecine populations could not have sustained themselves in the face of heavy predation without arboreal refugia. Their relatively slow speed and weak natural defenses, in combination with their lack of both controlled fire and manufactured stone weapons, would have created intolerable predation pressure. They would have been easy targets for the multispecies guild of large, group-hunting terrestrial predators that, in the Pliocene as today, would have sought out as preferred prey animals that were easiest to catch. Even a grove of trees that initially served well as a refuge could not have sufficed indefinitely. In time, a troop would have exhausted food resources within close range of any home base. It would then have been required to move across grassland at the risk of suffering predation.

Before 2.5 Ma, woodlands were widespread and numerous groves of trees were separated by narrow zones of grassland. When forests shrank and fragmented with the onset of the ice age, however, many populations of australopithecines must have suffered a devastating intensification of predation pressure. Shrinking groves of trees offered smaller stores of food, which necessitated more frequent migration, and expanding grasslands increased the risk of predation by lengthening dangerous journeys. Presumably, many populations suffered extinction. Others may have survived for a time in areas that continued to support woodlands of moderate extent.

Widespread replacement of woodland habitats by grasslands is also exactly the kind of environmental forcing factor that could be expected to have obliged some populations to abandon habitual arboreal activity. Such a restriction of behavior automatically opened the way for encephalization through evolutionary extension of Phase I growth into the postnatal interval: physically helpless infants, though ecologically problematic, were now tolerable because mothers no longer climbed trees. Overriding the problems of raising highly dependent offspring, coping with predators, and losing arboreal food resources were the profound advantages of brain expansion—especially the ability to offset relatively weak physical attributes with innate cunning, advanced cooperative behavior, and sophisticated weaponry. These advantages of encephalization applied not only to avoidance of predators but also to development of hunting prowess that expanded trophic resources on the ground.

An important aspect of this scenario is that the first step was a simple change in behavior—one that amounted to a reduction of the preexisting behavioral repertoire. The result was that powerful natural selection pressures were brought to bear, so that major morphological changes ensued. Furthermore, the evolutionary retardation of development that produced encephalization was a relatively simply change, in that it represented only a modification of timing, not the origin of an entirely new pattern of development. This is not to say that the brain changed only by expanding. There was also a reorganization of brain anatomy, which we are only beginning to understand (Deacon, 1990).

The general evolutionary scenario outlined here entailed a shift to a new adaptive zone, not by an entire populous species but by a relatively small population of such a species that survived an environmental crisis. Other populations may have survived for a time with little change, in areas where environmental deterioration was less extensive. At least one fossil individual dated at about 1.6 Ma had a relatively small brain and more apelike proportions than individuals assigned unequivocally to early *Homo* (Leakey *et al.*, 1989). In addition, two robust australopithecine species persisted well into Pleistocene time. The enormous molars and powerful jaw muscles of these forms endowed them with the ability to process a wider variety of plant foods than gracile forms, however, and this may have increased their chances for survival by reducing the need to migrate to new food supplies. Even these forms died out at about 1 Ma. This was approximately the time when glacial maxima and minima became more extreme (Stanley and Ruddiman, Chapter 7, this volume) and when carbon isotopes show that true savannas appeared (Cerling, 1992). Perhaps the increased severity of droughts during glacial maxima caused the extinction of the robust australopithecines.

There is evidence that *Australopithecus africanus* persisted to about 2.3 Ma (Delson, 1988), but we do not now know for sure that it survived beyond the origin of *Homo* at about 2.4 Ma. Thus, we cannot know for sure whether *Homo* emerged from the entire surviving population of the decimated ancestral australopithecine species or whether the ancestral species gave rise to *Homo* by the evolutionary divergence of just one of its populations and then survived for a time alongside it, though possibly in other geographic regions. Discovery of a temporal overlap within the poorly documented interval between 2.5 and 2.0 Ma would settle the issue in favor of evolutionary branching, as opposed to the bottlenecking of an entire species.

The mechanism of climatic forcing that I have described is compatible with either possibility, in that environmental deterioration must have been a complex process in time and space, and different populations were undoubtedly subjected to different patterns of environmental change. In any event, by mid-Pleistocene time, only the fully terrestrial genus *Homo* remained.

We tend to think of the environmental changes associated with the onset of the Plio-Pleistocene ice age as constituting a deterioration of habitats. Thus, it might seem a great irony that the origin of our genus, which we inevitably view as a positive event, was wrought by what, from a different perspective, has been widely viewed as an environmental crisis.

REFERENCES

Aiello, L., and C. Dean (1990). *An Introduction to Human Evolutionary Anatomy*, Academic Press, London, 596 pp.

Bonnefille, R. (1976). Implications of pollen assemblage from the Koabi Fora Formation, East Rudolf, Kenya, *Nature 264*, 487-491.

Bonnefille, R. (1983). Evidence for a cooler and drier climate in the Ethiopian uplands towards 2.5 Myr ago, *Nature 303*, 487-491.

Brain, C. K. (1981). *The Hunters or the Hunted? An Introduction to African Cave Taphonomy*, University of Chicago Press, Chicago, 365 pp.

Cartmill, M. (1974). Pads and claws in arboreal locomotion, in *Primate Locomotion*, F. A. Jenkins, ed., Academic Press, New York, pp. 45-83.

Cerling, T. (1992). Development of grasslands and savannahs in East Africa during the Neogene, *Palaeogeography, Palaeoclimatology, Palaeoecology 97*, 241-247.

Conrad, G. (1968). *Evolution Continental Post-Hercynienne du Sahara Algerien*, CNRS, Paris.

Coque, R. (1962). *La Tunisie Présaharienne*, Amrand Colin, Paris, 476 pp.

Count, E. W. (1947). Brain and body weight in man: Their antecedents in growth and evolution, *Annals of the New York Academy of Sciences 46*, 993-1122.

Deacon, T. W. (1990). Problems of ontogeny and phylogeny in brain-size evolution, *Internat. J. Primatol. 11*, 237-282.

Delson, E. (1988). Chronology of South African australopith site units, in *Evolutionary History of the "Robust" Australopithecines*, F. E. Grine, ed., Aldine de Gruyter, New York, pp. 317-324.

Devore, I., and S. L. Washburn (1963). Baboon ecology and human evolution, in *African Ecology and Human Evolution*, F. C. Howell and F. Bourliere, eds., Aldine, Chicago, pp. 335-367.

Harris, J. W. K. (1983). Cultural beginnings: Plio-Pleistocene archaeological occurrences from the Afar, Ethiopia, *African Archaeol. Rev. 1*, 3-31.

Hill, A. S., A. Deino, G. Curtiss, and R. Drake (1992). Earliest *Homo*, *Nature 355*, 719-722.

Holt, A. B., D. B. Cheek, E. D. Mellits, and D. E. Hill (1975). Brain size and the relation of the primate to the nonprimate, in *Fetal and Postnatal Cellular Growth*, D. B. Cheek, ed., John Wiley & Sons, New York, pp. 23-44.

Jungers, W. L., and J. T. Stern (1983). Body proportions, skeletal allometry and locomotion in the Hadar hominids: A reply to Wolpoff, *J. Human Evol. 12*, 673-684.

Kay, R. F. (1985). Dental evidence for the diet of Australopithecus, *Ann. Rev. Phys. Anthrop. 14*, 315-341.

Kennedy, G. E. (1983). A morphometric and taxonomic assessment of a hominine femur from the lower member, Koobi Fora, Lake Turkana, *Amer. J. Phys. Anthrop. 61*, 429-436.

Krogman, W. G. (1972). *Child Growth*, University of Michigan Press, Ann Arbor, 231 pp.

Kruuk, H. (1972). *The Spotted Hyena*, University of Chicago Press, Chicago, 335 pp.

Latimer, B., and C. O. Lovejoy (1990). Hallucal tarsometatarsal joint in *Australopithecus afarensis*, *Amer. J. Phys. Anthrop. 82*, 125-133.

Latimer, B., J. C. Ohman, and C. O. Lovejoy (1987). Talocrural joint in African hominoids: Implications for *Australopithecus afarensis*, *Amer. J. Phys. Anthrop. 74*, 155-175.

Leakey, M. D., and R. L. Hay (1979). Pliocene footprints in the Laetolil beds at Laetoli, northern Tanzania, *Nature 202*, 7-9.

Leakey, R. E. F., A. Walker, C. V. Ward, and H. M. Grausa (1989). A partial skeleton of a gracile hominid from the upper Burgi member of the Koobi Fora Formation, east Lake Turkana, Kenya, in *Hominidae*, G. Giacobini, ed., Jaca, Milan, pp. 209-215.

Marean, C. W. (1989). Saber-tooth cats and their relevance for early hominid diet and evolution, *J. Human Evol. 18*, 559-582.

McHenry, H. M. (1986). The first bipeds: A comparison of the A. afarensis and A. africanus postcranium and implications for the evolution of bipedalism, *J. Human Evol. 15*, 177-191.

McHenry, H. M. (1991). Sexual dimorphism in *Australopithecus afarensis*, *J. Human Evol. 20*, 21-32.

Rodman, P. S., and H. M. McHenry (1980). Bioenergetics and the origin of hominoid bipedalism, *Amer. J. Phys. Anthropol. 63*, 371-378.

Schaller, G. B. (1972). *The Serengeti Lion*, University of Chicago Press, Chicago, 480 pp.

Servant, M., and S. Servant-Vildary (1980). L'environment quaternaire du bassin du Tchad, in *The Sahara and the Nile*, M. A. J. Williams and H. Faure, eds., Balkema, Rotterdam, pp. 133-162.

Skeat, W. W., and C. O. Blagden (1906). *Pagan Races of the Malay Penninsula*, MacMillan, New York, 724 pp.

Stanley, S. M. (1992). An ecological theory for the origin of *Homo*, *Paleobiology 18*, 237-257.

Stern, J. T., and R. L. Susman (1983). The locomotory anatomy of *Australopithecus afarensis*, *Amer. J. Phys. Anthrop. 60*, 279-317.

Susman, R. L., J. T. Stern, and W. L. Jungers (1984). Arboreality and bipedality in the Hadar hominids, *Folia Primatol. 43*, 113-156.

Tague, R. G., and C. O. Lovejoy (1986). The obstetric pelvis of A. L. 288-1 (Lucy), *J. Human Evol. 15*, 237-255.

Vrba, E. S. (1974). Chronological and ecological implications of the fossil Bovidae at the Sterkfontein australopithecine site, *Nature 250*, 19-23.

Vrba, E. S. (1975). Some evidence of chronology and palaeoecology of Sterkfontein, Swartkrans and Kromdraai from the fossil Bovidae, *Nature 254*, 301-304.

Vrba, E. S. (1979). A new study of the scapula of *Australopithecus africanus* from Sterkfontein, *Amer. J. Phys. Anthrop. 51*, 117-130.

Vrba, E. S. (1980). The significance of bovid remains as indicators of environment and predation patterns, in *Taphonomy and Paleoecology*, A. K. Behrensmeyer and A. P. Hill, eds., University of Chicago Press, Chicago, pp. 247-271.

Vrba, E. S. (1985a). African Bovidae: Evolutionary events since the Miocene, *S. African J. Sci. 81*, 263-266.

Vrba, E. S. (1985b). Ecological and adaptive changes associated with early hominid evolution, in *Ancestors: The Hard Evidence*, E. Delson, ed., Alan R. Liss, New York, pp. 63-71.

Vrba, E. S. (1988). Late Pliocene climatic events and hominid evolution, in *Evolutionary History of the "Robust" Australopithecines*, F. E. Grine, ed., Aldine de Gruyter, New York, pp. 405-426.

Vrba, E. S., G. H. Denton, and M. L. Prentice (1989). Climatic influences on early hominid behavior, *Ossa 14*, 127-156.

Wesselman, H. B. (1985). Fossil micromammals as indicators of climatic change about 2.4 Myr ago in the Omo Valley, Ethiopia, *S. African J. Sci. 81*, 260-261.

Wood-Jones, F. (1900). *Arboreal Man*, E. Arnold, London, 230 pp.

Index

A

Abiotic stress, 136
Acacia, 176
Acritarchs, 26
Adaptation
    human genus, 234-241
    land mammals, 185-205
Adaptive radiation, 10-15, 42
Aerobic metabolism, 26
Aerobic organisms, 21-31
Africa, 123, 130, 189
    human genus origin, 234-241
Alaska, 121, 157-158
    North Slope, 159
Albedo, 9
Algae, 26
Allocasuarina, 181
Allopatry, 226
Amebomastigotes, 28
American West, 127
Ammonites, 50, 60
Amphicyonids, 192
Anaerobic organisms, 21-31
Angiosperms, 8, 150, 151, 157-158, 166, 167
Anoxia, 39-42
Antarctic Circumpolar Current, 5-6, 168, 178, 201
Antarctica, 5-6, 8-9, 97, 103, 111, 164-168, 175, 185, 201
Antelopes, 124, 240
Antilocapridae, 192
Araucarians, 164
Arboreal habitat, 234-241
Arboreal mammals, 189
Arctic, 121
Arid regions, 176-177
Aridity, 123, 191, 194
Arkansas River, 58
Armadillos, 195
Artiodactyls, 96, 190, 191, 200
Ash, 139
Asia, 130, 192
Atlantic Coastal Plain, 129
Australia, 5, 13, 24, 164
    flora and climatic changes, 174-182
Australian Plate, 174
Australopithecus, 12, 234-241
Avatars, 135
Axial Basin, 58

B

Baboons, 239
Bahamas, 121, 123
Baltoscania, 36-39
Banded iron formations, 22
Bears, 192, 195
Beavers, 192
Bennettitaleans, 164
Benthic foraminifera, 94-105, 121, 200, 211
Bering land bridge, 188
Bering Strait, 121
Beringia, 188
Biomass, 95, 141
Biomes
    land mammal biomes, 189-195
Biostratigraphy, 74, 96, 119, 185, 198
Biotas, *see* Ecosystems; Fauna; Flora
Biozones, 49, 50-53, 59-60
Birch, 224
Bivalves, 10, 50, 120
Black Sea, 212
Bogota, Colombia, 123
Bolide impacts, 8-9, 57, 63-64, 95, 101
    Cretaceous-Tertiary mass extinction, 73-91, 97
Boulder, Colorado, 57
Bovids, 200
Brachiopods, 13, 15, 35, 42-43
Brain anatomy, 241
Brain size and development
    human genus origin, 234-241
Brazos, Texas, 83-84

Breakpoints, 142, 148
Bridge Creek Limestone, 60
Browsers, 13, 190-193
Bryophytes, 157
Buoyancy flux, 111

C

Calcareous nannoplankton, 78-79
Calcium carbonate tests, 212
California, 40, 193
Camels, 192
Capybaras, 125, 195
Caravaca, Spain, 82-83
Carbon, 8, 56, 129
Carbon capture, 169
Carbon dioxide, 26, 43, 112, 127
Carbon isotopes, 3, 30, 87-88, 95, 100, 102, 240
Carbonate dissolution, 75-76
Cardiocarpus, 141
Caribbean, 54, 58, 121
Carnivores, 190-195
Cascade Range, 119
Caspian Sea, 212
Casuarina, 181
Catastrophism, 8-9
Cats, 193
Cenogram, 189
Cheetahs, 239
Chelation, 136
Chemical weathering, 103, 129
Chenopod scrub, 176
Chert, 24
Chimpanzees, 234, 236, 239
China, 41-42, 123, 130, 199
Chitinozoans, 43
Chlamytheres, 195
Chronofaunas, 185-205
Chronostratigraphy, 49, 58, 59, 74, 185
Circulatory systems, 29
Clay, 103
Climacograptids, 41
CLIMAP, 126
Climate curves, 148-149
Climatic changes, 4-15
  Australia flora and climatic changes, 174-182
  coal swamp ecosystems, 140, 147-149
  Cretaceous-Tertiary mass extinction, 72-91
  Eocene-Oligocene, 5-8
  human genus origin, 240-241
  Neogene ice age, biotic effects, 119-123
  terminal Paleocene mass extinction, 94-105
Climatic forcing factors, 110, 127

Climatic stability, 108-115
Cloud, Preston, 21
Coal, 134-152, 157, 166
Coal balls, 135-147
Coal seams, 136-138
Coal swamp, 10, 11
  climatic change effects, 134-152
Coelenterates, 29
Coevolution, 21
Collagen manufacture, 29
Colombia, 53, 54, 57
Colorado Plateau, 127
Complex Mesophyll Vine Forest, 176
Compression floras, 149
Concretions, 136
Condylarthra, 190
Conifers, 157-158, 164, 167
Conodonts, 43
Continental growth, 24
Continental margins, 201
Continental red beds, 22-23
Continental shelf, 73-91, 101
Convergence, 193
Corals, 10, 111, 113
Cordaites, 141, 145
Coriolis deflection, 36
Coryphodontidae, 190
Craton, 147-148
Creodonta, 190
Cretaceous-Tertiary mass extinction, 73-91, 97
Cryosphere, 9-10
Cyanobacteria, 24
Cycads, 157-159, 164
Cyclostratigraphy, 50, 58
Cypress, 124, 190

D

Dark shale, 39
Darwin, Charles, 182
Deep dwellers, 87-90
Deep mantle outgassing, 49, 62, 63
Deep sea
  Cretaceous-Tertiary mass extinction, 73-91, 97
  Paleocene mass extinction, 94-105
Deep Sea Drilling Project (DSDP), 84, 95, 167
Deep water, 39-42, 102, 111, 130
  see also North Atlantic deep water
Deep-sea cores, 119
Deer, 194, 199
Delayed recovery, 13-15
Demes, 229
Dermopterans, 190
Desert, 119

Detrivores, 139
Diagenesis, 23, 56
Diamictites, 35
Dinocerata, 190
Discoasters, 97
Disseminules, 140
Dob's Linn, Scotland, 39
Domed swamps, 136
Drake Passage, 201
Dust, 120, 127, 128
Dysaerobism, 102

E

Ecomorphs, 139, 140, 146, 189
Ecosystems, 10-15, 80
  coal swamp ecosystems, 134-152
  high-resolution stratigraphy, research needs, 16
  land mammals (Cenozoic), 185-205
  planktonic foraminifera, 211-212
  plant communities (Quaternary), 221-229
Ecotones, 137, 141, 224
Edaphic constraints, 138, 151, 178
Ediacaran fauna, 28, 29
El Kef, Tunisia, 4, 75, 76, 77, 80-82
Elephas, 199
Ellesmere Island, 190
Emergent properties, 135
England, 53, 143
Entamebas, 28
Environmental tolerances of organisms, 113-113, 212
Eolian transport, 192
Epicontinental sea, 58
Epifaunal forms, 96
Equisetites, 159, 164
Equus, 199
Ericridaceae, 176
Ethiopia, 123
Eubrontothere, 191
Eucalyptus, 176-178
Euglenids, 28
Eukaryotes, 24-25, 26, 28
Europe, 54, 120, 123, 124, 188, 190, 198
Euryhaline, 218
Eustasy, 148
Evaporation, 110
Evolution, 10, 151-152, 179
  Australia flora and climatic change, 174-182
  high-latitude vegetation (Cretaceous/Cenozoic), 157-169
  human genus, 234-241
  late Paleocene, 95-96
  oxygen and proterozoic evolution, 21-31

Quaternary, 226-229
  research needs, 16
Exercise metabolism, 29
Extinction, 11-12, 123, 168
  coal swamp ecosystems, 141-146
  see also Mass extinction
Extra-terrestrial impacts, see Bolide impacts

F

Fauna, 10-15, 78, 95-96
  Gulf of Mexico biotic response to deglaciation, 210-218
  land mammal chronofaunas, 185-205
  Ediacaran, 28, 29
  Ordovician glaciation and marine fauna, 35-44
Ferns, 136, 137, 157-158, 164-167
Floods, 136, 139
Flora, 8, 11, 80, 123, 136, 235, 240-241
  Australia flora and climatic changes, 174-182
  compression, 149
  cryptogamic, 168
  deciduous versus evergreen, 167-168
  fauna-flora feedback loop, 192
  grassland savanna in North America, 193-194
  vegetation change in late Quaternary, 221-229
Florida, 120, 194
Flowering plants, 136
Foliar physiognomy, 178
Forest, 10, 11, 123, 136, 157, 240-241
  closed forest system, 175-176
  land mammal habitat, North America, 189-190
  open forest system, 176
  Quaternary, 224
  see also Rain forest
France, 53
Functional morphology, 3
Fungi, 157, 166
Fusain, 139, 141

G

Gastropods, 120
Gauss Chron, 123
Gazelles, 191
Gebel Zelten, Libya, 200
Gene pool, 175
General Circulation Model (GCM), 3-4
  atmospheric, 110, 114
  oceanic, 111, 114, 127
Geochronometry, 49, 186

Giardia, 26-27
Ginkgophytes, 157-159
Glaciation and deglaciation, 4-8, 29, 148, 201, 203
  Gulf of Mexico biotic response to deglaciation, 210-218
  Ordovician glaciation and marine fauna, 35-44
  Quaternary vegetation change, 221-229
  see also Ice ages
Global cooling, 43
Global warming, 48-64, 89
  terminal Paleocene mass extinction, 94-105
Globigerina falconensis, 215
Globigerinoides ruber, 213-218
Globorotalia inflata, 214-215
Globorotalia menardii, 215
Glyptodonts, 195
Gondwanaland, 7, 176
  Ordovician glaciation and marine fauna, 35-44
Grande Coupure, 198
Granite, 23
Graptolites, 4, 8, 15, 35-44
Grassland, 10, 123, 125, 166, 240-241
Grazers, 189, 193
Great Basin, 50, 58, 119, 194
Great Plains, 125
Greenhouse world, 5, 48-64, 103, 156, 178, 200
Greenland, 9, 119, 166, 212
Ground sloths, 195
Gulf of California, 212
Gulf of Mexico, 210-218
Gulf of Aden, 211
Gulf Stream, 37, 120
Gymnosperms, 136, 141
Gymnostoma, 181

H

Habitat, 141, 143-147
Habitat destruction, 87-90
Heath scrub, 176
Hepatics, 164
Herbivores, 191
Heterotrophs, 27, 28
Hiatus patterns, 76, 90, 95
High latitudes, 8, 80, 86, 90, 103, 119
  vegetation and climate (Cretaceous/Cenozoic), 156-169
High-resolution stratigraphy, 2, 4, 49, 58, 74, 95, 97, 185, 212
  research needs, 16
Himalayan Plateau, 199
Hipparion, 199

Hippopotamus, 193
Hirnantia, 7, 15, 37-44
Hominoids, 200
Homo, 12, 48, 234-241
Horses, 191, 192, 195, 198-199, 203
Humans, 234-241
Humboldt Current, 36
Humerofemoral index, 236
Hyaenids, 194
Hyaenodontidae, 190
Hydrogenosomes, 27
Hydrothermal activity, 103
Hyenas, 239
Hypselodonty, 189
Hypsodonty, 189, 191

I

Ice ages, 109-110, 240-241
  Neogene, biotic effects, 118-130
  see also Glaciation and deglaciation
Icehouse world, 5, 156, 186, 200
Iceland, 121
India, 199-200
Indian Ocean, 211
Individualism, 135
Insectivores, 190, 191
Insolation, 129
Interdisciplinary research, 2-3
  research needs, 15-16
Intermediate dwellers, 87-90
Iridium, 54, 56, 62, 80, 83
Iron, 24, 29, 39-42
Isospores, 151
Isotopes, 29-30
Isthmus of Panama, 120, 126, 127

J

James Ross Island, 164
Jaramillo magnetic reversal, 123

K

Karsts, 35
Kinetoplastids, 27
King George Island, 167

L

Lagarostrobus, 164, 165, 166
Lagomorphs, 191
Land bridges, 188, 199-200
Landscape, 139, 141-147
Lauraceae, 181
Leaf physiognomy, 5, 157, 164
  roseaceous, 165

Leopards, 239
Lepidodendrids, 141, 146
Leptomeryx, *191*
Lichens, 168
Light, 157
Lignotubers, 176
Limestone, 60
Lions, 239
Locomotion, 238
Loess, 122, 130
Low latitudes, 80, 86, 90
Lucy skeleton, 236
Lycopods, 10, 157
Lycopsids, 136, 137, 139, 141, 143, 146
Lycospora, 146
Lyginopteris, 143-145

M

Macedonia, 123
Macroscopic animals, 28
Magnetochron 24 R, 96
Magnetostratigraphy, 49, 123
Malaysia, 236
Mallee vegetation, 176
Mammals, 10, 12-13, 95-96, 123
    Cenozoic faunas, 185-205
Mammoths, 195
Manganese, 39-42
Mangroves, 111
Marattialeans, 151
Mass extinction, 4-5, 7-8, 9
    Cenomanian-Turonian, 4, 48-64, 69-71
    Cretaceous-Tertiary mass extinction, 72-91
    land mammal rapid turnover episodes, 186-188
    Ordovician glaciation and marine fauna, 35-44
    Pennsylvanian coal swamps, 146
    research needs, 16
    terminal Paleocene, 94-105
Mats, 164
Matuyama Chron, 123
Maud Rise, Antarctica, 99, 167
Mediterranean, 120, 123, 127, 211
Meltwater, 7, 213-214
Merychippus, 192
Mesohippus, 191
Mesophytic flora, 149
Metabolites, 168
Metasequoia, 165
Micromammals, 124
Microsporidia, 26
Migration, 10, 221, 227
    North America land mammal immigration (Cenozoic), 185-205
    research needs, 16

Milankovitch climate cycles, 4-5, 50, 58, 60
Mineral charcoal, 139
Miospores, 137
Mississippi River, 214
Mitochondria, 26, 27
Mitrospermum, 141
Mixing, 226
Models, 4
    climatic changes, 109, 110, 222
    "conveyor belt" ocean circulation, 211
    environmental/biological coevolution, 31
    mass extinctions, 64
    research needs, 16
Mollusks, 12, 62, 120, 126
Monkeys, 124, 200
Monsoons, 36-39, 119, 127
Morozovellids, 97
Moschidae, 192
Mosses, 168
Mountain glaciers, 119
Multituberculata, 190
Musk oxen, 195
Mustelids, 193

N

Nannoplankton, 51
Natural selection, *see* Evolution
Nebraska, 193
Nekton, 39
Neoecology, 151
Neogloboquadrina dutertrei, 215
Neritic environment, *see* Continental shelf
Netherlands, 123
Neutron activation, 39
Nevada, 193
New Zealand, 165
Nitrates, 26
NOAA Geophysical Fluid Dynamics Laboratory, 112
North America, 119, 178, 222-223
    Cenozoic mammal faunas, 185-205
North Atlantic deep water (NADW), 6-7, 102, 127, 130, 211, 214
North Atlantic
    Neogene ice age, biotic effects, 118-130
North Pacific, 212
North Sea, 120, 127
Northern Hemisphere, 119-130, 156-169
Northern Plains, 127, 128
Nothofagus, 165, 166-168, 179-180
Nuttallides truempi, 97

O

Ocean circulation, 5-7, 12, 36-39, 102
Ocean Drilling Program (ODP), 95
Ocean Drilling Program Site 119, 164
Ocean Drilling Program Site 738C, 84-86
Ocean floor, *see* Deep sea
Ocean temperature, 10, 43, 54, 56, 63, 97-105, 109, 122, 201, 212
Ocean ventilation, 42
Ocean warming, 97-105
Ocean-climate system, 50, 56, 62
Oceanic anoxic events, 56, 62, 102
Omo Valley, Ethiopia, 240
Omomyidae, 190
Orangeburg scarp, 120
Orbital cycles, 4-5, 129, 226
Ordinations, 143
Oreodonts, 191, 192
Ostracods, 100, 126
Oxygen, 8, 56
    proterozoic evolution and oxygen, 21-31
Oxygen isotopes, 3, 87-88, 100, 102, 108-112, 122, 130, 178, 186-188, 200-205, 211, 213-217
Oxygen minimum zone (OMZ), 58
Oxygen-depleted water, 39-42
Ozone screen, 26

P

Pacific, 121, 211
Pacific Northwest, 188
Pacific superplume, 57, 58, 62
Pakistan, 199-200
Palaeomerycidae, 192
Paleo-Antarctic Circle, 168
Paleo-Arctic Circle, 168
Paleobathymetry, 58
Paleobiology, 3, 94-105
    Australia flora and climatic changes, 174-182
    Gulf of Mexico biotic response to deglaciation, 210-218
    high-latitude vegetation and climate, 156-169
    research needs, 16
    vegetation change in late Quaternary, 221-229
Paleoecology
    coal swamp ecosystems, 134-152
Paleogeography, 3, 112, 138, 148, 157
    Ordovician, 36-39
    pollen records, late Quaternary, 221-229
    research needs, 16
Paleomagnetic chronology, 185

Paleosols, 23, 101, 148, 191
Palms, 124
Palynology, 147, 157, 164, 166, 178
　late Quaternary vegetation, 221-229
Panama land bridge, 188
Pandas, 194
Pantodonta, 190
Parahippus, 192
Paralycopodites, 143-145
Patagonia, 168
Peat-forming swamps, *see* Coal swamps
Peccaries, 125, 195
Peripheral isolates, 151
Perissodactyls, 96, 190, 198
Petrography, 139
Petroleum source rock, 40, 56
Phosphorus, 24
Photosynthesis, 26, 103
Photosynthetic bacteria, 23
Phylogeny, 26-28
Planar swamps, 136
Plankton, 39
Planktonic foraminifera, 12, 15, 50-53, 97-100, 109, 120, 121
　Cretaceous-Tertiary mass extinction, 73-91, 97
　Gulf of Mexico, response to deglaciation, 210-218
Plant community ecology, 135
Plate rafting, 175
Plenus Marls, 53
Podocarps, 164, 166, 167
Poland, 124
Poleward heat transport, 110-111
Pollen, 119, 123, 135, 157
　late Quaternary vegetation, 221-229
　triaperturate, 164
Porcupines, 195
Portugal, 35
Prague Basin, Czechoslovakia, 35
Precipitation, 104, 129, 157, 178
Predators, 238-239
Present atmospheric level (PAL), 23, 27, 29
Primates, 96, 190
Proboscideans, 193, 198
Prokaryotes, 31
Proteaceae, 165, 167, 181
Proteus, 102, 103
Protists, 25, 30
Protoceratids, 195
Psaronius, 141, 145
Pteridosperms, 137, 141, 142, 146, 150, 164
Ptilodus, 190
Pueblo, Colorado, 58
Pulleniata obliquiloculata, 215
Pyrite, 30, 40

Q
Queensland, 176

R
Rabbits, 191, 199
Raccoons, 192
Radiative energy input, 110
Radiometric dating, 54-55, 59, 60, 185, 211
Rain forest, 11, 64, 175, 179
Rapid turnover episodes (RTE), 186-188, 197, 198-200
Rare earth elements, 24
Red Sea, 211
Research needs, 15-16
Resource depletion, 64
Reversed magnetic interval, 96, 190
Rhinocerotids, 191, 193
Rhizovores, 191
RNA, 27
Rock accumulation rates, 49
Rock Canyon Anticline, 58
Rocky Mountains, 191
Rodents, 13-15, 190, 191
Ross Sea, 167
Rotalipora foraminifer, 57
Rudists, 15, 54, 114
Ruminants, 198, 200

S
S. beccariformis, 96
Sabertooth cats, 239
Sagebrush, 178
Sahara, 120
Salinity, 100, 112, 214-217
Salix, 166
San Salvador, 123
Savanna, 190-194, 201, 233
Scales, 227-229
Scandinavia, 119
Scansorial mammals, 189
Sclerophyll plants, 13, 178, 182
Sea otters, 121
Sea-floor spreading, 129
Sea-level changes, 11, 54, 56, 201, 204, 212
　Cretaceous-Tertiary mass extinction, 88-91
　Ordovician glaciation and marine fauna, 35-44
Sediment traps, 212
Sedimentation rates, 4, 40, 75, 102
Seed ferns, 143
Seed plants, 136

Seismic stratigraphy, 200-201
Semiarid regions, 176-177
Sessile plants, 221
Seymour Island, 164, 185
Shales, 137
Shatsky Rise, 109
Shocked quartz grains, 62
Shrub eaters, 192
Siberia, 157
Sigillarians, 151
Signor-Lipps effect, 9, 52-53
Silicate rocks, 129
Siwalik Hills, 199-200
Slime molds, 28
Snakes, 15
Songbirds, 13-15
South Atlantic, 211
South Shetland Islands, 165
Southern Hemisphere, 156-169
Space-time box, 229
Spatial and temporal scales, 227-229
Speciation, 124, 152, 178, 188
Species turnover, 12-13, 146-147, 149, 151, 186, 198-200
Sphenopsids, 141, 142, 151
Spinels, 80
Spruce, 224
St. Lawrence River, 214
Stable isotope stratigraphy, 96-97
Stenothermal species, 125
Steppe, 188
Sterols, 27
Straits of Florida, 121
Stratigraphic anomaly 24, 190
Stromatolites, 24
Stromatoporoids, 114
Strontium isotopes, 30
Subcanopy, 176
Succession, 185, 221, 222
　land mammals in North America, 189-197
Sulfates, 23
Sulfur isotopes, 30
Sulu Sea, 211
Sundra Plate, 175
Surface dwellers, 87-90
Svalbard, 166
Symbiosis, 26
Sympatry, 226
Synchroneity, 102
Syncopated equilibrium, 197

T
Taiga, 156, 166
Taphonomic criteria, 139, 189
Tasmania, 168, 175

Tectonic activity, 62, 119, 123, 127, 148, 168, 175, 194, 226
Teeth, 101, 189
Tegelen faunas, 124
Teleoceras, 193
Temperature, 157, 178
Terminations, 122
Terrestrial habitat, 234-241
Tethyan Ocean, 80, 114
Tethys Seaway, 103
Thermocline, 39, 89, 101, 212
Thermohaline circulation, 6-7, 111
Thresholds, 9-10
Thulean land bridge, 188, 201
Tibetan Plateau, 127
Tillodontia, 190
Time scale, 2
Titanotheres, 13
Tools, 240
Toxic waters, 39
Trace elements, 56, 62
Tree ferns, 141, 142, 145, 150
Tree rings, 159
Trees, 140, 141
Trichomonads, 26
Trilobites, 15, 43
Trochospiral forms, 100
Tropics, 36-39, 104, 211
climatic stability, 108-115
Tundra, 166, 188
Turnover-pulse hypothesis, 188

U

Ultraviolet (UV) radiation, 26
Ungulates, 189, 191, 193, 204
Uraninite, 23

V

Vapor pressure, 104, 110
Vegetation, *see* Flora
Vertical advection, 39-42
Virginia, 120
Volcanic activity, 103, 129

W

Warm saline deep water, 102, 104, 111
Water column, 87-90
Western Interior Seaway, 58
Wetland habitats
coal swamp ecosystems, 134-152
White River chronofauna, 190-194
Winds, 127
Woody shrubs, 192
Wyoming, 194

X

Xeromorphy, 141, 147

Y

Yorktown Formation, 120-121
Younger Dryas cooling, 7, 210-218
Yucatan Channel, 121